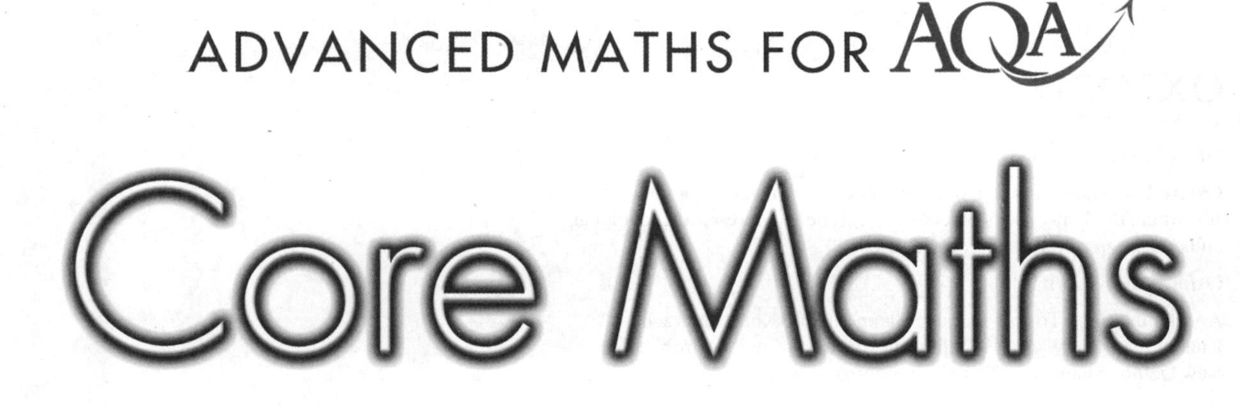

Garry Wiseman

Course consultant: Jeff Searle

C3

C4

OXFORD
UNIVERSITY PRESS

Great Clarendon Street, Oxford OX2 6DP

Oxford University Press is a department of the University of Oxford.
It furthers the University's objective of excellence in research, scholarship,
and education by publishing worldwide in

Oxford New York

Auckland Cape Town Dar es Salaam Hong Kong Karachi
Kuala Lumpur Madrid Melbourne Mexico City Nairobi
New Delhi Shanghai Taipei Toronto

With offices in

Argentina Austria Brazil Chile Czech Republic France Greece
Guatemala Hungary Italy Japan South Korea Poland Portugal
Singapore Switzerland Thailand Turkey Ukraine Vietnam

First published 2005

British Library Cataloguing in Publication Data

Data available

ISBN 978 0 19 914987 2

10 9

Typeset by Tech-Set Ltd, Gateshead, Tyne and Wear
Printed and bound in Great Britain by Bell and Bain Ltd., Glasgow

Acknowledgements
The publishers would like to thank AQA for their kind permission to reproduce past
paper questions. AQA accept no responsibility for the answers to the past paper
questions which are the sole responsibility of the publishers.

The publishers would also like to thank James Nicholson for his authoritative guidance
in preparing this book.

Paper used in the production of this book is a natural, recyclable product made from
wood grown in sustainable forests. The manufacturing process conforms to the
environmental regulations of the country of origin.

About this book

This Advanced level book is designed to help you get your best possible grade in the AQA MPC3 and MPC4 modules for first examination in 2006. These two modules can contribute to an award in GCE A level Mathematics.

The book is divided into the two modules, C3 and C4, and you can use the tabs at the edge of the page for quick reference.

Each chapter starts with an overview of what you are going to learn and a list of what you should already know. The 'Before you start' section contains 'Check in' questions, which will help to prepare you for the topics in the chapter.

You should know how to ...	Check in
1 Find the derivative of x^n for any rational number n.	**1** Find $\frac{dy}{dx}$ when $y = 2x^{\frac{3}{2}} - x^2$.

Key information is highlighted in the text so you can see the facts you need to learn.

$$\operatorname{cosec}\theta = \frac{1}{\sin\theta} \qquad \sec\theta = \frac{1}{\cos\theta} \qquad \cot\theta = \frac{1}{\tan\theta}$$

Worked examples showing the key skills and techniques you need to develop are shown in boxes. Also hint boxes show tips and reminders you may find useful.

Example 4

Solve the equation $|x - 1| = 4$.

$$|x-1|^2 = 4^2$$
$$(x-1)^2 = 4^2$$
$$x^2 - 2x + 1 = 16$$
$$x^2 - 2x - 15 = 0$$
$$(x-5)(x+3) = 0$$

Solving gives $x = 5$ and $x = -3$.

Square both sides to ensure that the LHS is positive.

The questions are carefully graded, with lots of basic practice provided at the beginning of each exercise. At the end of an exercise, you will sometimes find underlined questions. These are optional questions that go beyond the requirements of the specification and are provided as a challenge.

At the end of each chapter there is a summary. The 'You should now know' section is useful as a quick revision guide, and each 'Check out' question identifies important techniques that you should remember.

You should know how to ...	Check out
1 Locate the roots of an equation between two values of x.	**1** a) Show that $3x^2 + 2x - 7 = 0$ has a root α $1 < \alpha < 2$. b) Show that $x^2 - 1 = 3\sqrt{x}$ has a root β $2.1 < \beta < 2.5$.

Following the summary you will find a revision exercise with past paper questions from AQA. These will enable you to become familiar with the style of questions you will see in the exam.

At the end of each module you have a Practice Paper. These will directly help you to prepare for your exams.

At the end of the book you will find numerical answers, a list of formulae you need to learn, and a list of useful mathematical notation.

Contents

1 Functions

This chapter will show you how to

- ✦ Determine the domain and range of a function
- ✦ Recognise and use the modulus function
- ✦ Find composite and inverse functions
- ✦ Use inverse trigonometric functions

Before you start

You should know how to ...	Check in
1 Use function notation $f(x)$.	**1** a) Evaluate each of these functions at the given value of x. i) $f(x) = 3x - 2$; $x = 4$ ii) $g(x) = \sin 2x$; $x = 50°$ iii) $h(x) = 4^{2x}$; $x = 3$ b) Find the value of these functions at the value stated. i) $f(-1)$ where $f(x) = 2x^2 - 3x + 4$ ii) $g(1.5)$ where $g(x) = \cos^2 x$ (x in radians) iii) $h(0.25)$ where $h(x) = \frac{2}{x}$
2 Recognise graphs of functions.	**2** Match each of these functions to the sketch of its graph. a) $y = 3 - x$ b) $y = x^3 - 9x$ c) $y = \cos 3x$ d) $y = \frac{3}{x}$ A 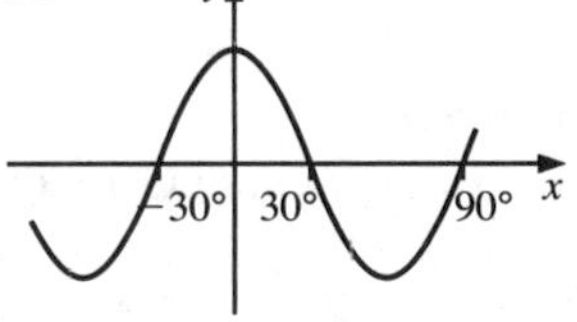 B 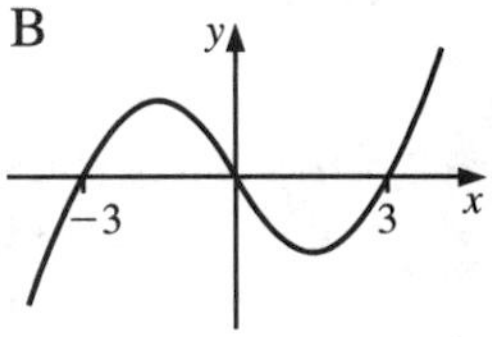 C 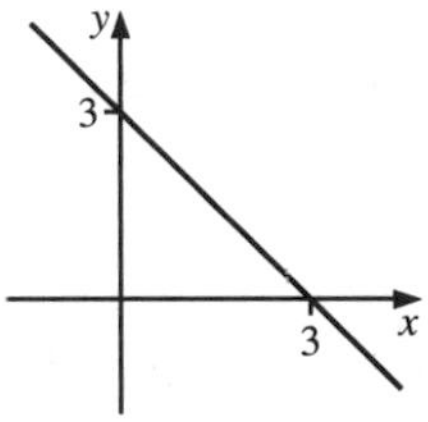 D 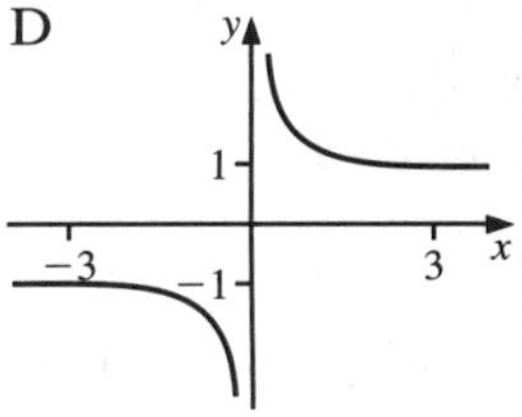 *(continued)*

3 Solve linear equations.	**3** Solve: a) $3x - 2 = 7$ b) $3(2x - 1) = 4 - 2(1 + 2x)$
4 Solve quadratic equations.	**4** Solve: a) $2x^2 + 5x - 3 = 0$ b) $x^2 - 3x - 7 = 0$
5 Manipulate algebraic formulae.	**5** Make x the subject of these formulae. a) $y = \dfrac{3}{2x - 1}$ b) $y = \sqrt{9 - 4x^2}$

C3

1.1 Domain and range of a function

The equation of a curve can be written in the form

$y =$ 'some expression in x'

or, alternatively, using **functional notation**, as:

$y = \text{f}(x)$

Remember:
f(x) means 'a function of x'.

For example, you could write $y = x^2$ as $\text{f}(x) = x^2$. To evaluate the function when $x = 3$, say, write

$$\text{f}(3) = 3^2$$
$$\therefore \quad \text{f}(3) = 9$$

You can say that 9 is the image of 3 under the function f. This is the same as saying when $x = 3$, $y = 9$.

In general:

- f(x) is called the **image** of x.
- The set of permitted x values is called the **domain** of the function.
- The set of all images is called the **range** of the function.

When a function is defined for all real values, you can write the domain of f as

$\{x : x \in \mathbb{R}\}$

$\{x : x \in \mathbb{R}\}$ means 'the set of x such that x belongs to the set of real numbers. Put more simply, 'x is a real number.'

or simply $x \in \mathbb{R}$, where $\mathbb{R}$ is the set of all real numbers.

If a function f is defined for all real values except one particular value, say c, then you write the domain of f as $x \in \mathbb{R}, x \neq c$.

Example 1

The function f is defined by $f(x) = x + \frac{3}{x}$, for $x \geqslant 2$.

a) Evaluate f(2). b) Find the value of x for which $f(x) = 4$.

a) $f(2) = 2 + \frac{3}{2} = \frac{7}{2}$

b) If $f(x) = 4$, then

$$x + \frac{3}{x} = 4$$

$$\therefore \quad x^2 - 4x + 3 = 0$$

$$\therefore \quad (x-1)(x-3) = 0$$

Solving gives $x = 1$ and $x = 3$.

Since the domain of f is $\{x : x \geqslant 2\}$ the only value of x that is required is $x = 3$.

Range of a function

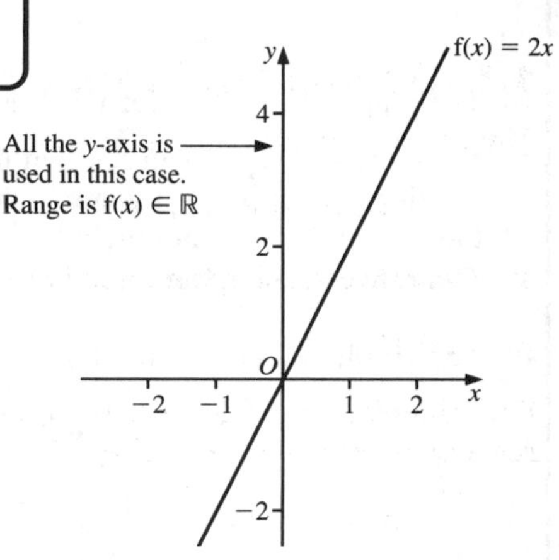

To identify the range of a function, it is very useful to have the graph of the function. For example, if the function $f(x) = 2x$ is defined for all real values of x then the graph of f is as shown. The range of the function is the set of all images of the function. (In other words, 'that part of the y-axis which is used up by the function'.) Therefore, the range of the function is the set of all real values, written as:

$$\{f(x) : f(x) \in \mathbb{R}\}$$

or simply $f(x) \in \mathbb{R}$.

A function may be defined on a restricted domain, that is, only a limited set of x values. For example, consider the function f defined by

$$f(x) = 2x, \quad -1 < x < 4$$

The graph of f is as shown, and the range of the function is the set of real values from -2 to 8, excluding -2 and 8, since -1 and 4 are excluded in the domain. Write this as $-2 < f(x) < 8$.

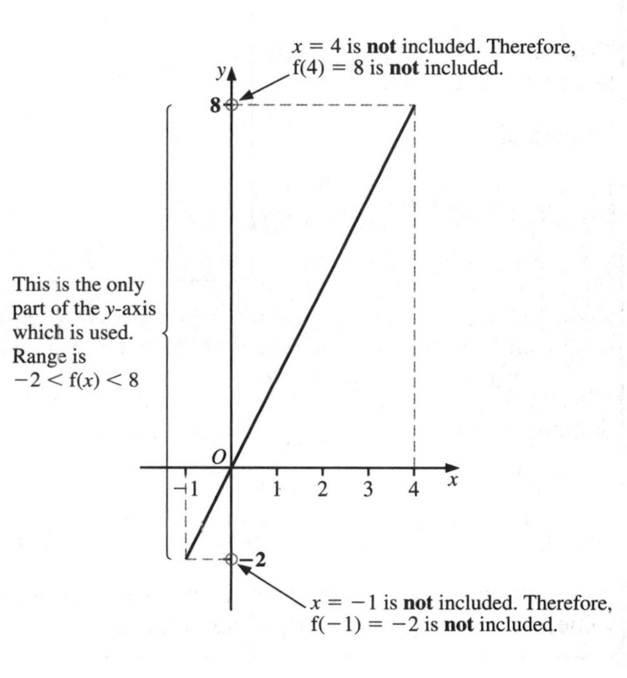

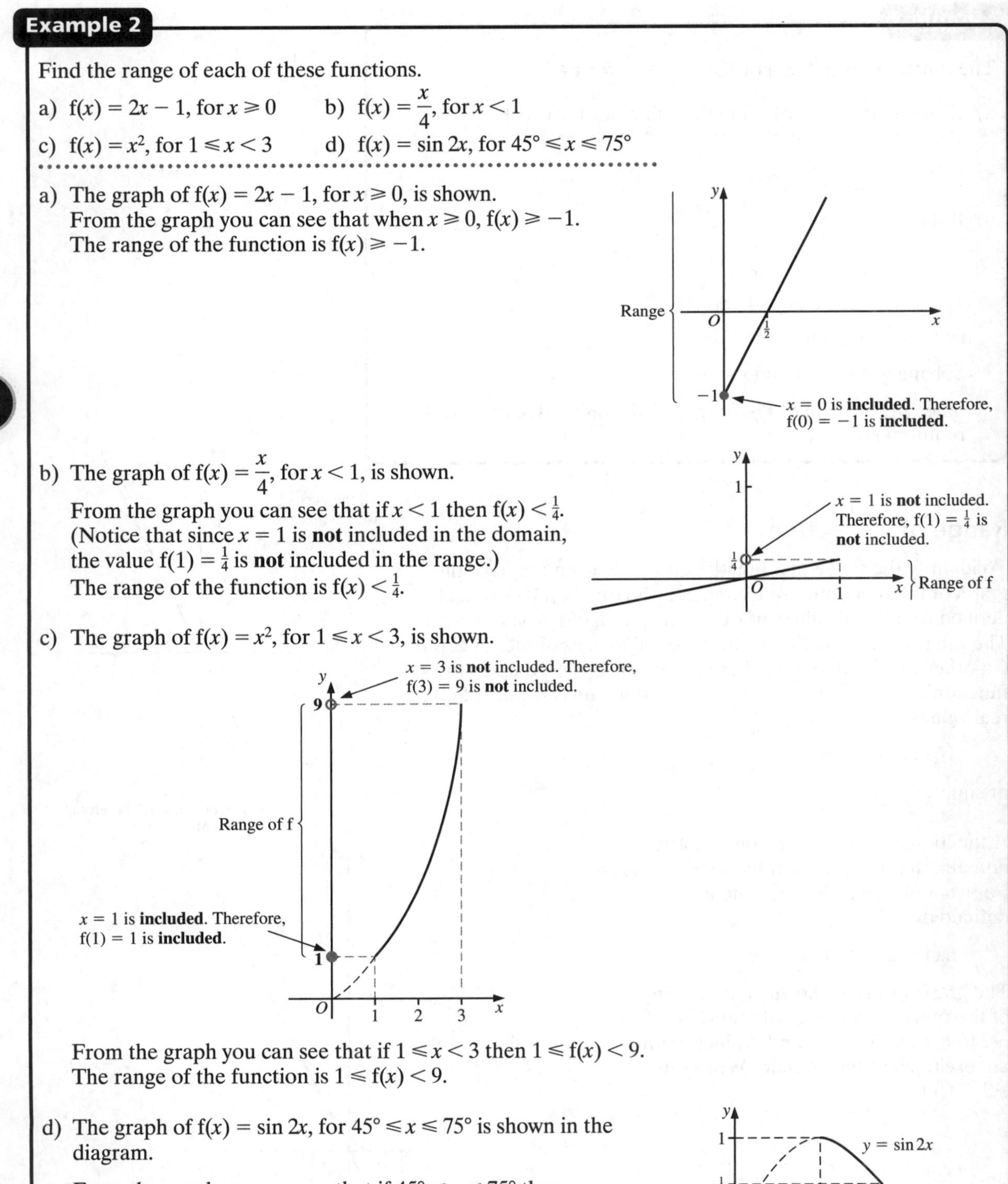

Example 2

Find the range of each of these functions.

a) $f(x) = 2x - 1$, for $x \geqslant 0$ b) $f(x) = \frac{x}{4}$, for $x < 1$

c) $f(x) = x^2$, for $1 \leqslant x < 3$ d) $f(x) = \sin 2x$, for $45° \leqslant x \leqslant 75°$

a) The graph of $f(x) = 2x - 1$, for $x \geqslant 0$, is shown.
From the graph you can see that when $x \geqslant 0$, $f(x) \geqslant -1$.
The range of the function is $f(x) \geqslant -1$.

b) The graph of $f(x) = \frac{x}{4}$, for $x < 1$, is shown.
From the graph you can see that if $x < 1$ then $f(x) < \frac{1}{4}$.
(Notice that since $x = 1$ is **not** included in the domain, the value $f(1) = \frac{1}{4}$ is **not** included in the range.)
The range of the function is $f(x) < \frac{1}{4}$.

c) The graph of $f(x) = x^2$, for $1 \leqslant x < 3$, is shown.

From the graph you can see that if $1 \leqslant x < 3$ then $1 \leqslant f(x) < 9$.
The range of the function is $1 \leqslant f(x) < 9$.

d) The graph of $f(x) = \sin 2x$, for $45° \leqslant x \leqslant 75°$ is shown in the diagram.

From the graph you can see that if $45° \leqslant x \leqslant 75°$ then

$\frac{1}{2} \leqslant f(x) \leqslant 1$

The range of the function is $\frac{1}{2} \leqslant f(x) \leqslant 1$.

Exercise 1A

1 The function f(x) is defined by $f(x) = \frac{4}{x-1}$ for $x > 1$.

a) Evaluate f(3).

b) Find the value of x for which $f(x) = 8$.

2 The function f(x) is defined by $f(x) = 2x - \frac{4}{x}$ for $x > 0$.

a) Evaluate f(5).

b) Find the value of x for which $f(x) = 7$.

3 The function f(x) is defined by $f(x) = \frac{x^2}{x+3}$ for $x > 2$.

a) Find the value of x for which $f(x) = 4$.

b) Show that there are no values of x for which $f(x) = \frac{1}{4}$.

C3

4 The function f(x) is defined by $f(x) = \cos x$ for $-\pi < x < \pi$.

a) Evaluate $f(\frac{\pi}{3})$.

b) Find the two values of x for which $f(x) = 0$.

5 Determine the range of each of these functions.

a) $f(x) = x + 4,\ x \in \mathbb{R},\ 0 < x < 5$

b) $f(x) = x^2 + 7,\ x \in \mathbb{R}$

c) $f(x) = 2x - 3,\ x \in \mathbb{R},\ 2 < x \leqslant 6$

d) $f(x) = \frac{1}{x^2 + 2},\ x \in \mathbb{R},\ 1 \leqslant x \leqslant 4$

e) $f(x) = (x^2 + 3)^2,\ x \in \mathbb{R}$

f) $f(x) = 5x^3 - 1,\ x \in \mathbb{R},\ 1 < x < 3$

g) $f(x) = x^2 - 6x,\ x \in \mathbb{R},\ 0 \leqslant x \leqslant 6$

h) $f(x) = \frac{1}{x + 1},\ x \in \mathbb{R},\ 1 \leqslant x < 9$

i) $f(x) = 3\sqrt{x} - 4,\ x \in \mathbb{R},\ 0 < x < \infty$

j) $f(x) = \sqrt{3x - 2},\ x \in \mathbb{R},\ 2 \leqslant x \leqslant 9$

k) $f(x) = x^4 + x^2,\ x \in \mathbb{R},\ 0 < x \leqslant 2$

l) $f(x) = \frac{1}{3 + x^4},\ x \in \mathbb{R}$

6 Determine the range of each of these functions.

a) $f(x) = \cos 2x,\ x \in \mathbb{R},\ 0 < x < \frac{\pi}{6}$

b) $f(x) = \tan x,\ x \in \mathbb{R},\ -\frac{\pi}{4} \leqslant x < \frac{\pi}{4}$

c) $f(x) = \sin\left(\frac{x}{2}\right),\ x \in \mathbb{R},\ \frac{\pi}{3} \leqslant x \leqslant \pi$

d) $f(x) = 2 + \sin x,\ x \in \mathbb{R},\ 0 < x \leqslant \frac{\pi}{6}$

e) $f(x) = 3 - \cos x,\ x \in \mathbb{R},\ \frac{\pi}{2} \leqslant x \leqslant \pi$

f) $f(x) = 4 \tan 3x,\ x \in \mathbb{R},\ \frac{\pi}{12} < x < \frac{\pi}{6}$

1.2 Mappings

This section will show you a different way of looking at functions.

Consider two non-empty sets A and B. A **mapping** from A to B is a rule which associates with each element of A an element of B.

You can represent a mapping by a mapping diagram. Suppose the set $A = \{-2, -1, 0, 1, 2\}$ is mapped to the set $B = \{0, 1, 2, 3, 4, 5, 6\}$.

Case 1

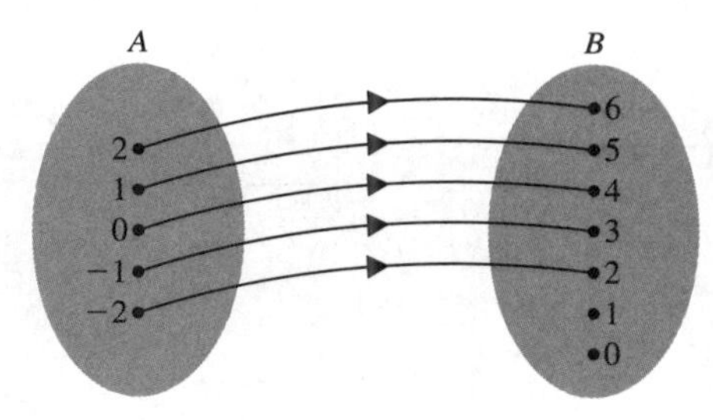

Note: You will not be examined on the use of mapping diagrams. They are included for clarity only.

In case 1, each element of A maps to one and only one element of B. This is called a **one-to-one mapping**. It doesn't matter that no element of A maps to either of the elements 0 or 1 in B.

Case 2

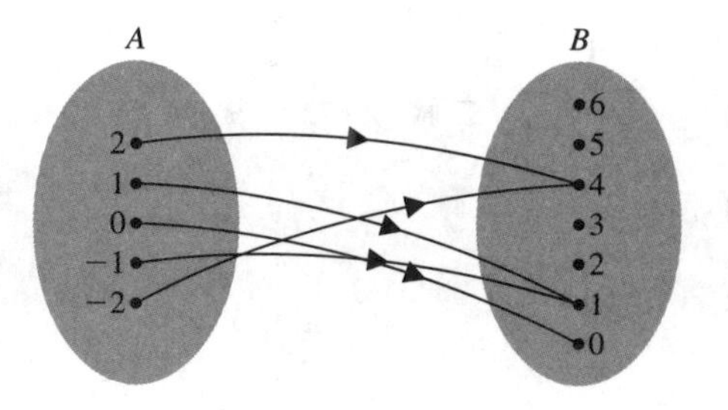

In case 2, two elements of A map to one element of B. This is called a **two-to-one-mapping**, or a **many-to-one mapping**.

A one-to-one mapping or a many-to-one mapping is called a **function**. You can denote the rule which associates each element of A to an element B by f.

As long as a mapping is one-to-one or many-to-one, it is just another way of representing a function.

For example, in case 1 the rule is 'add 4'. Using 'functional notation', you would write

$$f(x) = x + 4$$

In case 2 the rule is

$$f(x) = x^2$$

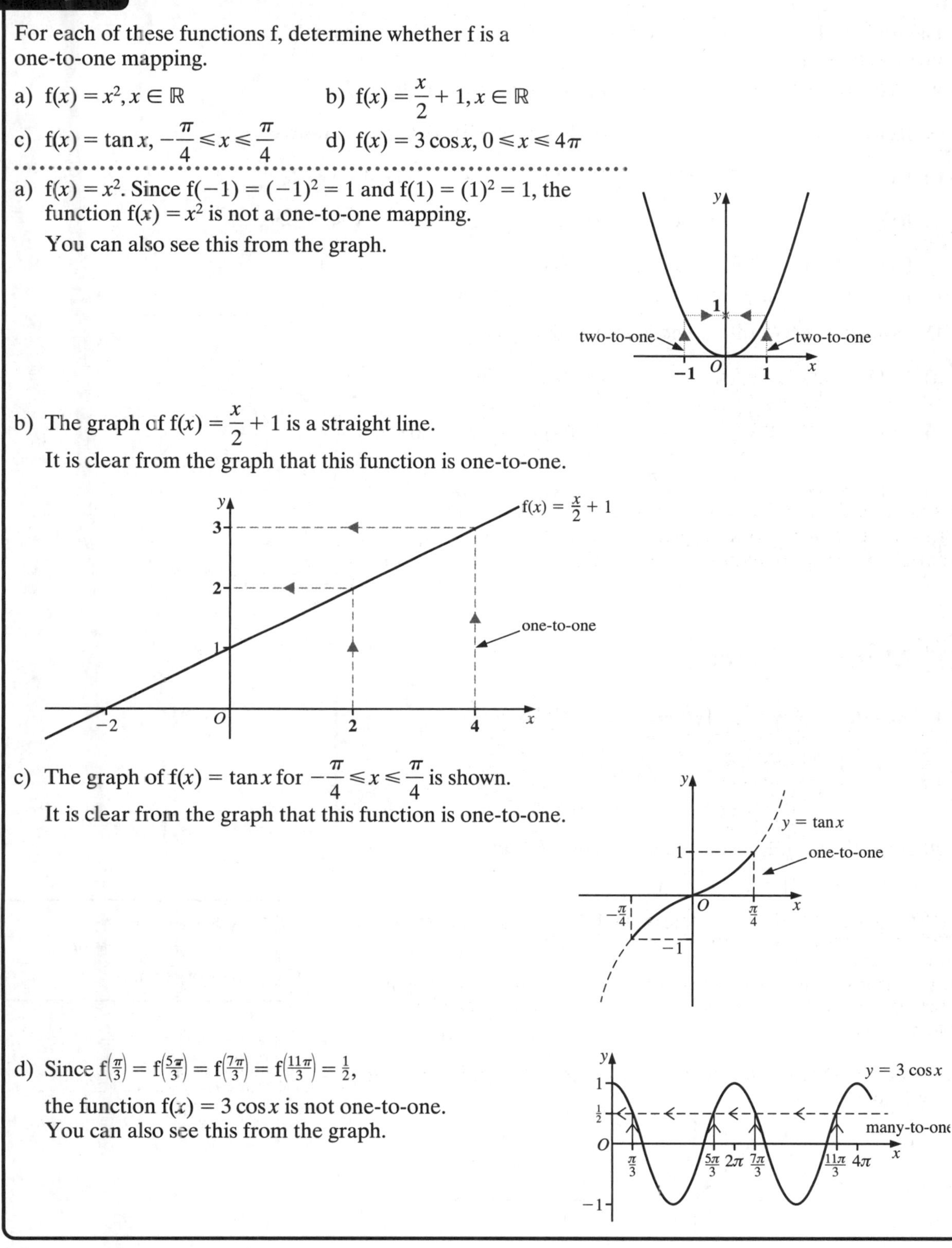

Example 3

For each of these functions f, determine whether f is a one-to-one mapping.

a) $f(x) = x^2, x \in \mathbb{R}$

b) $f(x) = \dfrac{x}{2} + 1, x \in \mathbb{R}$

c) $f(x) = \tan x, -\dfrac{\pi}{4} \leqslant x \leqslant \dfrac{\pi}{4}$

d) $f(x) = 3\cos x, 0 \leqslant x \leqslant 4\pi$

a) $f(x) = x^2$. Since $f(-1) = (-1)^2 = 1$ and $f(1) = (1)^2 = 1$, the function $f(x) = x^2$ is not a one-to-one mapping.
You can also see this from the graph.

b) The graph of $f(x) = \dfrac{x}{2} + 1$ is a straight line.
It is clear from the graph that this function is one-to-one.

c) The graph of $f(x) = \tan x$ for $-\dfrac{\pi}{4} \leqslant x \leqslant \dfrac{\pi}{4}$ is shown.
It is clear from the graph that this function is one-to-one.

d) Since $f\left(\frac{\pi}{3}\right) = f\left(\frac{5\pi}{3}\right) = f\left(\frac{7\pi}{3}\right) = f\left(\frac{11\pi}{3}\right) = \frac{1}{2}$,
the function $f(x) = 3\cos x$ is not one-to-one.
You can also see this from the graph.

C3

Exercise 1B

1 Determine which of these functions are one-to-one and which are not one-to-one.

a) $f(x) = x + 3,\ x \in \mathbb{R}$

b) $f(x) = x^2 + 3,\ x \in \mathbb{R}$

c) $f(x) = \frac{1}{x},\ x \in \mathbb{R}, x \neq 0$

d) $f(x) = (x - 4)^2,\ x \in \mathbb{R}, 2 \leqslant x \leqslant 6$

e) $f(x) = x^2 - 4x,\ x \in \mathbb{R}, 0 < x < 4$

f) $f(x) = x^2 - 4x,\ x \in \mathbb{R}, 0 < x < 2$

g) $f(x) = x^4 - 3,\ x \in \mathbb{R}, 3 \leqslant x \leqslant 6$

h) $f(x) = \frac{2}{x - 3},\ x \in \mathbb{R},\ -1 < x < 2$

i) $f(x) = x^3 - x^2,\ x \in \mathbb{R}, 0 \leqslant x \leqslant 1$

j) $f(x) = x^6,\ x \in \mathbb{R}, -2 < x < 0$

k) $f(x) = x^6,\ x \in \mathbb{R}, -2 < x < 2$

l) $f(x) = (x^4 + 1)^2 - 3,\ x \in \mathbb{R}$

2 Determine which of these functions are one-to-one.

a) $f(x) = 2\sin x,\ 0 < x < \pi$

b) $f(x) = 2\sin x,\ \frac{\pi}{2} < x < \frac{3\pi}{2}$

c) $f(x) = \tan 3x,\ -\frac{\pi}{2} < x < \frac{\pi}{2}$

d) $f(x) = 2 + \cos x,\ 0 < x < \frac{\pi}{3}$

e) $f(x) = 3 - \sin x,\ -\frac{\pi}{2} < x < \frac{\pi}{2}$

f) $f(x) = 1 + \tan 4x,\ 0 < x < \frac{\pi}{8}$

g) $f(x) = 1 + 2\cos x,\ 0 < x < 4\pi$

h) $f(x) = 6 + 5\sin 3x,\ \frac{\pi}{6} < x < \frac{\pi}{3}$

1.3 Modulus function

The **modulus** of x, written $|x|$, is defined as

$$|x| = \begin{cases} x & \text{for } x \geqslant 0 \\ -x & \text{for } x < 0 \end{cases}$$

In other words, $|x|$ means the magnitude of x. For example,

$$|-2| = 2 \quad |2| = 2 \quad \text{and} \quad |-\tfrac{1}{2}| = \tfrac{1}{2}$$

The modulus function is sometimes called the **absolute value function**.

From the graph of $f(x) = |x|$ you can see that $|-1| = |1| = 1$. In other words, the mapping $x \to |x|$ is not a one-to-one mapping.

The modulus of x is always positive, regardless of whether x is positive or negative.

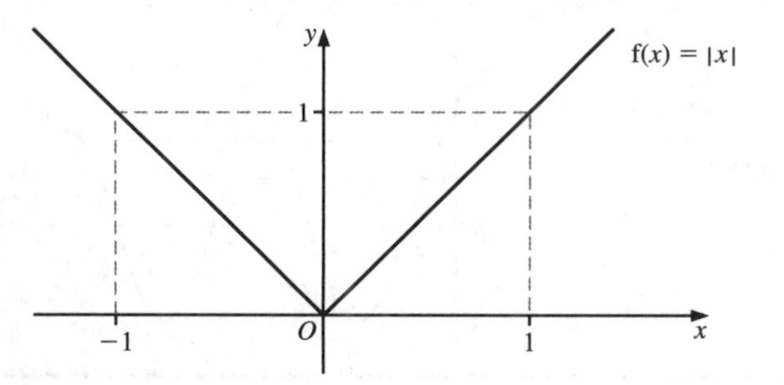

To get the graph of $f(x) = |x|$ from the graph of $f(x) = x$, reflect the part of the graph for which $f(x) < 0$ in the x-axis, as shown.

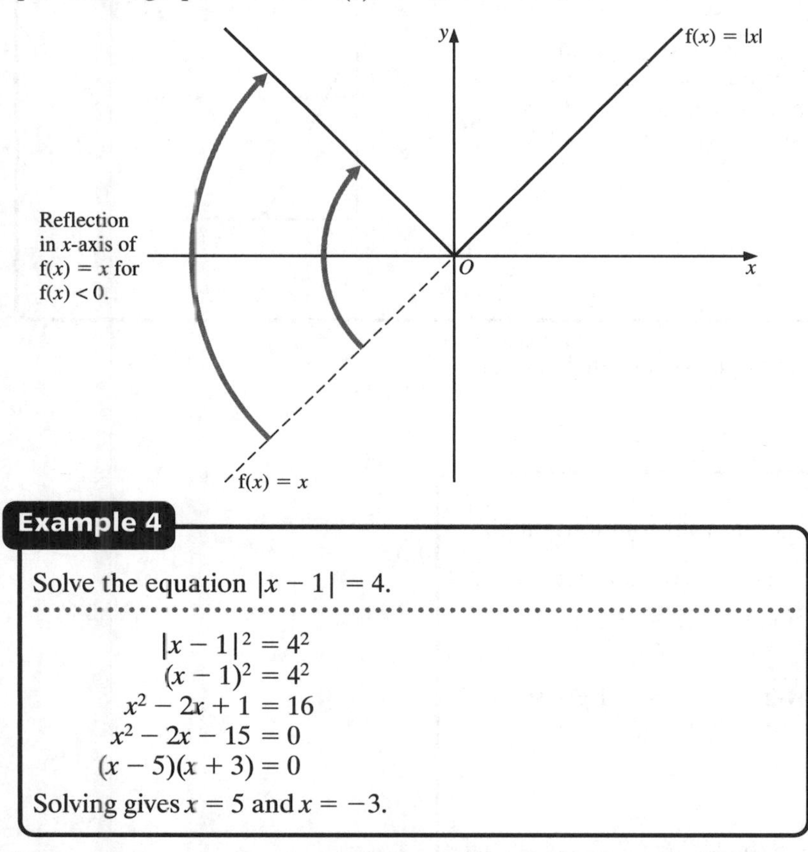

Example 4

Solve the equation $|x - 1| = 4$.

$$|x - 1|^2 = 4^2$$
$$(x - 1)^2 = 4^2$$
$$x^2 - 2x + 1 = 16$$
$$x^2 - 2x - 15 = 0$$
$$(x - 5)(x + 3) = 0$$

Solving gives $x = 5$ and $x = -3$.

Square both sides to ensure that the LHS is positive.

You will often be asked to sketch a graph of a modulus function and use it to solve an equation.

Example 5

Sketch the graph of $f(x) = |2x - 3|$ and hence solve the equation $|2x - 3| = 2$.

The graph of $f(x) = 2x - 3$ is shown.

Reflect the part of the graph for which $f(x) < 0$ in the x-axis.

This gives you the graph of $f(x) = |2x - 3|$.

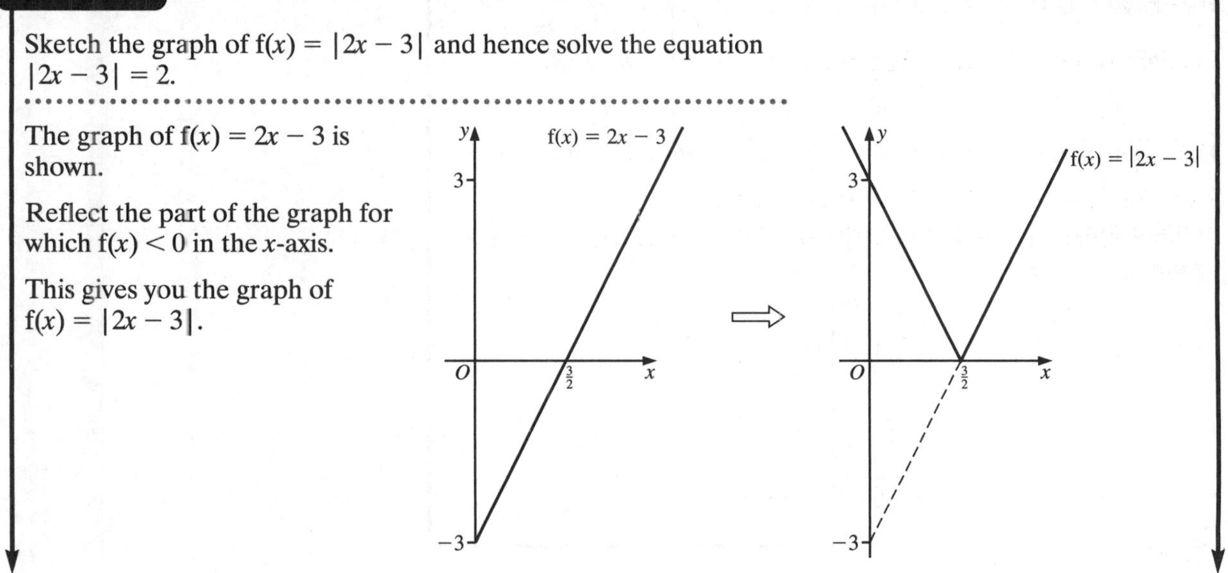

To solve the equation $|2x - 3| = 2$, draw the line $f(x) = 2$ on the graph of $|2x - 3|$, as shown.

The x-coordinates of the intersection points of $f(x) = 2$ and $f(x) = |2x - 3|$ give the solutions of the equation $|2x - 3| = 2$. The x-coordinates of the intersection points are $x = \frac{1}{2}$ and $x = \frac{5}{2}$. Therefore, the solutions are $x = \frac{1}{2}$ and $x = \frac{5}{2}$.

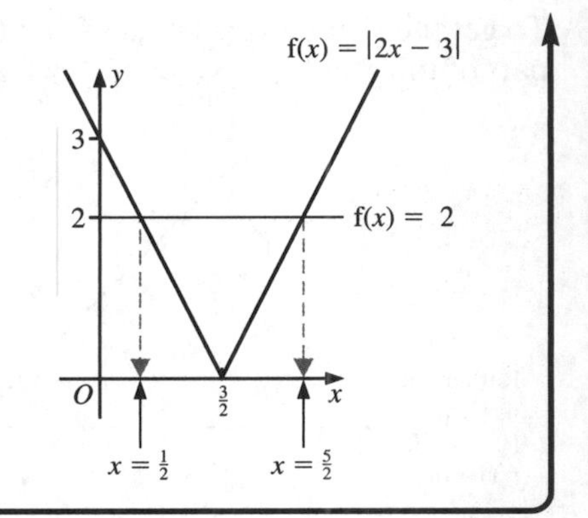

Example 6 illustrates an alternative technique for solving a modulus equation.

C3

Example 6

Solve the equation $|9 - 2x^2| = 1$. Illustrate your solutions on a graph.

Since $|9 - 2x^2| = 1$ you know that either $9 - 2x^2 = +1$ or $9 - 2x^2 = -1$.

Solving $9 - 2x^2 = +1$ gives:

$$2x^2 = 8$$
$$x^2 = 4$$
$$\therefore \quad x = \pm 2$$

Solving $9 - 2x^2 = -1$ gives:

$$2x^2 = 10$$
$$x^2 = 5$$
$$\therefore \quad x = \pm\sqrt{5}$$

The graph of $f(x) = 9 - 2x^2$ is shown. Reflect in the x-axis the part of the graph for which $f(x) < 0$, to get the graph of $f(x) = |9 - 2x^2|$.

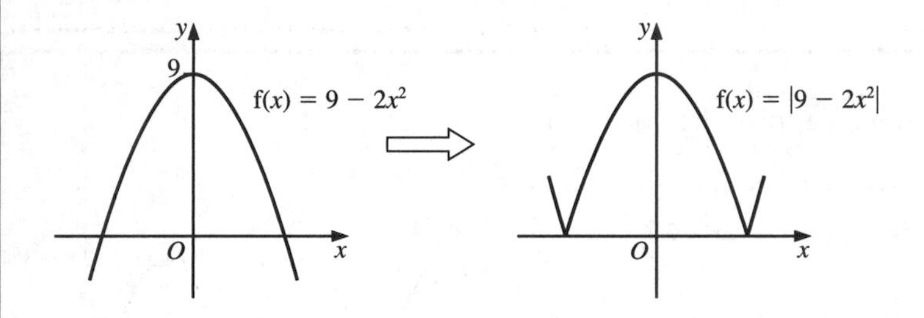

To solve the equation $|9 - 2x^2| = 1$, draw the line $f(x) = 1$ on the graph of $|9 - 2x^2|$.

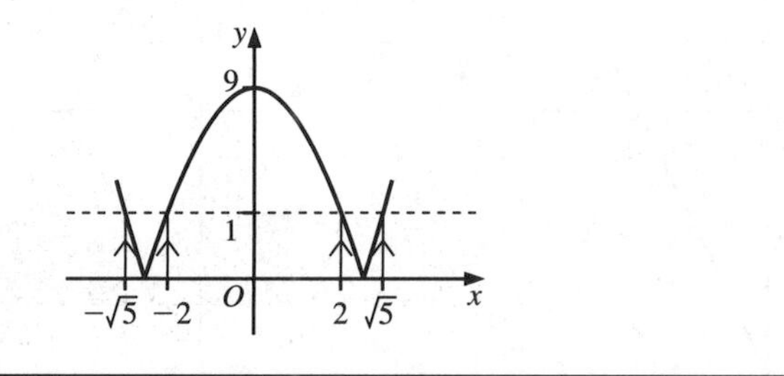

Inequalities

You can also use the graph of a modulus function to solve an inequality.

Example 7

Sketch the graph of $f(x) = |3x + 1|$ and hence solve the inequality $|3x + 1| \leqslant 2$.

The graph of $f(x) = |3x + 1|$ is shown.

Draw the line $f(x) = 2$ on the same set of axes.

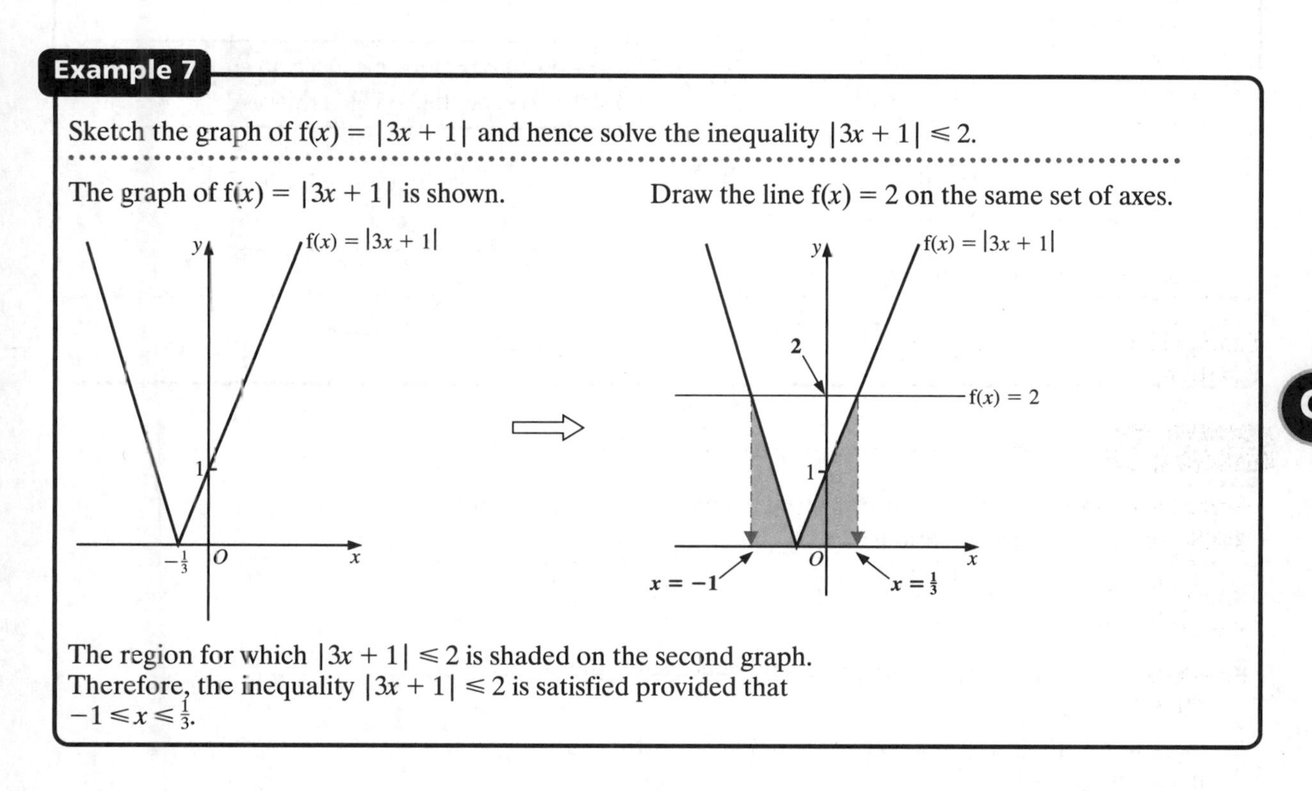

The region for which $|3x + 1| \leqslant 2$ is shaded on the second graph. Therefore, the inequality $|3x + 1| \leqslant 2$ is satisfied provided that $-1 \leqslant x \leqslant \frac{1}{3}$.

Modulus functions may also apply to trigonometric functions.

Example 8

Sketch the graph of $f(x) = |\sin 2x|$ for $0 \leqslant x \leqslant 360°$ and hence solve the inequality $|\sin 2x| \geqslant \frac{1}{2}$.

The graph of $f(x) = |\sin 2x|$ is:

Draw the line $f(x) = \frac{1}{2}$ on the same set of axes:

The region for which $|\sin 2x| \geqslant \frac{1}{2}$ is shaded on the graph. Therefore, the inequality $|\sin 2x| \geqslant \frac{1}{2}$ is satisfied for $15° \leqslant x \leqslant 75°$, $105° \leqslant x \leqslant 165°$, $195° \leqslant x \leqslant 255°$ and $285° \leqslant x \leqslant 345°$.

Example 9

Solve the inequality $|2x+1| \geqslant |x+3|$.

Square both sides of the inequality to ensure that both the LHS and the RHS are positive:

$$|2x+1|^2 \geqslant |x+3|^2$$
$$\therefore \quad (2x+1)^2 \geqslant (x+3)^2$$
$$4x^2+4x+1 \geqslant x^2+6x+9$$
$$\therefore \quad 3x^2-2x-8 \geqslant 0$$
$$\therefore \quad (3x+4)(x-2) \geqslant 0$$

You can show this on a number line.

The solution regions are $x \leqslant -\frac{4}{3}$ and $x \geqslant 2$.

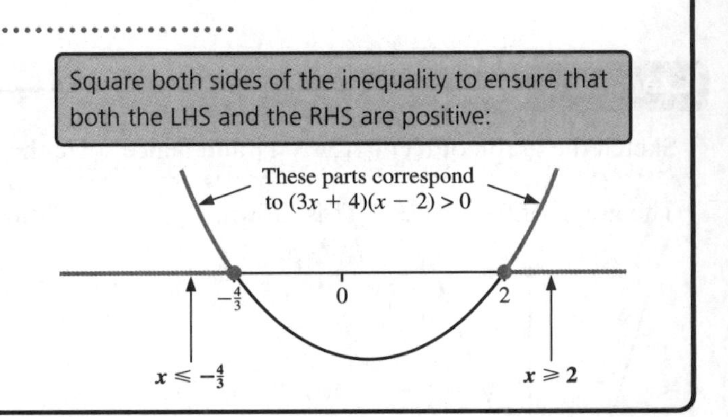

C3

Exercise 1C

1 Solve each of these equations.

a) $|x-2|=4$ b) $|x+4|=5$ c) $|3-x|=6$
d) $|4-x|=2$ e) $|3x+1|=4$ f) $|5x-3|=7$

2 Sketch the graph of $f(x)=|3x-2|$, and hence solve the equation $|3x-2|=5$.

3 Sketch the graph of $f(x)=|2x-1|$, and hence solve the equation $|2x-1|=3$.

4 By first sketching the graph of $y=|\frac{1}{2}x-3|$, find the solutions to the equation $|\frac{1}{2}x-3|=2$.

5 Solve each of these equations.

a) $|x^2+1|=5$ b) $|x^2-5|=4$ c) $|2x^2-7|=1$
d) $|12-x^2|=3$ e) $|9-4x^2|=5$ f) $|14-3x^2|=2$

6 Sketch the graph of $f(x)=|\cos x|$, for $0 \leqslant x \leqslant 360°$, and hence solve the equation $|\cos x|=\frac{1}{2}$.

7 Sketch the graph of $f(x)=|2\sin x|$, for $0 \leqslant x \leqslant 360°$, and hence solve the equation $|2\sin x|=1$.

8 Sketch the graph of $f(x)=|\tan 2x|$, for $-90° \leqslant x \leqslant 90°$, and hence solve the equation $|\tan 2x|=1$.

9 Sketch the graph of $f(x)=|4\cos 3x|$, for $0 \leqslant x \leqslant 180°$, and hence solve the equation $|4\cos 3x|=2$.

10 Solve each of these equations.

a) $|x+1|=|x-3|$ b) $|x-4|=|6-x|$
c) $|2x-1|=|x|$ d) $|3x+1|=|x+4|$
e) $|2x-5|=|2-3x|$ f) $|2x-1|=|4x+3|$

11 On the same set of axes sketch the graphs of the functions $f(x) = |x - 2|$, and $g(x) = |x - 6|$. Hence solve the equation $|x - 2| = |x - 6|$.

12 Sketch the graph of $y = |2x + 5|$, and hence solve the inequality $|2x + 5| < 7$.

13 Use a graph to solve the inequality $|8x - 3| > 9$.

14 By first sketching the graph of the function $f(x) = |\frac{1}{4}x + 3|$, find the solution to the inequality $|\frac{1}{4}x + 3| \geqslant 3$.

15 On one set of axes sketch the graphs of the functions $f(x) = |x| - 4$, and $g(x) = \frac{1}{2}x$. Hence solve the inequality $|x| - 4 \leqslant \frac{1}{2}x$

16 Solve each of these inequalities.

a) $|x + 1| > |x - 3|$
b) $|2x + 3| \leqslant |2x - 1|$
c) $|x + 4| \geqslant |2x - 3|$
d) $|x - 5| \leqslant |3x + 2|$
e) $|x - 1| > |2x + 5|$
f) $|3 - x| < |x - 4|$

17 Sketch the graph of $f(x) = |12 - x^2|$, and hence solve the inequality $|12 - x^2| < 8$.

18 Sketch the graph of $f(x) = |2x^2 - 3|$, and hence solve the inequality $|2x^2 - 3| \geqslant \frac{7}{2}$.

19 Sketch the graph of $f(x) = |\cos x|$, for $0 \leqslant x \leqslant 360°$ and hence solve the inequality $|\cos x| \geqslant \frac{\sqrt{3}}{2}$.

20 Sketch the graph of $f(x) = |\tan 2x|$, for $-180° \leqslant x \leqslant 180°$ and hence solve the inequality $|\tan 2x| < 1$.

C3

1.4 Composite functions

Consider the two functions $f(x) = x - 3$ and $g(x) = 2x + 5$, where the domain of g is {1, 2, 3, 4} and the domain of f is the range of g. You can show this on a mapping diagram.

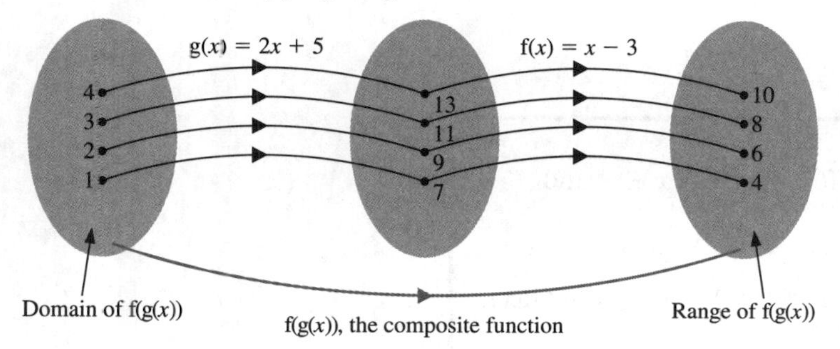

The function indicated on the diagram with domain {1, 2, 3, 4} and range {4, 6, 8, 10} is called the **composite function**. This function is written as $f(g(x))$ or $fg(x)$.

You can work out a single 'rule' for the composite function fg(x) in terms of x.

f(x) = $x - 3$
g(x) = $2x + 5$

Notice that g is written nearest to the variable x since g is the first function to operate on the set {1, 2, 3, 4}. The rule for the composite function is given by

$$\begin{aligned} fg(x) &= f(2x+5) \\ fg(x) &= (2x+5) - 3 \\ &= 2x + 2 \end{aligned}$$

The composite function is defined by fg(x) = $2x + 2$ with domain {1, 2, 3, 4} and range {4, 6, 8, 10}.

If the function f were to operate first on the set {1, 2, 3, 4} and then g were to operate on the range of f, the mapping diagram would be:

The range of f is the domain of g.

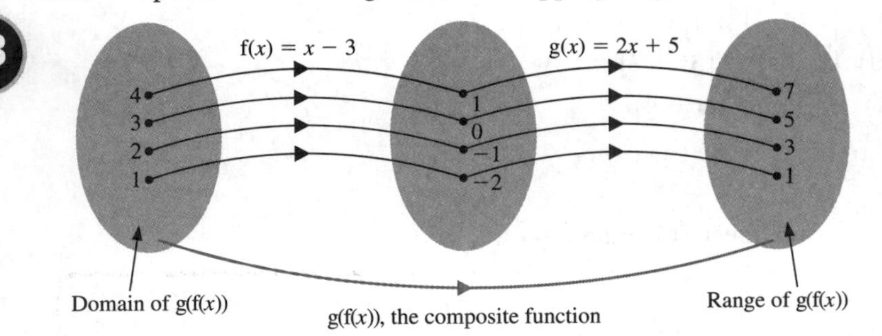

In this case, the composite function which has domain {1, 2, 3, 4} and range {1, 3, 5, 7} is written g(f(x)) or gf(x).

The rule for gf(x) is given by

$$\begin{aligned} gf(x) &= g(x-3) \\ &= 2(x-3) + 5 \\ &= 2x - 6 + 5 \\ \therefore\ gf(x) &= 2x - 1 \end{aligned}$$

The composite function is defined by gf(x) = $2x - 1$ with domain {1, 2, 3, 4} and range {1, 3, 5, 7}.

Example 10

The functions f and g are defined by $f(x) = 3x - 5, x \in \mathbb{R}$ and $g(x) = 3 - 2x, x \in \mathbb{R}$.

a) Evaluate i) f(2), ii) fg(3).

b) The composite function h is defined by h(x) = gf(x). Find h(x).

a) i) Since $f(x) = 3x - 5$,

$$f(2) = 3(2) - 5$$

$$\therefore\ f(2) = 1$$

ii) To find fg(3), first evaluate g(3). Since $g(x) = 3 - 2x$,

$$g(3) = 3 - 2(3)$$
$$\therefore \quad g(3) = -3$$

Therefore,

$$fg(3) = f(-3)$$
$$= 3(-3) - 5$$
$$\therefore \quad fg(3) = -14$$

b) You are given that $h(x) = gf(x)$. Therefore

$$h(x) = g(3x - 5)$$
$$= 3 - 2(3x - 5)$$
$$= 3 - 6x + 10$$
$$\therefore \quad h(x) = 13 - 6x$$

C3

Example 11

The functions f and g are defined by

$$f(x) = x^2, \qquad 0 \leqslant x \leqslant 4$$

and $g(x) = x + 3, \qquad x \in \mathbb{R}$

Find the composite function gf(x) and state the range of this function.

$$gf(x) = g(x^2)$$
$$= x^2 + 3$$
$$\therefore \quad gf(x) = x^2 + 3$$

Since f is the first function to operate in the composite function gf, you need the range of f, as this will be the domain of g.

Sketching the graph of $f(x) = x^2$, for $0 \leqslant x \leqslant 4$, gives the diagram shown.

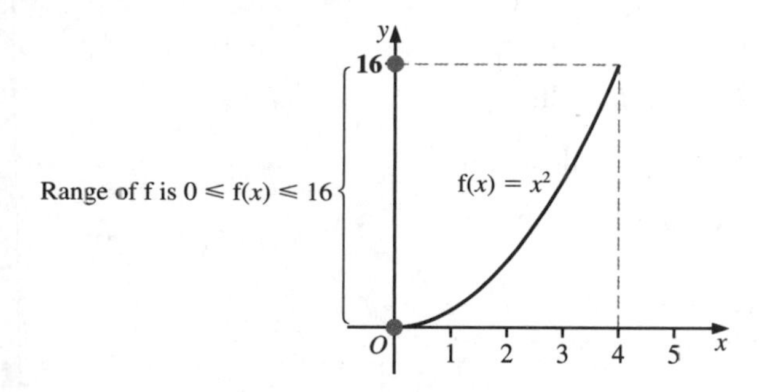

You can see from the graph of f that the range of f is $0 \leqslant f(x) \leqslant 16$. Therefore, the domain of g is $0 \leqslant x \leqslant 16$, giving the graph of g as shown.

From the graph of g you can see that the range of g (when its domain is $0 \leqslant x \leqslant 16$) is $3 \leqslant x \leqslant 19$. Therefore, the composite function gf has range $3 \leqslant \mathrm{gf}(x) \leqslant 19$.

Alternatively, since you know that $\mathrm{gf}(x) = x^2 + 3$ and it has domain $0 \leqslant x \leqslant 4$, you can sketch the graph of $\mathrm{gf}(x)$.

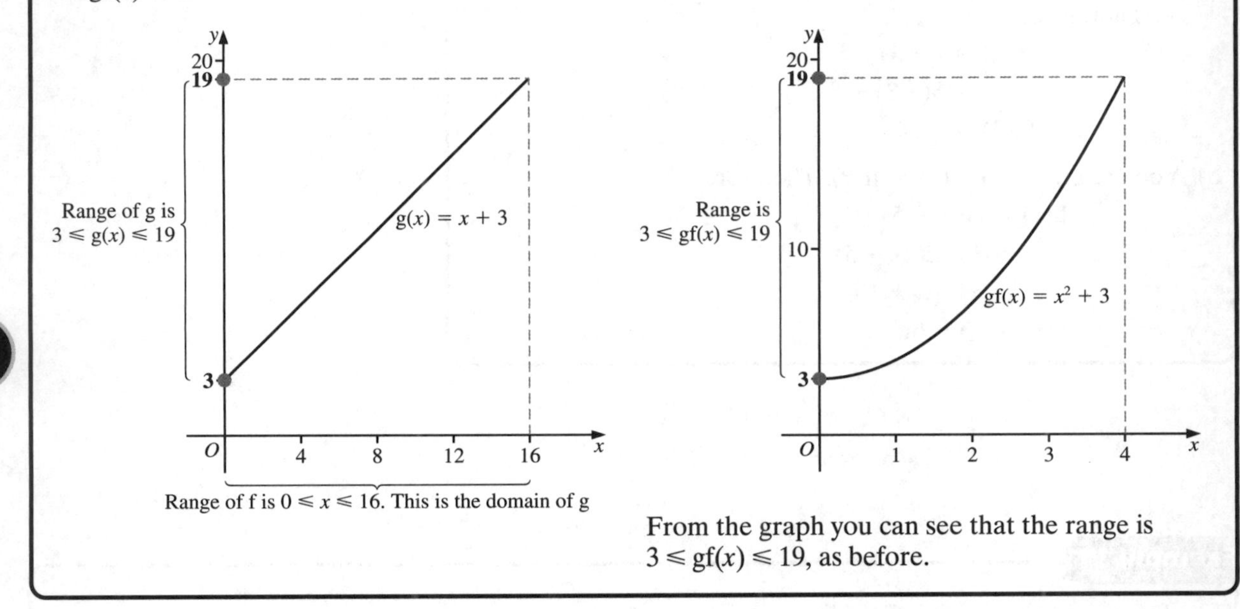

From the graph you can see that the range is $3 \leqslant \mathrm{gf}(x) \leqslant 19$, as before.

Note that sometimes the composite function is more complicated and therefore the alternative method shown in Example 11 is not quite as straightforward.

Exercise 1D

Throughout this exercise, the domain of each function is the set of real numbers unless specifically stated otherwise.

1 Given $\mathrm{f}(x) = 2x + 1$, $\mathrm{g}(x) = x^2$ and $\mathrm{h}(x) = \frac{1}{x}$ evaluate each of these functions.

a) f(3) b) g(2) c) hg(2) d) fg(−3)
e) gf(1) f) gh(−2) g) hf(4) h) ff(5)
i) gg(−3) j) hh(12) k) fgh(2) l) hfg(4)

2 Given $\mathrm{f}(x) = 3x - 1$, $\mathrm{g}(x) = x^2$ and $\mathrm{h}(x) = \frac{2}{x}$, write down and simplify expressions for each of these composite functions.

a) fg(x) b) gf(x) c) fh(x)
d) hg(x) e) gg(x) f) ff(x)

3 Functions f and g are defined by

$$\mathrm{f}(x) = x^2 + 3 \quad \mathrm{g}(x) = x + 5$$

a) Write down and simplify expressions for i) fg(x), ii) gf(x).
b) Hence solve the equation fg(x) = gf(x).

4 Functions f and g are defined by

$$f(x) = \frac{3}{x} \quad g(x) = x + 5$$

a) Write down an expression for fg(x), and hence solve the equation fg(x) = 1.

b) Write down an expression for gf(x), and hence solve the equation gf(x) = 6.

5 Given $f(x) = x^2$ and $g(x) = 2x + 5$, solve these equations.

a) $fg(x) = 9$ b) $gg(x) = 21$

6 Given

$$f(x) = x^2,\ x \in \mathbb{R},\ 1 \leqslant x \leqslant 5 \quad \text{and} \quad g(x) = 2x + 5,\ x \in \mathbb{R}$$

find an expression for the composite function gf(x). State the domain and range of gf(x).

7 Functions f and g are defined by

$$f(x) = 3x^2 + 1,\ x \in \mathbb{R},\ 0 \leqslant x \leqslant 2 \quad \text{and} \quad g(x) = x^2 - 2,\ x \in \mathbb{R}$$

Find the composite function gf(x) and state its range.

8 Given

$$f(x) = x^2 + 4,\ x \in \mathbb{R} \quad \text{and} \quad g(x) = \frac{1}{x - 3},\ x \in \mathbb{R},\ x \geqslant 4$$

find an expression for the composite function gf(x) and state its range.

9 Functions f and g are defined by

$$f(x) = x^2 + 3,\ x \in \mathbb{R} \quad \text{and} \quad g(x) = |x| - 5,\ x \in \mathbb{R}$$

a) Write down an expression for fg(x) and state its range.

b) Write down an expression for ff(x) and state its range.

10 Given

$$f(x) = \sqrt{x + 1},\ x \in \mathbb{R},\ x > 0 \quad \text{and} \quad g(x) = x^2,\ x \in \mathbb{R},$$

a) find an expression for fg(x) and state its range

b) find an expression for gf(x) and state its range.

1.5 Inverse functions

Consider the function f defined by $f(x) = x + 3$ with domain {1, 2, 3}. The range of f is {4, 5, 6}. We can define a function f^{-1}, called the **inverse function**, which has domain {4, 5, 6} and range {1, 2, 3} such that

$$f^{-1}(4) = 1 \quad f^{-1}(5) = 2 \quad \text{and} \quad f^{-1}(6) = 3$$

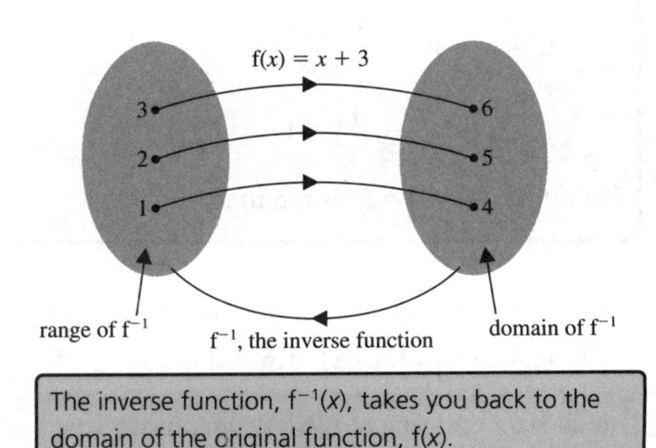

The inverse function, $f^{-1}(x)$, takes you back to the domain of the original function, f(x).

In this case, it is easy to see that the inverse function f^{-1}, is given by

$$f^{-1}(x) = x - 3$$

However, in some examples it is not quite so easy to identify a formula for f^{-1}. So you need a technique for finding such a formula. Consider the function $y = f(x)$ whose mapping diagram is shown.

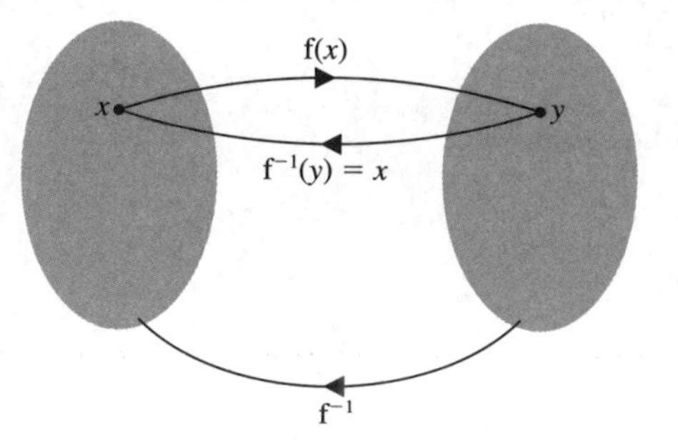

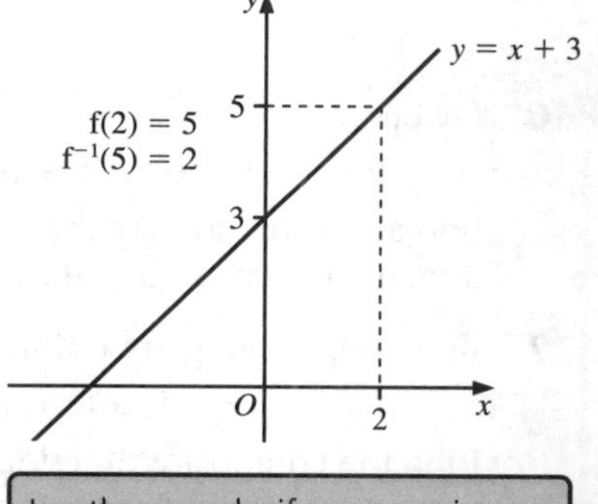

You want the function f^{-1} such that $f^{-1}(y) = x$. In other words, you require x to be expressed as a function of y. So a useful technique for finding the formula for an inverse function is to let $y = f(x)$ and rearrange for x.

In this example, let $y = x + 3$. Rearranging for x gives $x = y - 3$. Therefore, the inverse function is given by $f^{-1}(x) = x - 3$.

In other words, if you are given y, the corresponding x value can be found using $x = y - 3$.

If a function f has an inverse f^{-1}, then the composite ff^{-1} is given by $ff^{-1}(x) = x$, and similarly the composite function $f^{-1}f$ is given by $f^{-1}f(x) = x$.

C3

Example 12

The function f is defined by $f(x) = 5x + 4, x \in \mathbb{R}$. Find $f^{-1}(x)$ and verify that $ff^{-1}(x) = x$.

To find $f^{-1}(x)$, let $y = 5x + 4$. Then rearranging for x gives $x = \dfrac{y-4}{5}$.

Therefore, the inverse function is given by $f^{-1}(x) = \dfrac{x-4}{5}$.

The composite function $ff^{-1}(x)$ is given by

$$ff^{-1}(x) = f\left(\frac{x-4}{5}\right)$$
$$= 5\left(\frac{x-4}{5}\right) + 4 = x$$

Therefore, $ff^{-1}(x) = x$, as required.

You can also find an inverse function by constructing a function chart.

The function chart for $f(x) = 5x + 4$ is: $x \rightarrow \boxed{\times 5} \xrightarrow{5x} \boxed{+4} \rightarrow 5x + 4$

This shows x entering at the left side. In the first box x is multiplied by 5. In the second box 4 is added to that answer.

To find the inverse, start at the right side and invert each of the functions:

- the inverse of $+4$ is -4
- the inverse of $\times 5$ is $\div 5$

$\leftarrow \boxed{\div 5} \leftarrow \boxed{-4} \leftarrow$

Entering x at the right side gives the inverse function chart: $\frac{x-4}{5} \leftarrow \boxed{\div 5} \xleftarrow{x-4} \boxed{-4} \leftarrow x$

So $f^{-1}(x) = \dfrac{x-4}{5}$

Example 13

Two functions f and g are defined by

$$f(x) = 7x + 1, \ x \in \mathbb{R}$$

and $g(x) = \dfrac{x}{3} - 1, \ x \in \mathbb{R}$

Find the inverse functions f^{-1} and g^{-1} and verify that $(fg)^{-1} = g^{-1}f^{-1}$.

To find the inverse function of f, let $y = 7x + 1$. Then rearrange for x:

$$x = \frac{y-1}{7}$$

Therefore, $f^{-1}(x) = \dfrac{x-1}{7}$.

To find the inverse function of g, let $y = \dfrac{x}{3} - 1$. Then rearrange for x:

$$x = 3y + 3$$

Therefore, $g^{-1}(x) = 3x + 3$.

To show that $(fg)^{-1} = g^{-1}f^{-1}$, first look at the LHS. You need the composite function $fg(x)$, which is given by

$$\begin{aligned} fg(x) &= f\left(\frac{x}{3} - 1\right) \\ &= 7\left(\frac{x}{3} - 1\right) + 1 \\ &= \frac{7x}{3} - 7 + 1 \end{aligned}$$

$$\therefore \quad fg(x) = \frac{7x}{3} - 6$$

To find the inverse of $fg(x) = \dfrac{7x}{3} - 6$, let $y = \dfrac{7x}{3} - 6$. Then rearranging for x gives

$$x = \frac{3y + 18}{7}$$

Therefore,

$$fg^{-1}(x) = \frac{3x + 18}{7}$$

Next, look at the RHS. You need the composite function $g^{-1}(f^{-1}(x))$, which is given by

$$g^{-1}(f^{-1}(x)) = g^{-1}\left(\frac{x-1}{7}\right)$$
$$= 3\left(\frac{x-1}{7}\right) + 3$$
$$= \frac{3x-3}{7} + 3$$
$$= \frac{3x-3+21}{7}$$

Therefore $g^{-1}(f^{-1}(x)) = \frac{3x+18}{7} = (fg)^{-1}(x)$ as required.

C3

The function charts for $f(x) = 7x + 1$, $g(x) = \frac{x}{3} - 1$ and

$fg(x) = \frac{7x}{3} - 6$, along with their inverses, are:

$f(x) = 7x + 1$ $x \rightarrow$ [×7] $\xrightarrow{7x}$ [+1] $\rightarrow 7x + 1$

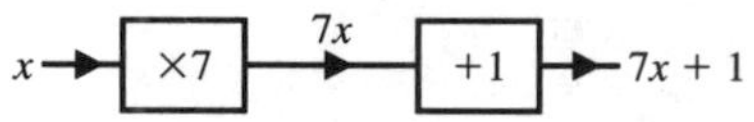

$f^{-1}(x) = \frac{x-1}{7}$ $\frac{x-1}{7} \leftarrow$ [÷7] $\xleftarrow{x-1}$ [−1] $\leftarrow x$

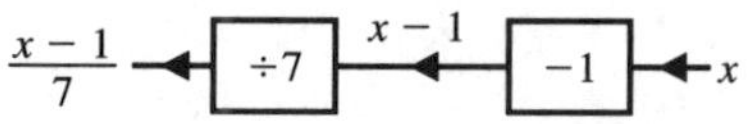

$g(x) = \frac{x}{3} - 1$ $x \rightarrow$ [÷3] $\xrightarrow{\frac{x}{3}}$ [−1] $\rightarrow \frac{x}{3} - 1$

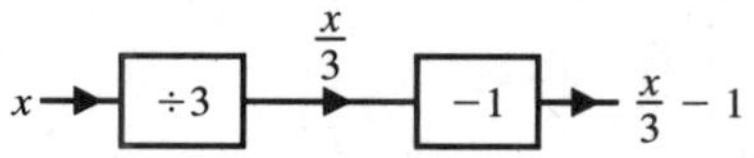

$g^{-1}(x) = 3(x + 1)$ $3(x+1) \leftarrow$ [×3] $\xleftarrow{x+1}$ [+1] $\leftarrow x$

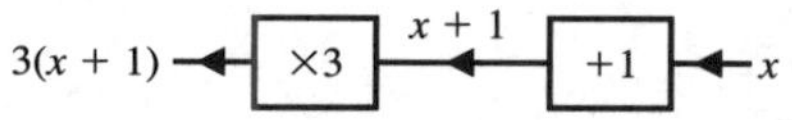

$fg(x) = \frac{7x}{3} - 6$ $x \rightarrow$ [×7] $\xrightarrow{7x}$ [÷3] $\xrightarrow{\frac{7x}{3}}$ [−6] $\rightarrow \frac{7x}{3} - 6$

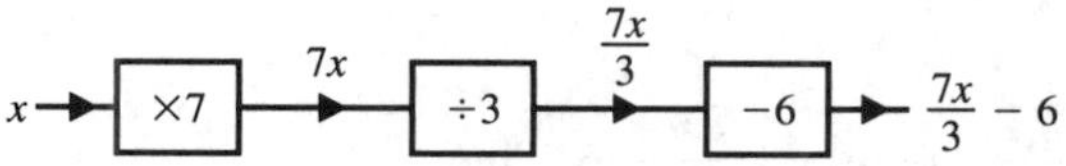

$fg^{-1}(x) = \frac{3(x+6)}{7}$ $\frac{3(x+6)}{7} \leftarrow$ [÷7] $\xleftarrow{3(x+6)}$ [×3] $\xleftarrow{x+6}$ [+6] $\leftarrow x$

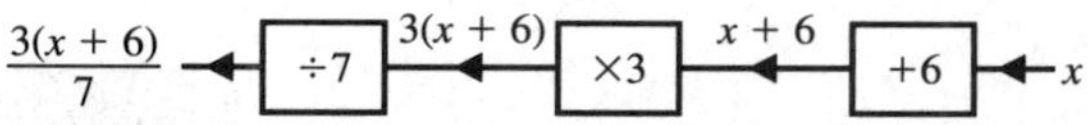

So far all the examples have involved one-to-one functions. This has been a good thing, as the next two diagrams show.

This function is one-to-one. Therefore for any value, b, on the y-axis there is a unique value of x such that $f(x) = b$.

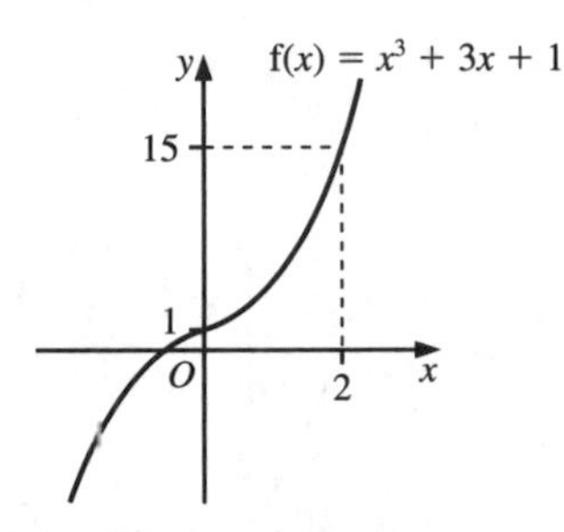

For example, $x = 2$ is the only solution to the equation $f(x) = 15$.

Here there are as many as three values of x such that $g(x) = b$.

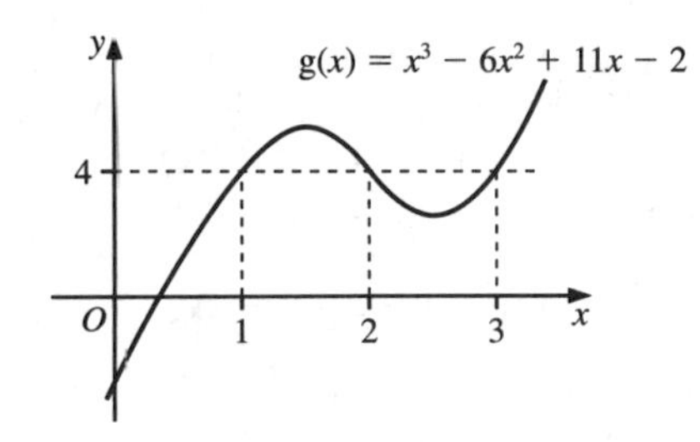

So $g(x)$ does not have an inverse. This is because $g(x)$ is not one-to-one.

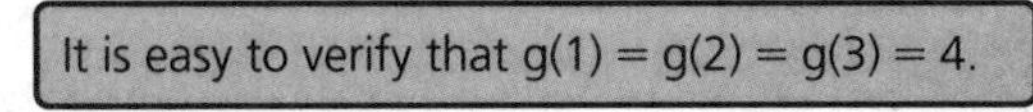
It is easy to verify that $g(1) = g(2) = g(3) = 4$.

For a graph to have an inverse there must always be a unique value in the domain which maps to a specific value in the range. This means that only graphs which are one-to-one can have inverses. However, there is a way round this.

You can find an inverse of a many-to-one function if you restrict the domain so that the new function *is* one-to-one. For example, the function $f(x) = x^2$ defined for all real x is a two-to-one function and has the graph shown.

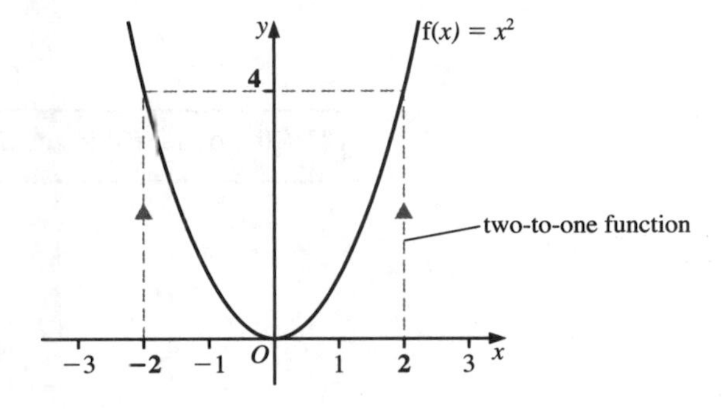

If you restrict the domain to $x \geqslant 0$, you now have a one-to-one function which *will* have an inverse. The inverse $f^{-1}(x) = +\sqrt{x}$.

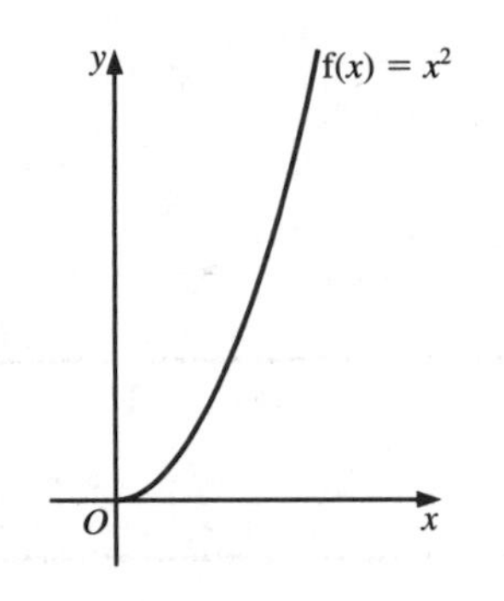

When the domain is restricted to $x \geqslant 0$, $f(x) = x^2$ becomes a one-to-one function.

Example 14

The function f is defined by $f(x) = x^2 - 2x$, for $x \in \mathbb{R}$. Explain why f^{-1} does not exist. Restrict the domain to $x \geqslant 1$, and explain why $f^{-1}(x)$ now exists. Find $f^{-1}(x)$ and state the range of the function f^{-1}.

The graph of $f(x) = x^2 - 2x$ is sketched in the diagram.

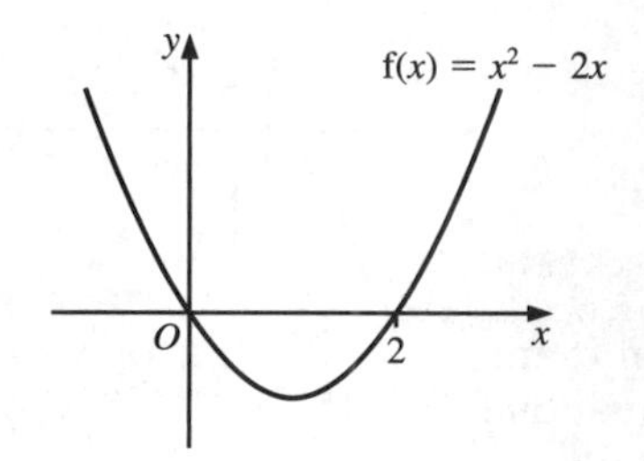

Since f(x) is two-to-one, $f^{-1}(x)$ does not exist. Now restrict the domain to $x \geqslant 1$.

From the second graph you can see that the range of f is now $f(x) \geqslant -1$. You can also see that f is a one-to-one function and therefore f^{-1} exists.

Range of f is $f(x) \geqslant -1$

To find $f^{-1}(x)$, let $y = x^2 - 2x$. Then rearranging for x gives a quadratic in x. That is,

$$x^2 - 2x - y = 0$$

So:

$$x = \frac{-(-2) \pm \sqrt{(-2)^2 - 4(1)(-y)}}{2(1)}$$

Use the quadratic formula.

$$= \frac{2 \pm \sqrt{4 + 4y}}{2}$$

$$= \frac{2 \pm 2\sqrt{1 + y}}{2}$$

$$\therefore \quad x = 1 \pm \sqrt{1 + y}$$

You want the positive square root. Therefore,

$$f^{-1}(x) = 1 + \sqrt{1 + x}$$

The range of f^{-1} is the domain of f. Therefore, the range of f^{-1} is the set $\{x : x \geqslant 1\}$.

Alternatively, you could use the method of completing the square. Starting with $y = x^2 - 2x$,

$$y = (x - 1)^2 - 1$$

$$y + 1 = (x - 1)^2$$

$$\therefore \quad x - 1 = \pm\sqrt{y + 1}$$

$$\therefore \quad x = 1 \pm \sqrt{y + 1}$$

as before.

Graph of an inverse function

Consider the function $f(x) = x + 3$ with inverse $f^{-1}(x) = x - 3$. Plotting graphs of both functions on the same set of axes gives the lines shown.

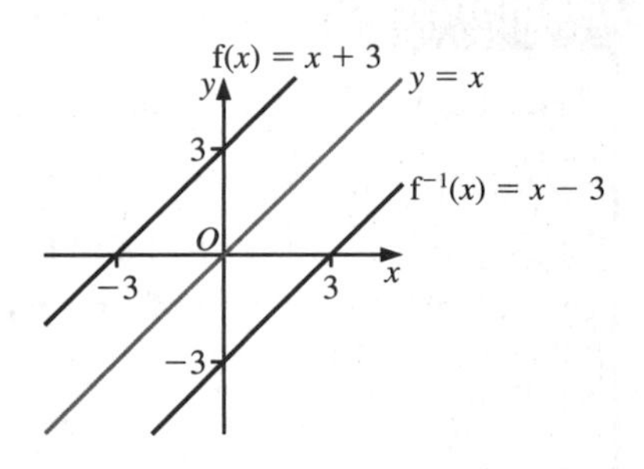

The graph of the inverse function is a reflection of the graph of f in the line $y = x$. This is because for every point (x, y) on the graph of the function f there is a point (y, x) on the graph of the function f^{-1}.

Example 15

The function f is defined by $f(x) = 3x - 6$ for all real values of x. Find the inverse function f^{-1}. Sketch the graphs of f and f^{-1} on the same set of axes and hence find the coordinates of the point of intersection of the graphs of f and f^{-1}.

To find f^{-1}, let $y = 3x - 6$. Then rearranging for x gives $x = \dfrac{y+6}{3}$.

Therefore, $f^{-1}(x) = \dfrac{x+6}{3}$.

Sketching graphs of f and f^{-1} on the same set of axes gives the lines shown.

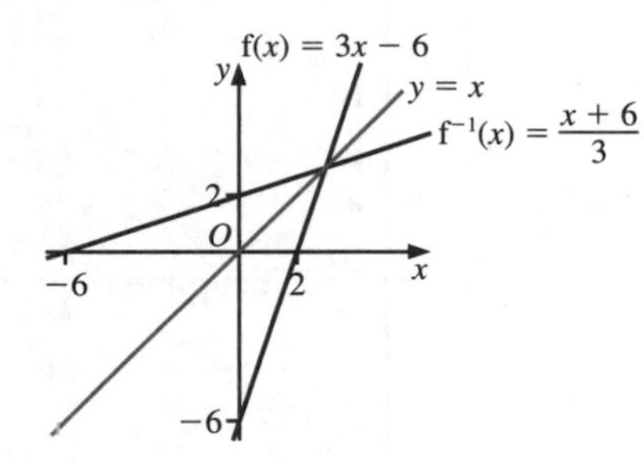

The point of intersection of the graphs of f and f^{-1} is also the point of intersection of the line $y = x$ with each of the graphs of f and f^{-1}. Therefore, to find the coordinates of this intersection point, you need to solve simultaneously the equations

$$y = x \quad \text{and} \quad y = 3x - 6$$

Eliminating y gives

$$\begin{aligned} x &= 3x - 6 \\ \therefore \quad 2x &= 6 \\ \therefore \quad x &= 3 \end{aligned}$$

Substituting $x = 3$ into $y = x$ gives $y = 3$.

The coordinates of the point of intersection of the graphs of f and f^{-1} are $(3, 3)$.

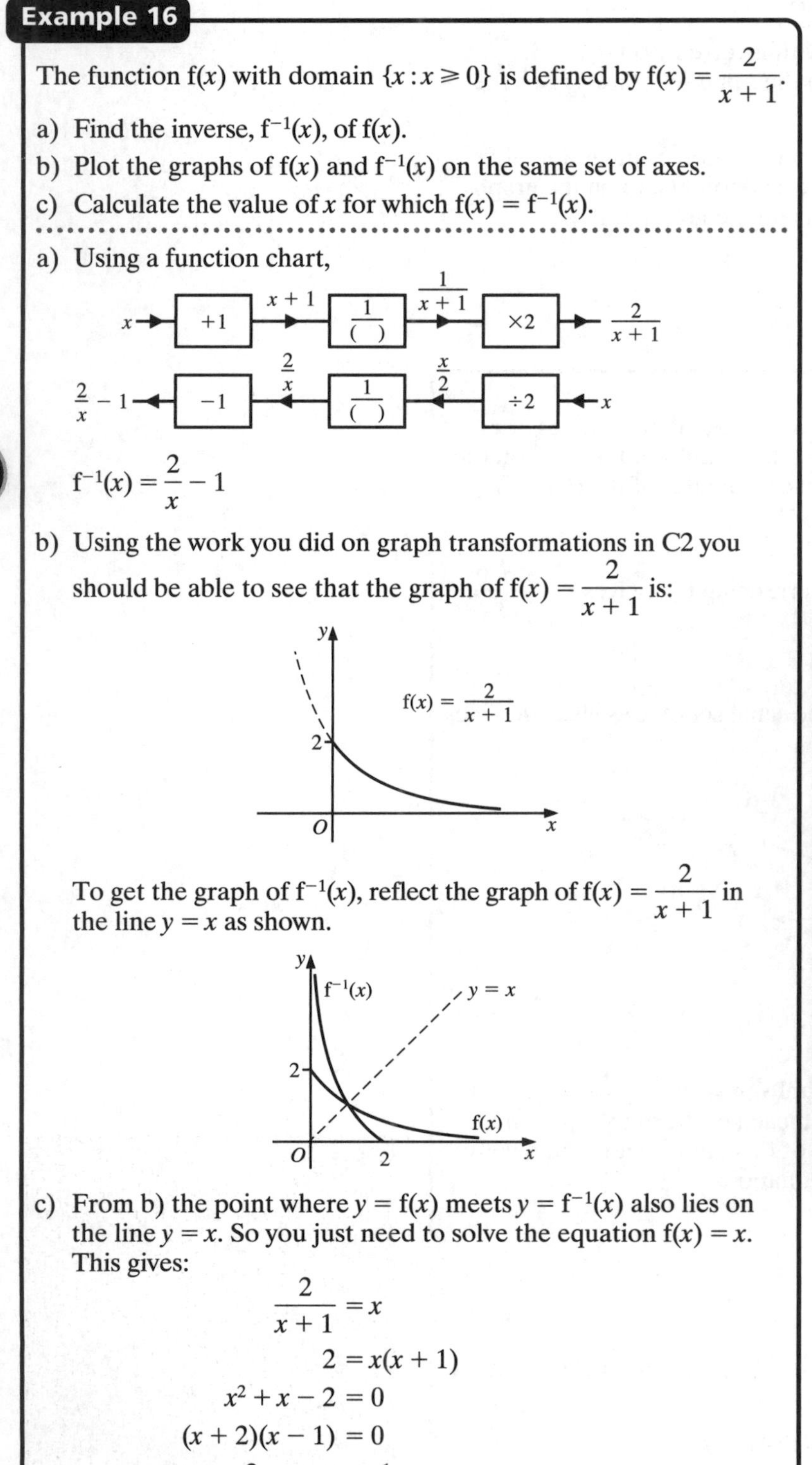

Example 16

The function f(x) with domain $\{x : x \geqslant 0\}$ is defined by $f(x) = \dfrac{2}{x+1}$.

a) Find the inverse, $f^{-1}(x)$, of f(x).

b) Plot the graphs of f(x) and $f^{-1}(x)$ on the same set of axes.

c) Calculate the value of x for which $f(x) = f^{-1}(x)$.

a) Using a function chart,

$$f^{-1}(x) = \frac{2}{x} - 1$$

b) Using the work you did on graph transformations in C2 you should be able to see that the graph of $f(x) = \dfrac{2}{x+1}$ is:

To get the graph of $f^{-1}(x)$, reflect the graph of $f(x) = \dfrac{2}{x+1}$ in the line $y = x$ as shown.

c) From b) the point where $y = f(x)$ meets $y = f^{-1}(x)$ also lies on the line $y = x$. So you just need to solve the equation $f(x) = x$. This gives:

$$\frac{2}{x+1} = x$$

$$2 = x(x+1)$$

$$x^2 + x - 2 = 0$$

$$(x+2)(x-1) = 0$$

$$\therefore \quad x = -2 \quad \text{or} \quad x = 1$$

Since $x \geqslant 0$ the solution is $x = 1$.

The graph of $f(x) = \dfrac{1}{x}$, $x \geqslant 0$, is

- The graph of $f(x) = \dfrac{1}{x+1}$ is translated -1 unit parallel to the x-axis.
- The graph of $f(x) = \dfrac{2}{x+1}$ is stretched parallel to the y-axis by a scale factor of 2.

Example 17

The function f is defined by

$$f(x) = \frac{2x+1}{x+2}, \text{ for } x > -2$$

Find the inverse function f^{-1} and find the coordinates of the points of intersection of the graphs of f and f^{-1}.

To find f^{-1}, let $y = \frac{2x+1}{x+2}$. Then rearranging for x gives

$$y(x+2) = 2x+1$$
$$yx + 2y = 2x + 1$$
$$yx - 2x = 1 - 2y$$
$$x(y-2) = 1 - 2y$$
$$x = \frac{1-2y}{y-2}$$

Therefore, $f^{-1}(x) = \frac{1-2x}{x-2}$.

The graphs of f and f^{-1} intersect at the points where the graphs of $y = x$ and $y = f(x)$ intersect. To find the x-coordinates of the points of intersection, solve simultaneously the equations

$$y = x \quad \text{and} \quad y = \frac{2x+1}{x+2}$$

Eliminating y gives

$$x = \frac{2x+1}{x+2}$$

$$\therefore \quad x(x+2) = 2x+1$$
$$x^2 - 1 = 0$$
$$\therefore \quad (x-1)(x+1) = 0$$

Solving gives $x = 1$ or $x = -1$.

When $x = 1, y = 1$ and when $x = -1, y = -1$. Therefore, the coordinates of the points of intersection of the graphs of f and f^{-1} are $(1, 1)$ and $(-1, -1)$.

Two useful techniques for sketching the graph of an inverse function are as follows.

- Reflect the graph of the function f in the line $y = x$.
- Sketch the graph of $y = f(x)$, turn the page over and then turn it through 90° clockwise. What you see through the page is the graph of the inverse function.

Note that
- a reflection in the y-axis followed by a rotation through 90° clockwise is equivalent to a reflection in the line $y = x$, and
- a reflection in the x-axis followed by a rotation through 90° anticlockwise is equivalent to a reflection in the line $y = -x$.

Exercise 1E

Throughout this exercise, the domain of each function is the set of real numbers unless specifically stated otherwise.

1 Find the inverse of each of these functions.

a) $f(x) = 3x + 2$

b) $f(x) = 5x - 1$

c) $f(x) = 4 - 3x$

d) $f(x) = \dfrac{2}{x}, \ x \neq 0$

e) $f(x) = \dfrac{3}{x-1}, \ x \neq 1$

f) $f(x) = \dfrac{5}{2-3x}, \ x \neq \frac{3}{2}$

g) $f(x) = \dfrac{x}{2+x}, \ x \neq -2$

h) $f(x) = \dfrac{2x}{5-x}, \ x \neq 5$

C3

2 Find the inverse of each of these functions, and state the domain on which each inverse is defined.

a) $f(x) = x^2, \ x \in \mathbb{R}, \ x > 2$

b) $f(x) = \dfrac{1}{2+x}, \ x \in \mathbb{R}, \ x > 0$

c) $f(x) = \sqrt{x-2}, \ x \in \mathbb{R}, \ x > 3$

d) $f(x) = 3x^2 - 1, \ x \in \mathbb{R}, \ 1 < x < 4$

e) $f(x) = \sqrt{2x+3}, \ x \in \mathbb{R}, \ x \geqslant 11$

f) $f(x) = \dfrac{1}{x} - 3, \ x \in \mathbb{R}, \ 2 < x < 5$

g) $f(x) = (x+2)^2 + 3, \ x \in \mathbb{R}, \ x \geqslant -2$

h) $f(x) = x^3 + 1, \ x \in \mathbb{R}$

3 Given $f(x) = 3x - 4, \ x \in \mathbb{R}$,

a) find an expression for the inverse function $f^{-1}(x)$

b) sketch the graphs of $f(x)$ and $f^{-1}(x)$ on the same set of axes

c) solve the equation $f(x) = f^{-1}(x)$.

4 a) Sketch the graph of the function defined by

$$f(x) = 10 - 2x, \ x \in \mathbb{R}, \quad x \geqslant 0$$

b) Find an expression for the inverse function $f^{-1}(x)$, and sketch the graph of $f^{-1}(x)$ on the same set of axes.

c) Calculate the value of x for which $f(x) = f^{-1}(x)$.

5 A function is defined by $f(x) = x^2 - 6, \ x \in \mathbb{R}, \quad x > 0$.

a) Find an expression for the inverse function $f^{-1}(x)$.

b) Sketch the graphs of $f(x)$ and $f^{-1}(x)$ on the same set of axes.

c) Calculate the value of x for which $f(x) = f^{-1}(x)$.

6 a) Sketch the graph of the function defined by

$$f(x) = (x-2)^2, \ x \in \mathbb{R}, \quad x \geqslant 2$$

b) Find an expression for the inverse function $f^{-1}(x)$, and sketch the graph of $f^{-1}(x)$ on the same set of axes.

c) Calculate the value of x for which $f(x) = f^{-1}(x)$.

7 The functions f and g are defined by

$$f(x) = 2x - 5, \ x \in \mathbb{R} \quad \text{and} \quad g(x) = 7 - 4x, \ x \in \mathbb{R}$$

a) Solve the equation $f(x) = g(x)$.

b) Write down expressions for $f^{-1}(x)$ and $g^{-1}(x)$.

c) Solve the equation $f^{-1}(x) = g^{-1}(x)$, and comment on your answer.

8 The function f with domain $\{x : x \geqslant 0\}$ is defined by
$\mathrm{f}(x) = \dfrac{4}{x+3}$.

a) Sketch the graph of f and state its range.
b) Find an expression for $\mathrm{f}^{-1}(x)$.
c) Calculate the value of x for which $\mathrm{f}(x) = \mathrm{f}^{-1}(x)$.

9 Functions f and g are defined by

$$\mathrm{f}(x) = 3x + 1,\ x \in \mathbb{R} \quad \text{and} \quad \mathrm{g}(x) = x - 2,\ x \in \mathbb{R}$$

a) Write down and simplify an expression for the composite function $\mathrm{fg}(x)$.
b) Find expressions for each of these inverse functions.
i) $\mathrm{f}^{-1}(x)$ ii) $\mathrm{g}^{-1}(x)$ iii) $(\mathrm{fg})^{-1}(x)$
c) Verify that $(\mathrm{fg})^{-1}(x) = \mathrm{g}^{-1}\mathrm{f}^{-1}(x)$.

10 The function g is defined by $\mathrm{g}(x) = 2x^2 - 3,\ x \in \mathbb{R}, x \geqslant 0$

a) State the range of g and sketch its graph.
b) Explain why the inverse function g^{-1} exists and sketch its graph.
c) Given also that h is defined by $\mathrm{h}(x) = \sqrt{5x+2},\ x \in \mathbb{R},\ x \geqslant -\frac{2}{5}$, solve the inequality $\mathrm{gh}(x) \geqslant x$.

11 Functions f and g are defined by

$$\mathrm{f}(x) = 2x + 3,\ x \in \mathbb{R} \quad \text{and} \quad \mathrm{g}(x) = \frac{1}{x-1},\ x \in \mathbb{R}, x \neq 1$$

a) Find an expression for the inverse function $\mathrm{f}^{-1}(x)$.
b) Find an expression for the composite function $\mathrm{gf}(x)$.
c) Solve the equation $\mathrm{f}^{-1}(x) = \mathrm{gf}(x) - 1$.

1.6 Inverse trigonometric functions

$\sin^{-1} x$

In C2 you met the graphs of the three basic trigonometric functions. For example, the graph of $\mathrm{f}(x) = \sin x$ for $-\pi \leqslant x \leqslant 2\pi$ is:

Throughout this section, angles are measured in radians.

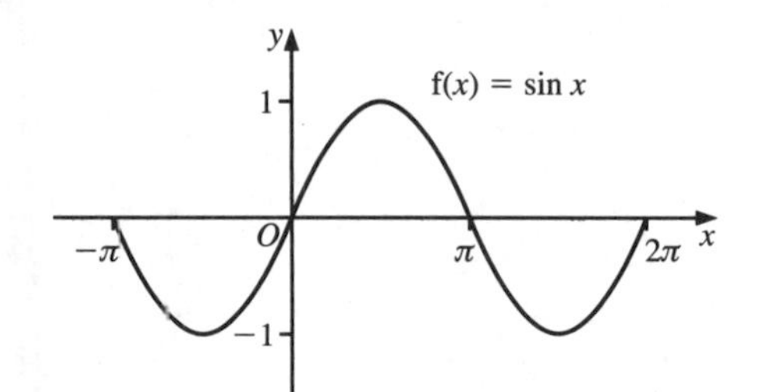

It is easy to see that $\mathrm{f}(x) = \sin x$ is not one-to-one on the given domain, and hence $\mathrm{f}(x) = \sin x$ does not have an inverse for $-\pi \leqslant x \leqslant 2\pi$. However, if you restrict the domain, you can make the graph one-to-one.

The next diagram shows the graph of the function $f(x) = \sin x$ for $-\frac{\pi}{2} \leqslant x \leqslant \frac{\pi}{2}$. For this domain the function $f(x) = \sin x$ is one-to-one, and hence it has an inverse.

The inverse function $f^{-1}(x) = \sin^{-1} x$ for $-1 \leqslant x \leqslant 1$ is also shown.

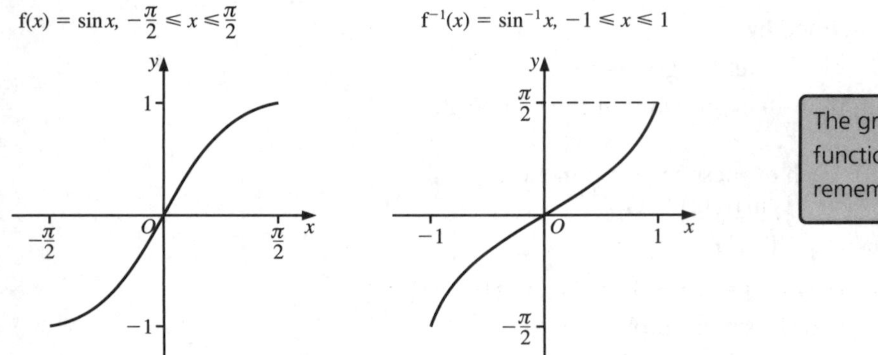

The graphs of these inverse trig functions need to be learnt and remembered.

$\cos^{-1} x$

Likewise, by restricting the domain, you can ensure that $f(x) = \cos x$ is one-to-one, and hence that $f^{-1}(x) = \cos^{-1} x$ exists.

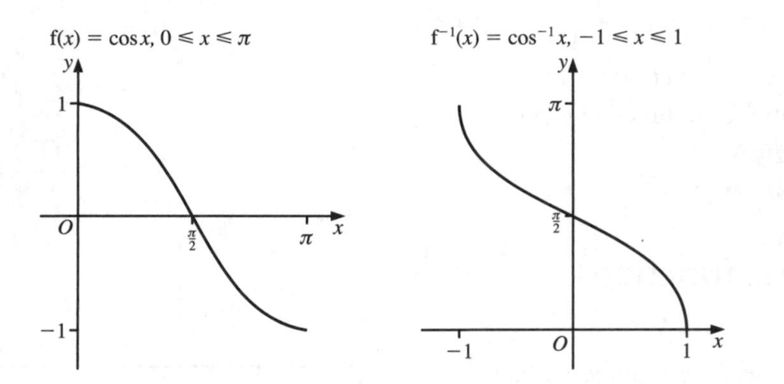

Remember the hints for sketching graphs of inverse functions (page 23).

$\tan^{-1} x$

The same treatment can be applied to $f(x) = \tan x$.

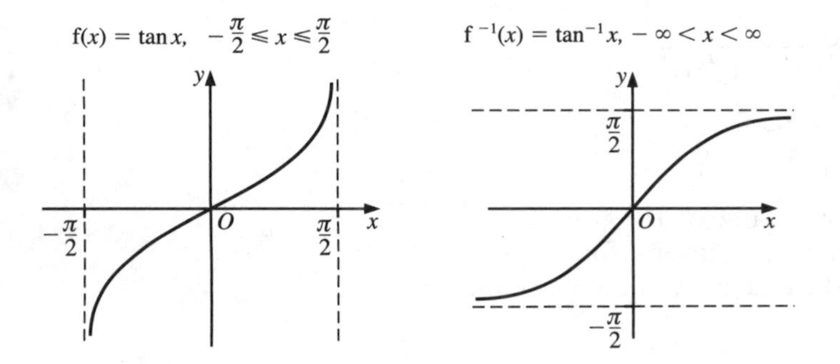

Example 18

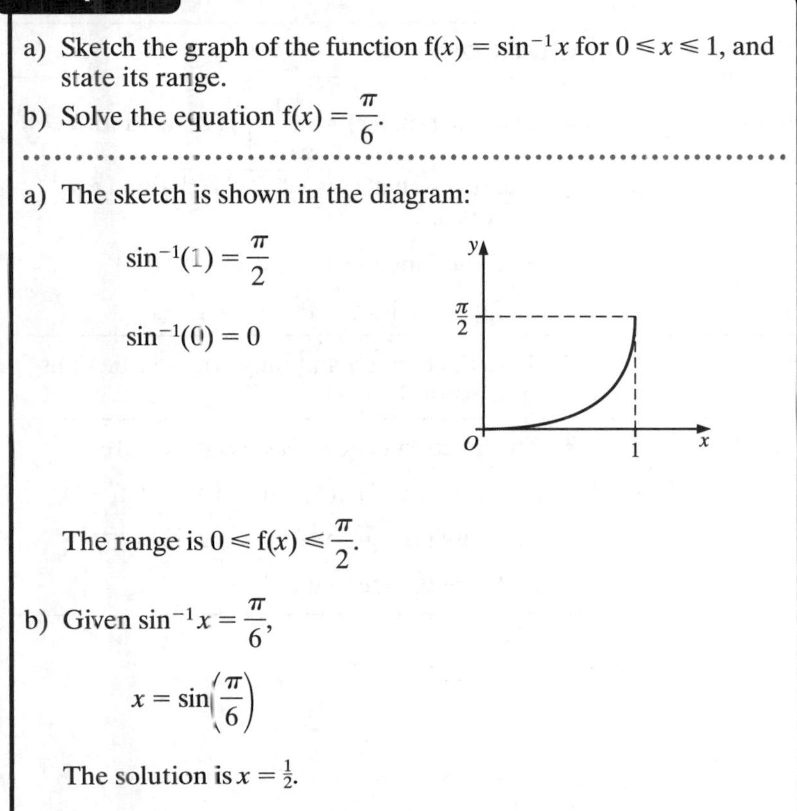

a) Sketch the graph of the function $f(x) = \sin^{-1} x$ for $0 \leqslant x \leqslant 1$, and state its range.

b) Solve the equation $f(x) = \frac{\pi}{6}$.

a) The sketch is shown in the diagram:

$$\sin^{-1}(1) = \frac{\pi}{2}$$

$$\sin^{-1}(0) = 0$$

The range is $0 \leqslant f(x) \leqslant \frac{\pi}{2}$.

b) Given $\sin^{-1} x = \frac{\pi}{6}$,

$$x = \sin\left(\frac{\pi}{6}\right)$$

The solution is $x = \frac{1}{2}$.

C3

Exercise 1F

1 a) Sketch the graph of the function $f(x) = \sin^{-1} x$ for $0 \leqslant x \leqslant 1$, and state its range.

b) Solve the equation $f(x) = \frac{\pi}{2}$.

2 a) Sketch the graph of the function $f(x) = \cos^{-1} x$ for $-1 \leqslant x \leqslant 1$, and state its range.

b) Solve the equation $f(x) = 0$.

3 a) Sketch the graph of the function $f(x) = \sin^{-1} x$ for $-1 \leqslant x \leqslant 1$, and state its range.

b) Solve the equation $f(x) = -\frac{\pi}{6}$.

4 a) Sketch the graph of the function $f(x) = \tan^{-1} x$ for $0 \leqslant x < \infty$, and state its range.

b) Solve the equation $f(x) = \frac{\pi}{4}$.

Summary

You should know how to ...	Check out
1 Find composite and inverse functions.	**1** The function $f(x) = \dfrac{1}{x^2 + 4}$ is defined for $x \geqslant 0$. The function $g(x) = \sqrt{4 - x^2}$ is defined for $0 \leqslant x \leqslant 2$. Find the functions: a) $fg(x)$ b) $f^{-1}(x)$ c) $g^{-1}(x)$
2 Find the domain and range of a function.	**2** State the domain and range of the functions in question 1.
3 Use the modulus function.	**3** The function f is given as $f(x) = \lvert x - 3 \rvert$. a) Evaluate i) $f(5)$ ii) $f(0)$ iii) $f(-2)$ b) Sketch the graph of $y = \lvert x - 3 \rvert$. c) Solve the inequality $\lvert x - 3 \rvert < 2$.

C3

Revision exercise 1

1 a) The functions f and g are defined by $f(x) = \sqrt{x}$ for $x \geqslant 0$; $g(x) = x - 1$ for all values of x.

i) Write down expressions for $fg(x)$ and $gf(x)$.

ii) Verify that $x = 1 \Rightarrow fg(x) = gf(x)$.

b) The diagram shows the graph of $y = h(x)$ where the function h is defined for the domain $1 \leqslant x \leqslant 5$ by $h(x) = \sqrt{x - 1}$.

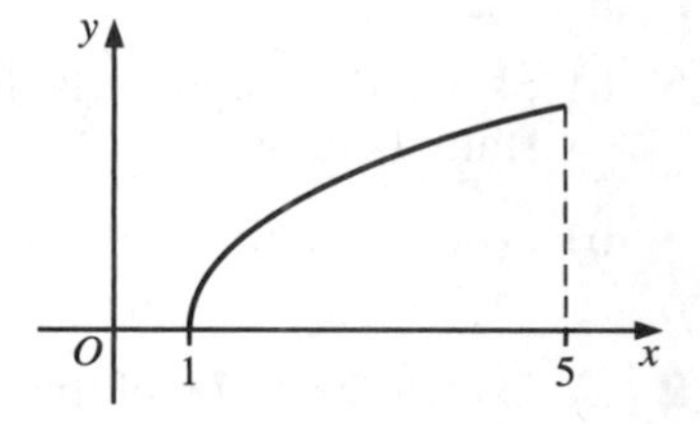

i) Describe the transformation by which the graph of $y = \sqrt{x - 1}$ can be obtained from the graph of $y = \sqrt{x}$.

ii) Write down the range of the function h.

iii) Write down the domain and range of the inverse function h^{-1}.

iv) Find an expression for $h^{-1}(x)$.

(AQA, 2004)

2 a) Sketch the graph of $y = \lvert 2x - 4 \rvert$. Indicate the points where the graph meets the coordinate axes.

b) i) The line $y = x$ intersects the graph of $y = \lvert 2x - 4 \rvert$ at two points P and Q. Find the x-coordinates of the points P and Q.

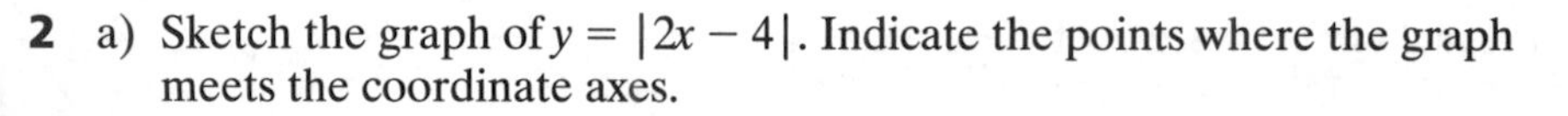

ii) Hence solve the inequality $\lvert 2x - 4 \rvert > x$.

c) The graph of $y = \lvert 2x - 4 \rvert + k$ touches the line $y = x$ at only one point. Find the values of the constant k.

(AQA, 2004)

3 The function f is defined for all real values of x by $f(x) = 3 - |2x - 1|$.

a) i) Sketch the graph of $y = f(x)$. Indicate the coordinates of the points where the graph crosses the coordinate axes.

ii) Hence show that the equation $f(x) = 4$ has no real roots.

b) State the range of f.

c) By finding the values of x for which $f(x) = x$, solve the inequality $f(x) < x$. *(AQA, 2003)*

4 The diagram shows the graphs of $y = x$ and $y = f(x)$.

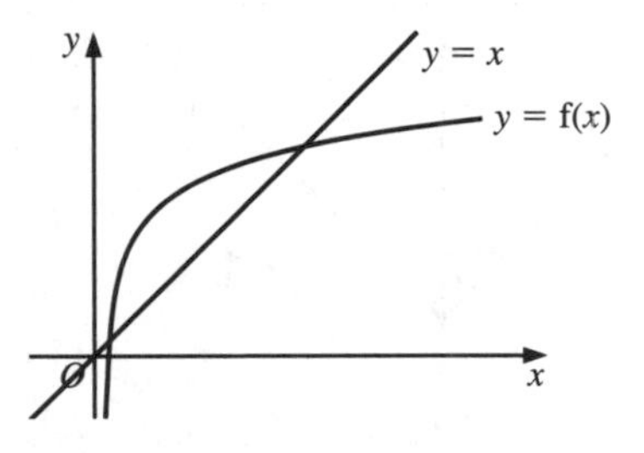

i) Describe the geometrical transformation by which the graph of $y = f^{-1}(x)$ can be obtained from the graph of $y = f(x)$.

ii) Copy the diagram and sketch on the same axes the graph of $y = f^{-1}(x)$. *(AQA, 2002)*

5 The diagram shows a sketch of the graph of $y = \cos 2x$ with a line of symmetry L.

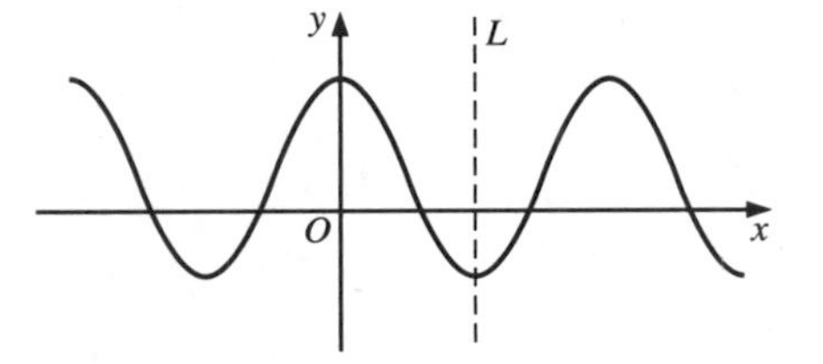

a) i) Describe the geometrical transformation by which the graph of $y = \cos 2x$ can be obtained from that of $y = \cos x$.

ii) Write down the equation of the line L.

The function f is defined for the restricted domain $0 \leqslant x \leqslant \frac{\pi}{2}$ by $f(x) = \cos 2x$.

b) i) State the range of the function f.

ii) Write down the domain and range of the inverse function f^{-1}, making it clear which is the domain of f^{-1} and which is its range.

iii) Sketch the graph of $y = f^{-1}(x)$.

The function g is defined for all real numbers by $g(x) = |x|$.

c) i) Write down an expression for $gf(x)$.

ii) Sketch the graph of $y = gf(x)$. *(AQA, 2002)*

6 The function f is defined for all real values of x by $f(x) = |2x - 3| - 1$.

a) Sketch the graph of $y = f(x)$. Indicate the coordinates of the points where the graph crosses the x-axis and the coordinates of the point where the graph crosses the y-axis.

b) State the range of f.

c) Find the values of x for which $f(x) = x$. *(AQA, 2002)*

7 The diagrams show the graphs of $y = f(x)$ and $y = g(x)$ where the functions f and g are defined on the domain of all real numbers by $f(x) = |x - 2|$ and $g(x) = |x| - 2$.

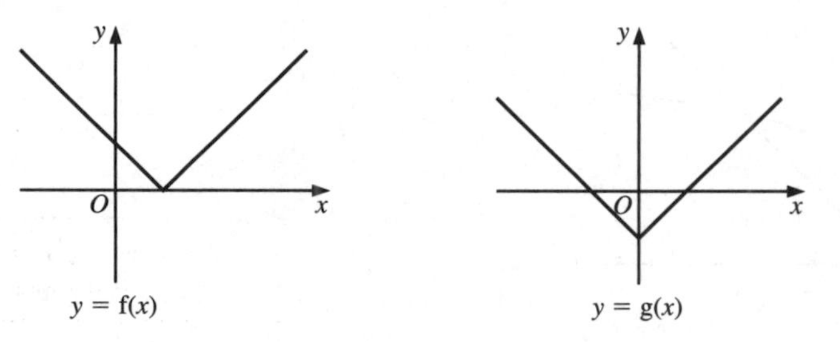

a) Describe the geometrical transformation by which each of these graphs can be obtained from the graph of $y = |x|$.

b) Sketch the graph of $f(x) - g(x)$.

c) i) State whether the function f has an inverse function.

ii) Give the range of the function h, where $h(x) = f(x) - g(x)$.

d) Solve the following inequalities:

i) $f(x) < 2$

ii) $g(x) < 2$

iii) $f(x) > g(x)$ *(AQA, 2001)*

2 Differentiation

This chapter will show you how to

- Differentiate composite functions
- Integrate using composite functions
- Differentiate the product of two functions
- Differentiate the quotient of two functions

Before you start

You should know how to ...	Check in
1 Find the derivative of x^n for any rational number n.	**1** Find $\frac{dy}{dx}$ when $y = 2x^{\frac{3}{2}} - x^2$.
2 Find the gradient of a curve at a given point on the curve.	**2** Find the gradient of the curve $y = 2x^{\frac{3}{2}} - x^2$ at the point with x-coordinate $x = 2$.
3 Use the first derivative to find stationary points on a curve.	**3** Find the coordinates of the stationary point on the curve $y = 2x^{\frac{3}{2}} - x^2$ $(x > 0)$.
4 Use the second derivative to determine the nature of stationary points.	**4** Determine whether the stationary point on the curve $y = 2x^{\frac{3}{2}} - x^2$ is a maximum or minimum.
5 Use the first derivative to determine where a function is increasing or decreasing.	**5** For what values of x is the function $f(x) = 2x^{\frac{3}{2}} - x^2$ an increasing function?

C3

2.1 Function of a function

If you write, for example, $y = (x + 2)^4$, you can say that y is a function of x. If you let $u = x + 2$ then

$$y = u^4 \text{ where } u = x + 2$$

In other words, y is now a function of u, and u is a function of x. The new variable, u, is the link between the two expressions.

In order to differentiate the expression $y = (x + 2)^4$ with respect to x, you would first need to expand the bracket. In this particular case, this would certainly be feasible. However, if you had an expression such as $y = (x + 2)^{12}$, this would involve more work. So you need a technique for differentiating such expressions more easily.

The term 'function of a function' is just another way of thinking of composite functions.
For example,
$f(x) = x^4$
$g(x) = x + 2$
Then $fg(x) = (x + 2)^4$
Composite functions are explained on page 13.

The **chain rule** states that:

If y is a function of u, and u is a function of x,

$$\frac{dy}{dx} = \frac{dy}{du}\frac{du}{dx}$$

Using the example of $y = (x + 2)^4$,

$$y = u^4 \quad \text{where } u = x + 2$$

Differentiating y with respect to u gives

$$\frac{dy}{du} = 4u^3$$

Differentiating u with respect to x gives

$$\frac{du}{dx} = 1$$

By the chain rule,

$$\frac{dy}{dx} = \frac{dy}{du}\frac{du}{dx} = (4u^3)(1)$$

$$\therefore \frac{dy}{dx} = 4u^3$$

Substituting $u = x + 2$ gives

$$\frac{dy}{dx} = 4(x + 2)^3$$

which is the required derivative.

C3

Example 1

If $y = (3x - 1)^7$ find $\frac{dy}{dx}$.

Let $u = 3x - 1$, so $y = u^7$.

Differentiating each expression gives

$$\frac{dy}{du} = 7u^6 \quad \text{and} \quad \frac{du}{dx} = 3$$

By the chain rule,

$$\frac{dy}{dx} = \frac{dy}{du}\frac{du}{dx}$$

$$= (7u^6)(3) = 21u^6$$

$$\therefore \frac{dy}{dx} = 21(3x - 1)^6$$

If you are using function notation then $y = f(g(x))$ and $\frac{dy}{dx} = f'(g(x))g'(x)$.

You can write this in a shorter way.

$$y = (3x - 1)^7$$

$$\frac{dy}{dx} = 7(3x - 1)^6 \times \frac{d}{dx}(3x - 1)$$

$$= 7(3x - 1)^6 \times 3$$

$$= 21(3x - 1)^6$$

Example 2

Find $\frac{dy}{dx}$ for each of these functions.

a) $y = 2(1 - x)^5$

b) $y = (x^2 + 3)^4$

c) $y = \frac{1}{(3 - 7x)}$

d) $y = \sqrt{6x + 1}$

a) When $y = 2(1 - x)^5$,

$$\frac{dy}{dx} = 2 \times 5(1 - x)^4 \times \frac{d}{dx}(1 - x)$$

$$= 10(1 - x)^4(-1)$$

$$\therefore \quad \frac{dy}{dx} = -10(1 - x)^4$$

b) When $y = (x^2 + 3)^4$,

$$\frac{dy}{dx} = 4(x^2 + 3)^3 \times \frac{d}{dx}(x^2 + 3)$$

$$= 4(x^2 + 3)^3(2x)$$

$$\therefore \quad \frac{dy}{dx} = 8x(x^2 + 3)^3$$

c) When $y = \frac{1}{3 - 7x} = (3 - 7x)^{-1}$,

$$\frac{dy}{dx} = -(3 - 7x)^{-2} \times \frac{d}{dx}[(3 - 7x)]$$

$$= -(3 - 7x)^{-2}(-7) = 7(3 - 7x)^{-2}$$

$$\therefore \quad \frac{dy}{dx} = \frac{7}{(3 - 7x)^2}$$

d) When $y = \sqrt{6x + 1} = (6x + 1)^{\frac{1}{2}}$,

$$\frac{dy}{dx} = \tfrac{1}{2}(6x + 1)^{-\frac{1}{2}} \times \frac{d}{dx}(6x + 1)$$

$$= \tfrac{1}{2}(6x + 1)^{-\frac{1}{2}}(6)$$

$$= 3(6x + 1)^{-\frac{1}{2}}$$

$$\therefore \quad \frac{dy}{dx} = \frac{3}{\sqrt{6x + 1}}$$

Exercise 2A

1 Find $\frac{dy}{dx}$ for each of these functions.

a) $y = (2x - 1)^3$ b) $y = (3x + 4)^2$ c) $y = (5x - 3)^4$
d) $y = (3 - x)^5$ e) $y = (4 - 3x)^6$ f) $y = (x^2 + 1)^4$
g) $y = (x^3 - 6)^2$ h) $y = (1 - 2x^2)^3$ i) $y = (4 - x^4)^2$
j) $y = (7 - 5x^3)^6$ k) $y = (6x^2 - 5)^4$ l) $y = (9 - 7x^2)^3$

2 Differentiate each of these expressions with respect to x.

a) $(2x - 5)^{-3}$ b) $(3x + 2)^{-1}$ c) $(x^2 + 3)^{-2}$
d) $(5 - 2x^3)^{-1}$ e) $\frac{1}{3 + 4x}$ f) $\frac{1}{4 - x^2}$
g) $\frac{5}{3 - 2x}$ h) $\frac{3}{(x + 1)^2}$ i) $\frac{7}{(2 - x^2)^5}$
j) $-\frac{1}{(3x^2 + 8)}$ k) $(5x^3 - 4)^{-4}$ l) $\frac{1}{2(5 - 3x^4)^2}$

3 Find $f'(x)$ for each of these functions.

a) $f(x) = (2x - 1)^{\frac{1}{2}}$ b) $f(x) = (6 - x)^{\frac{1}{3}}$ c) $f(x) = (x^3 - 2)^{\frac{2}{3}}$
d) $f(x) = (4 - x^5)^{-\frac{1}{5}}$ e) $f(x) = \sqrt{4x - 5}$ f) $f(x) = \sqrt[3]{x^2 + 3}$
g) $f(x) = \frac{1}{\sqrt{5 - 2x}}$ h) $f(x) = \frac{6}{\sqrt[3]{x^2 + 5}}$ i) $f(x) - \frac{3}{\sqrt[6]{4x - 7}}$
j) $f(x) = (5 - 4\sqrt{x})^5$ k) $f(x) = \sqrt{3 + \sqrt{x}}$ l) $f(x) = \frac{1}{4 - \sqrt[3]{x}}$

2.2 Inverse function of a function

If you have a function such as $f(x) = (x^2 + 1)^n$, you can multiply out the brackets and integrate each term separately. For example, if $n = 2$

$$\begin{aligned}\int (x^2 + 1)^2 \, dx &= \int (x^4 + 2x^2 + 1) dx \\ &= \int x^4 \, dx + 2\int x^2 \, dx + \int 1 \, dx \\ &= \tfrac{1}{5}x^5 + \tfrac{2}{3}x^3 + x + c\end{aligned}$$

If n is greater than 2 the algebra becomes more difficult. You need a better way to integrate functions of functions like this.

Consider the function $f(x) = (x^2 + 1)^4$.

$$f'(x) = 4(x^2 + 1)^3 \times \frac{d}{dx}(x^2 + 1)$$

$$\therefore \quad f'(x) = 8x(x^2 + 1)^3$$

So you can write

$$\int 8x(x^2 + 1)^3 \, dx = (x^2 + 1)^4 + c, \text{ and hence}$$

$$\int x(x^2 + 1)^3 \, dx = \tfrac{1}{8}(x^2 + 1)^4 + C$$

$\frac{c}{8}$ is still a constant, so for convenience you can just call it C

Consider the integral $\int x(3x^2 - 2)^5\,dx$. Notice first that the derivative of $(3x^2 - 2)$ is $6x$ and that there is an x term in the integral. So you should consider $(3x^2 - 2)^6$, which when differentiated gives $36x(3x^2 - 2)^5$. Therefore,

$$\int x(3x^2 - 2)^5\,dx = \tfrac{1}{36}(3x^2 - 2)^6 + c$$

Example 3

Find each of these integrals.

a) $\int (x-2)^2\,dx$ b) $\int x(3x^2 + 6)^4\,dx$ c) $\int 4x^2(x^3 - 3)^5\,dx$

d) $\int (x+2)(x^2 + 4x - 1)^3\,dx$ e) $\int \frac{x}{\sqrt{x^2+3}}\,dx$

a) Consider $(x-2)^3$, which when differentiated gives $3(x-2)^2$. So:

$$\int (x-2)^2\,dx = \frac{(x-2)^3}{3} + c$$

b) Notice that the derivative of $3x^2 + 6$ is $6x$ and there is an x term in the integral. Consider $(3x^2 + 6)^5$, which when differentiated gives $30x(3x^2 + 6)^4$. So:

$$\int x(3x^2 + 6)^4\,dx = \frac{(3x^2 + 6)^5}{30} + c$$

c) Notice that the derivative of $(x^3 - 3)$ is $3x^2$, and that there is an x^2 term in the integral. Consider $(x^3 - 3)^6$, which when differentiated gives $18x^2(x^3 - 3)^5$. Therefore,

$$\int 4x^2(x^3 - 3)^5\,dx = \tfrac{4}{18}(x^3 - 3)^6 + c$$
$$= \tfrac{2}{9}(x^3 - 3)^6 + c$$

d) Notice that the derivative of $(x^2 + 4x - 1)$ is $2x + 4 = 2(x + 2)$, and that there is the term $(x + 2)$ in the integral. So consider $(x^2 + 4x - 1)^4$, which when differentiated gives

$$4(2x + 4)(x^2 + 4x - 1)^3 = 8(x + 2)(x^2 + 4x - 1)^3$$

Therefore,

$$\int (x + 2)(x^2 + 4x - 1)^3\,dx = \tfrac{1}{8}(x^2 + 4x - 1)^4 + c$$

e) $\int \frac{x}{\sqrt{x^2+3}}\,dx = \int x(x^2 + 3)^{-\frac{1}{2}}\,dx$

Notice that the derivative of $(x^2 + 3)$ is $2x$ and that there is an x in the integral. Consider $(x^2 + 3)^{\frac{1}{2}}$, which when differentiated gives

$$2x \times \tfrac{1}{2}(x^2 + 3)^{-\frac{1}{2}} = x(x^2 + 3)^{-\frac{1}{2}}$$

as required. Therefore,

$$\int \frac{x}{\sqrt{x^2+3}}\,dx = (x^2 + 3)^{\frac{1}{2}} + c$$
$$= \sqrt{x^2 + 3} + c$$

Exercise 2B

1 Integrate each of these with respect to x.

a) $(2x-3)^4$ b) $(5x+8)^2$ c) $(3x-4)^5$

d) $3(x-7)^2$ e) $(4-x)^5$ f) $-(6-7x)^3$

g) $(3x-4)^{-3}$ h) $6(5-9x)^{-2}$ i) $\dfrac{1}{(2x-1)^7}$

j) $\dfrac{3}{(1-x)^2}$ k) $\sqrt{2x-3}$ l) $\dfrac{12}{\sqrt[3]{x-4}}$

Check your answers by differentiating.

2 Find each of these integrals.

a) $\int (2x-7)^5\,dx$ b) $\int \dfrac{1}{\sqrt{2x-1}}\,dx$ c) $\int x(x^2+2)^3\,dx$

d) $\int 2x(4-3x^2)^5\,dx$ e) $\int x^2(x^3-4)^2\,dx$ f) $\int \dfrac{4x}{(3-x^2)^2}\,dx$

g) $\int x^3\sqrt{x^4-1}\,dx$ h) $\int 4x\sqrt[3]{2-3x^2}\,dx$ i) $\int x^{\frac{1}{3}}(x^{\frac{4}{3}}-2)^2\,dx$

j) $\int \dfrac{25x^4}{(3-x^5)^2}\,dx$ k) $\int x(x^2+1)^2\,dx$ l) $\int (x^2+1)^2\,dx$

C3

2.3 Product rule

The **product rule** states that:

If $y = uv$, where u and v are both functions of x, then

$$\frac{d}{dx}(uv) = u\frac{dv}{dx} + v\frac{du}{dx}$$

An alternative form of this rule is: $(uv)' = uv' + u'v$

Example 4

If $y = x^2(x+2)^3$ find $\dfrac{dy}{dx}$.

In this example, y is the product of two functions u and v, where

$$u = x^2 \quad \text{and} \quad v = (x+2)^3$$

Differentiating each with respect to x gives

$$\frac{du}{dx} = 2x \quad \text{and} \quad \frac{dv}{dx} = 3(x+2)^2$$

Using the product rule gives

$$\frac{dy}{dx} = u\frac{dv}{dx} + v\frac{du}{dx}$$
$$= x^2[3(x + 2)^2] + (x + 2)^3 (2x)$$

Factorising gives

$$\frac{dy}{dx} = x(x + 2)^2[3x + 2(x + 2)]$$

$$\therefore \quad \frac{dy}{dx} = x(x + 2)^2(5x + 4)$$

In all the examples which follow, the answer will be fully factorised.

Example 5

Find $\frac{dy}{dx}$ for each of these functions.

a) $y = (2x + 3)(x + 5)^2$ b) $y = x\sqrt{3x - 1}$

a) When $y = (2x + 3)(x + 5)^2$, using the product rule gives:

$$\frac{dy}{dx} = (2x + 3) \times 2(x + 5) + (x + 5)^2 \times 2$$
$$= 2(x + 5)[2x + 3 + x + 5]$$
$$= 2(x + 5)(3x + 8)$$

$u = (2x + 3)$ $v = (x + 5)^2$

$\frac{du}{dx} = 2$ $\frac{dv}{dx} = 2(x + 5)$

b) When $y = x\sqrt{3x - 1}$, using the product rule gives:

$$\frac{dy}{dx} = x \times \frac{3}{2\sqrt{3x - 1}} + \sqrt{3x - 1} \times 1$$
$$= \frac{3x}{2\sqrt{3x - 1}} + \sqrt{3x - 1} \times \frac{2\sqrt{3x - 1}}{2\sqrt{3x - 1}}$$
$$= \frac{3x + 2(3x - 1)}{2\sqrt{3x - 1}}$$
$$= \frac{9x - 2}{2\sqrt{3x - 1}}$$

$u = x$ $v = \sqrt{3x - 1}$

$\frac{du}{dx} = 1$ $\frac{dv}{dx} = \frac{3}{2\sqrt{3x - 1}}$

C3

Exercise 2C

In each part of questions **1** to **4**, use the product rule to differentiate the given function with respect to x.

1 a) $(x + 3)(x - 4)$ b) $(2x + 5)(x - 7)$
c) $(3x - 4)(2x + 5)$ d) $(6 + x)(5 - x)$
e) $(3 - 2x)(7 + 3x)$ f) $(x + 4)(x^2 - 2)$
g) $(x^2 - 5)(4x - 1)$ h) $(x + 6)(x^2 - 3x + 3)$
i) $(3x - 5)(x^2 - 2x + 7)$ j) $(x^2 + 1)(x^3 - 5)$
k) $(5x^3 - 3)(x^2 + 4x - 1)$ l) $(x^6 - 2)(3x^3 - x^2 + 4)$

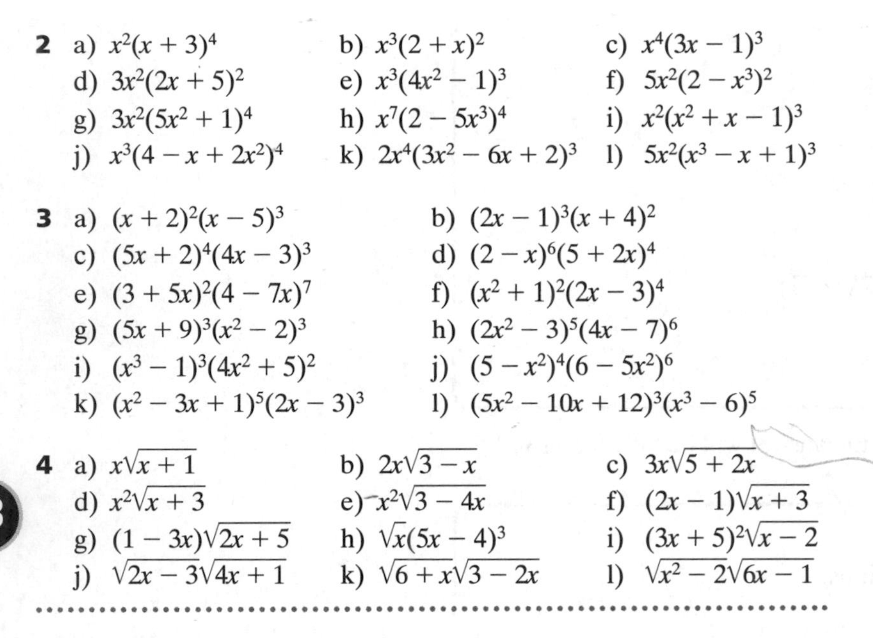

2 a) $x^2(x+3)^4$ b) $x^3(2+x)^2$ c) $x^4(3x-1)^3$
d) $3x^2(2x+5)^2$ e) $x^3(4x^2-1)^3$ f) $5x^2(2-x^3)^2$
g) $3x^2(5x^2+1)^4$ h) $x^7(2-5x^3)^4$ i) $x^2(x^2+x-1)^3$
j) $x^3(4-x+2x^2)^4$ k) $2x^4(3x^2-6x+2)^3$ l) $5x^2(x^3-x+1)^3$

3 a) $(x+2)^2(x-5)^3$ b) $(2x-1)^3(x+4)^2$
c) $(5x+2)^4(4x-3)^3$ d) $(2-x)^6(5+2x)^4$
e) $(3+5x)^2(4-7x)^7$ f) $(x^2+1)^2(2x-3)^4$
g) $(5x+9)^3(x^2-2)^3$ h) $(2x^2-3)^5(4x-7)^6$
i) $(x^3-1)^3(4x^2+5)^2$ j) $(5-x^2)^4(6-5x^2)^6$
k) $(x^2-3x+1)^5(2x-3)^3$ l) $(5x^2-10x+12)^3(x^3-6)^5$

4 a) $x\sqrt{x+1}$ b) $2x\sqrt{3-x}$ c) $3x\sqrt{5+2x}$
d) $x^2\sqrt{x+3}$ e) $x^2\sqrt{3-4x}$ f) $(2x-1)\sqrt{x+3}$
g) $(1-3x)\sqrt{2x+5}$ h) $\sqrt{x}(5x-4)^3$ i) $(3x+5)^2\sqrt{x-2}$
j) $\sqrt{2x-3}\sqrt{4x+1}$ k) $\sqrt{6+x}\sqrt{3-2x}$ l) $\sqrt{x^2-2}\sqrt{6x-1}$

C3

2.4 Quotient rule

The **quotient rule** states that:

If $y = \dfrac{u}{v}$, where u and v are functions of x, then

$$\frac{d}{dx}\left(\frac{u}{v}\right) = \frac{v\dfrac{du}{dx} - u\dfrac{dv}{dx}}{v^2}$$

An alternative form is

$$\left(\frac{u}{v}\right)' = \frac{u'v - uv'}{v^2}$$

Example 6

If $y = \dfrac{3x+2}{2x+1}$ find $\dfrac{dy}{dx}$.

In this example, y is the quotient of two functions u and v, where

$$u = 3x+2 \quad \text{and} \quad v = 2x+1$$

Differentiating each with respect to x gives

$$\frac{du}{dx} = 3 \quad \text{and} \quad \frac{dv}{dx} = 2$$

Using the quotient rule,

$$\frac{dy}{dx} = \frac{(2x+1)\times 3 - (3x+2)\times 2}{(2x+1)^2}$$

$$= \frac{6x+3-6x-4}{(2x+1)^2}$$

$$\therefore \quad \frac{dy}{dx} = -\frac{1}{(2x+1)^2}$$

$u = (3x+2)$ $v = 2x+1$

$\dfrac{du}{dx} = 3$ $\dfrac{dv}{dx} = 2$

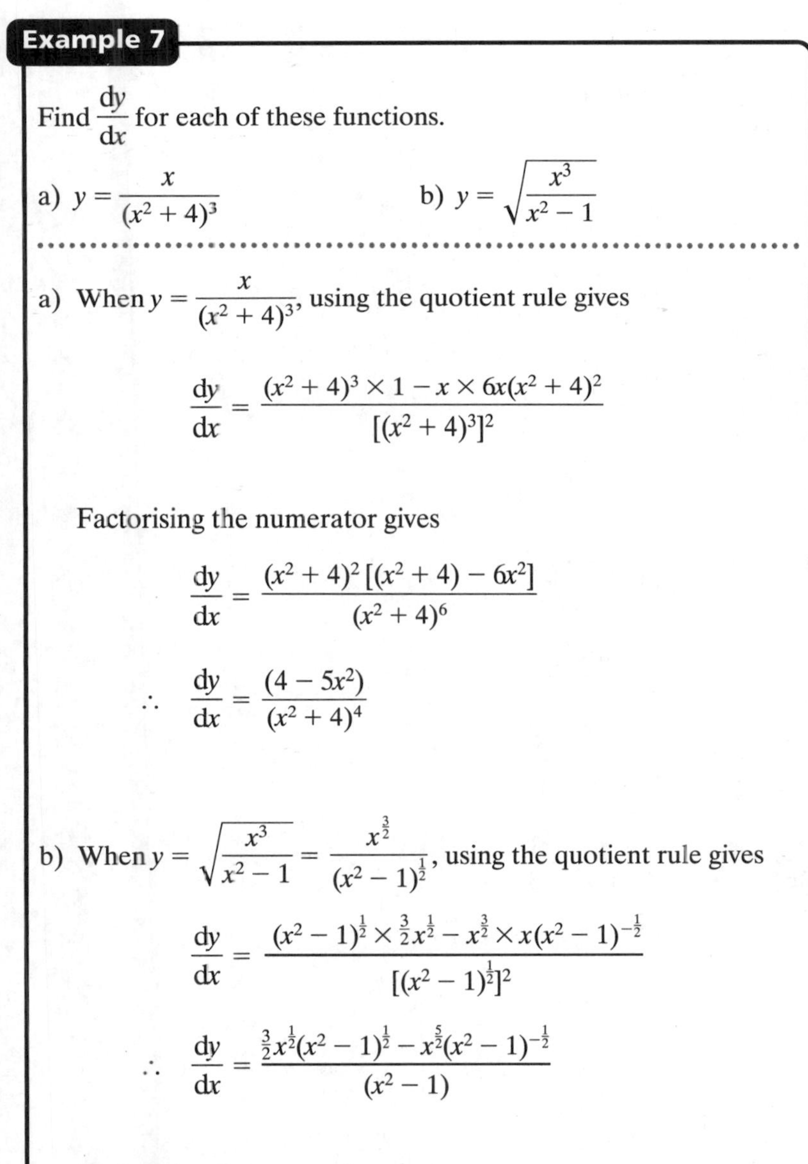

Example 7

Find $\dfrac{dy}{dx}$ for each of these functions.

a) $y = \dfrac{x}{(x^2 + 4)^3}$ b) $y = \sqrt{\dfrac{x^3}{x^2 - 1}}$

a) When $y = \dfrac{x}{(x^2 + 4)^3}$, using the quotient rule gives

$$\frac{dy}{dx} = \frac{(x^2 + 4)^3 \times 1 - x \times 6x(x^2 + 4)^2}{[(x^2 + 4)^3]^2}$$

$u = x$ $v = (x^2 + 4)^3$

$\dfrac{du}{dx} = 1$ $\dfrac{dv}{dx} = 6x(x^2 + 4)^2$

Factorising the numerator gives

$$\frac{dy}{dx} = \frac{(x^2 + 4)^2 [(x^2 + 4) - 6x^2]}{(x^2 + 4)^6}$$

$$\therefore \quad \frac{dy}{dx} = \frac{(4 - 5x^2)}{(x^2 + 4)^4}$$

b) When $y = \sqrt{\dfrac{x^3}{x^2 - 1}} = \dfrac{x^{\frac{3}{2}}}{(x^2 - 1)^{\frac{1}{2}}}$, using the quotient rule gives

$$\frac{dy}{dx} = \frac{(x^2 - 1)^{\frac{1}{2}} \times \frac{3}{2}x^{\frac{1}{2}} - x^{\frac{3}{2}} \times x(x^2 - 1)^{-\frac{1}{2}}}{[(x^2 - 1)^{\frac{1}{2}}]^2}$$

$u = x^{\frac{3}{2}}$ $v = (x^2 - 1)^{\frac{1}{2}}$

$\dfrac{du}{dx} = \frac{3}{2}x^{\frac{1}{2}}$ $\dfrac{dv}{dx} = x(x^2 - 1)^{-\frac{1}{2}}$

$$\therefore \quad \frac{dy}{dx} = \frac{\frac{3}{2}x^{\frac{1}{2}}(x^2 - 1)^{\frac{1}{2}} - x^{\frac{5}{2}}(x^2 - 1)^{-\frac{1}{2}}}{(x^2 - 1)}$$

Factorising the numerator gives

$$\frac{dy}{dx} = \frac{\frac{1}{2}x^{\frac{1}{2}}(x^2 - 1)^{-\frac{1}{2}}[3(x^2 - 1) - 2x^2]}{(x^2 - 1)}$$

$$= \frac{\frac{1}{2}x^{\frac{1}{2}}(x^2 - 1)^{-\frac{1}{2}}(x^2 - 3)}{(x^2 - 1)}$$

$$\therefore \quad \frac{dy}{dx} = \frac{\sqrt{x}(x^2 - 3)}{2\sqrt{(x^2 - 1)^3}}$$

Exercise 2D

In each part of questions **1** and **2**, use the quotient rule to differentiate the given fraction with respect to x.

1 a) $\dfrac{x}{x-2}$ b) $\dfrac{x+3}{x-1}$ c) $\dfrac{3-x}{4+x}$

d) $\dfrac{4x-3}{x+2}$ e) $\dfrac{2x-5}{x+4}$ f) $\dfrac{5x}{x+2}$

g) $\dfrac{1+3x}{2-5x}$ h) $\dfrac{4x+3}{2x-1}$ i) $\dfrac{x^2}{x+3}$

j) $\dfrac{x^2}{4-x}$ k) $\dfrac{x^3}{2x-3}$ l) $\dfrac{x^5}{3-x}$

C3

2 a) $\dfrac{(3x-2)^2}{\sqrt{x}}$ b) $\dfrac{(5x+1)^3}{\sqrt{x}}$ c) $\dfrac{(x^2-4)^5}{\sqrt{x}}$

d) $\dfrac{\sqrt{x}}{2x-1}$ e) $\dfrac{3-\sqrt{x}}{(2+x)^2}$ f) $\dfrac{5+2\sqrt{x}}{(5-4x)^3}$

g) $\dfrac{(3x^2+2)^4}{\sqrt{2x-1}}$ h) $\dfrac{(2-3x)^2}{\sqrt{1-x^2}}$ i) $\sqrt{\dfrac{x-2}{x+1}}$

j) $\sqrt{\dfrac{x-3}{2x+5}}$ k) $\sqrt{\dfrac{3+x}{2-3x}}$ l) $\sqrt{\dfrac{x^2+1}{x^3-3}}$

3 *Mixed exercise*

Find $\dfrac{dy}{dx}$ for each of these.

a) $y = x^3(3-x)^2$ b) $y = \dfrac{x}{2x-1}$

c) $y = \sqrt{x}(5x-1)^2$ d) $y = \dfrac{2x}{\sqrt{x}+1}$

e) $y = (5x+3)^3(x-2)^2$ f) $y = \sqrt{\dfrac{3x-2}{x-3}}$

g) $y = x^3\sqrt{7-2x}$ h) $y = \dfrac{x^2}{2-x}$

i) $y = \dfrac{\sqrt{x}+1}{\sqrt{x}-1}$ j) $y = (3-x)^4(2+x)^5$

k) $y = \dfrac{x^2+1}{3x-1}$ l) $y = \sqrt{(x-1)(2x+1)}$

2.5 x as a function of y

Suppose you have a function in which x is a function of y, for example, $x = y^3$. There are two ways of finding an expression for $\frac{dy}{dx}$.

One way is to rearrange the expression to get y as a function of x.

$$x = y^3$$
$$\therefore \quad y = x^{\frac{1}{3}}$$

Differentiating with respect to x gives:

$$\frac{dy}{dx} = \tfrac{1}{3}x^{-\frac{2}{3}}$$

So $\frac{dy}{dx} = \frac{1}{3\sqrt[3]{x^2}}$

Another way of finding $\frac{dy}{dx}$ is to differentiate the original expression with respect to y.

$$x = y^3$$
$$\therefore \quad \frac{dx}{dy} = 3y^2$$

Now, in general:

$$\frac{dy}{dx} = \frac{1}{\frac{dx}{dy}}$$

So taking the reciprocal of each side gives:

$$\frac{dy}{dx} = \frac{1}{3y^2}$$

Since $y = x^{\frac{1}{3}}$ you know that $y^2 = x^{\frac{2}{3}}$, and hence $\frac{dy}{dx} = \frac{1}{3\sqrt[3]{x^2}}$ as before.

In this example you could rearrange the expression $x = y^3$ to get y as a function of x. So it was easy to find an expression for $\frac{dy}{dx}$ without needing to work out $\frac{dx}{dy}$ first. However, it is not always possible to rearrange an expression in this way – and this is when this method really comes into its own. The following example illustrates this.

Example 8

Given $x = y^4 - 3y^3$ find the value of $\frac{dy}{dx}$ at the point where $y = 3$.

$$x = y^4 - 3y^3$$

$$\therefore \quad \frac{dx}{dy} = 4y^3 - 9y^2$$

Taking the reciprocal of each side:

$$\frac{dy}{dx} = \frac{1}{4y^3 - 9y^2}$$

Substituting $y = 3$ gives

$$\frac{dy}{dx} = \frac{1}{4(3)^3 - 9(3)^2}$$

$$= \frac{1}{108 - 81} = \frac{1}{27}$$

At $y = 3$, $\frac{dy}{dx} = \frac{1}{27}$.

The technique described here will be further explored in the section on **implicit functions**, on page 212.

C3

Exercise 2E

1 For each of these find an expression for $\frac{dy}{dx}$, leaving your answer in terms of y.

a) $x = y^2 - 2y$ b) $x = 3 - 5y^3$

c) $x = 4y - 6y^3$ d) $x = 1 + \sqrt{y}$

e) $x = (2y - 1)^3$ f) $x = y + \frac{1}{y}$

g) $x = y^2(1 + y)^3$ h) $x = \frac{y}{1 + y}$

i) $x = \frac{1 - \sqrt{y}}{1 + \sqrt{y}}$

2 For each of these find the value of $\frac{dy}{dx}$ at the point indicated by the given value of y.

a) $x = y^3 - 2y$, at $y = 3$ b) $x = 3y - 2y^4$, at $y = -1$

c) $x = y - \sqrt{y}$, at $y = 4$ d) $x = \frac{1}{3 + y}$, at $y = -2$

e) $x = y - \frac{1}{y}$, at $y = 5$ f) $x = \sqrt{y^2 - 5}$, at $y = 3$

g) $x = y^3(y + 3)^2$, at $y = -2$ h) $x = \frac{2y + 3}{3y + 2}$, at $y = 1$

3 a) Given $x = \frac{y-1}{y+1}$, show that $\frac{dy}{dx} = \frac{(y+1)^2}{2}$.

b) By rearranging the expression $x = \frac{y-1}{y+1}$ show that $y = \frac{1+x}{1-x}$.

c) Differentiate $y = \frac{1+x}{1-x}$ to show that $\frac{dy}{dx} = \frac{2}{(1-x)^2}$.

d) Show that your answers to a) and c) are the same.

2.6 Applications

Example 9

Find the equation of the tangent to the curve $y = x^2(x+1)^4$ at the point $P(1, 16)$.

Use the product rule and the chain rule.

When $y = x^2(x+1)^4$, using the product rule gives:

$$\frac{dy}{dx} = x^2 \times 4(x+1)^3 + (x+1)^4 \times 2x$$
$$= 4x^2(x+1)^3 + 2x(x+1)^4$$
$$= 2x(x+1)^3[2x + (x+1)]$$
$$\therefore \quad \frac{dy}{dx} = 2x(x+1)^3(3x+1)$$

$u = x^2$ $\quad v = (x+1)^4$

$\frac{du}{dx} = 2x$ $\quad \frac{dv}{dx} = 4(x+1)^3$

At $P(1, 16)$,

$$\left.\frac{dy}{dx}\right|_{x=1} = 2(8)(4) = 64$$

Therefore, the equation of the tangent line is of the form

$$y = 64x + c$$

Since the tangent line passes through $P(1, 16)$,

$$16 = 64(1) + c$$
$$\therefore \quad c = -48$$

The equation of the tangent is $y = 64x - 48$.

Example 10

a) Find the coordinates of the stationary points on the curve $y = \frac{x}{x^2+4}$.

b) Show that $\frac{d^2y}{dx^2} = \frac{2x(x^2-12)}{(x^2+4)^3}$ and hence determine the nature of the stationary points.

a) At a stationary point $\frac{dy}{dx} = 0$.

Using the quotient rule gives:

$$\frac{dy}{dx} = \frac{(x^2+4)(1) - x(2x)}{(x^2+4)^2}$$
$$= \frac{x^2 + 4 - 2x^2}{(x^2+4)^2}$$
$$= \frac{4 - x^2}{(x^2+4)^2}$$

$u = x$ $\quad v = (x^2 + 4)$

$\frac{du}{dx} = 1$ $\quad \frac{dv}{dx} = 2x$

When $\frac{dy}{dx} = 0$: $\quad \frac{4 - x^2}{(x^2+4)^2} = 0$

$\therefore \quad 4 - x^2 = 0$

$\therefore \quad x = \pm 2$

C3

When $x = 2$: $\quad y = \frac{2}{2^2 + 4} = \frac{1}{4}$

When $x = -2$: $\quad y = \frac{-2}{(-2)^2 + 4} = -\frac{1}{4}$

The coordinates of the stationary points on the curve are $(2, \frac{1}{4})$ and $(-2, -\frac{1}{4})$.

b) Since $\frac{dy}{dx} = \frac{4 - x^2}{(x^2+4)^2}$, use the quotient rule to find $\frac{d^2y}{dx^2}$:

$$\frac{d^2y}{dx^2} = \frac{(x^2+4)^2(-2x) - (4 - x^2) \times 4x(x^2+4)}{(x^2+4)^4}$$
$$= \frac{-2x(x^2+4)^2 - 4x(x^2+4)(4 - x^2)}{(x^2+4)^4}$$
$$= \frac{-2x(x^2+4)[(x^2+4) + 2(4 - x^2)]}{(x^2+4)^4}$$
$$= \frac{-2x(x^2+4)(12 - x^2)}{(x^2+4)^4}$$
$$= \frac{2x(x^2 - 12)}{(x^2+4)^3}$$

$u = 4 - x^2$ $\quad v = (x^2 + 4)^2$

$\frac{du}{dx} = -2x$ $\quad \frac{dv}{dx} = 4x(x^2 + 4)$

as required.

When $x = 2$: $\quad \frac{d^2y}{dx^2} = \frac{-32}{512} = -\frac{1}{16} < 0$

Since $\frac{d^2y}{dx^2} < 0$, $\quad$ the stationary point $(2, \frac{1}{4})$ is a maximum.

When $x = -2$: $\quad \frac{d^2y}{dx^2} = \frac{32}{512} = \frac{1}{16} > 0$

Since $\frac{d^2y}{dx^2} > 0$, $\quad$ the stationary point $(-2, -\frac{1}{4})$ is a minimum.

Exercise 2F

1 Find the equation of the tangent and the normal to the curve $y = x(4 - x)^2$ at the point (2, 8).

2 Find the equation of the tangent and the normal to the curve $y = \dfrac{2x}{x - 1}$ at the point (3, 3).

3 Find the coordinates of the two points on the curve $y = \dfrac{x}{1 + x}$ where the gradient is $\frac{1}{9}$.

4 Show there is just one point, P, on the curve $y = x(x - 1)^3$, where the gradient is 7. Find the coordinates of P.

5 Given that $y = \dfrac{x^2}{2 - x}$, show that

$$\frac{dy}{dx} = \frac{x(4 - x)}{(2 - x)^2}$$

Hence find the coordinates of the two points on the curve $y = \dfrac{x^2}{2 - x}$ where the gradient of the curve is zero.

6 Given that $y = x\sqrt{3 + 2x}$, show that

$$\frac{dy}{dx} = \frac{3(1 + x)}{\sqrt{3 + 2x}}$$

Hence find the point on the curve $y = x\sqrt{3 + 2x}$ where the gradient is zero.

7 Find the coordinates of the stationary points on the curve $y = (x + 3)^2(2 - x)$, and determine their nature.

8 Show that the curve $y = \dfrac{x + 3}{(x + 4)^2}$ has a single stationary point.
Find the coordinates of that stationary point, and determine its nature.

9 Find the coordinates of the stationary point on the curve $y = \dfrac{x}{\sqrt{x - 5}}$, and determine its nature.

10 The total profit, y thousand pounds, generated from the production and sale of x items of a particular product is given by the formula

$$y = \frac{300\sqrt{x}}{100 + x}$$

Calculate the value of x which gives a maximum profit, and determine the maximum profit.

11 A rectangle is drawn inside a semicircle of radius 5 cm, in such a way that one of its sides lies along the diameter of the semicircle, as in the diagram on the right.

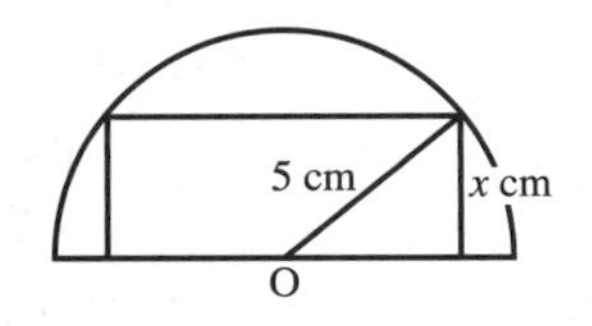

Given that the width of the rectangle is x cm, show that the area of the rectangle is $[2x\sqrt{25 - x^2}]$ cm². Calculate the maximum value of this area.

C3

Summary

You should know how to ...	Check out
1 Use the chain rule.	**1** Differentiate: a) $y = (x^2 + 5)^3$ b) $y = \frac{1}{\sqrt{3 + 2x}}$
2 Use the product rule.	**2** Differentiate: a) $y = x\sqrt{3 + 2x}$ b) $y = x^3(2 - x)^5$
3 Use the quotient rule.	**3** Differentiate: a) $y = \frac{x + 1}{x + 2}$ b) $y = \frac{x^3}{x^2 - 1}$

C3

Revision exercise 2

1 Differentiate each of the following with respect to x.

a) $(1 + x^2)^8$ b) $\frac{5x}{x^3 + 2}$ *(AQA, 2004)*

2 A model plane moves so that its height, y metres, above horizontal ground is given by

$$y = \frac{8x}{x^3 + 1} \quad x \geqslant 0$$

when its horizontal distance from the take off point on the ground is x metres.

a) Find the value of $\frac{dy}{dx}$ when $x = 1$.

b) i) Find the rate of change of y in m s^{-1} when $x = 1$ and x is increasing at a rate of 0.8 m s^{-1}.

ii) Interpret the sign in your answer to part i). *(AQA, 2004)*

3 The volume, V cm^3, of liquid in a container, with a depth of x cm, is given by

$$V = \frac{x^3}{x^2 + 1}$$

Find the value of $\frac{dV}{dx}$ when $x = 3$. *(AQA, 2003)*

4 The volume, V cm^3, of liquid in a container, when the depth is x cm is given by

$$V = (2x^2 + x^3)^{\frac{1}{2}}$$

a) Find the value of $\frac{dV}{dx}$ when $x = 2$. *(AQA, 2002)*

5 A curve has equation $y = \frac{x-2}{x^2+5}$

a) Determine the x-coordinates of the stationary points of the curve.

b) Find the equation of the tangent to the curve at the point where $x = 2$. *(AQA, 2001)*

6 a) Differentiate $(x^3 + 1)^{10}$ with respect to x.

b) Hence, or otherwise, evaluate $\int_0^1 x^2(x^3 + 1)^9 \, dx$ *(AQA/AEB, 1998)*

7 A curve is defined for $x \geqslant -2$ by the equation $y = (x^2 + 3)\sqrt{(x + 2)}$

a) Show that $\frac{dy}{dx} = 0$ when $x = -1$ and find the x-coordinate of the other stationary point.

b) Find the value of $\frac{d^2y}{dx^2}$ when $x = -1$. Hence determine whether your turning point when $x = -1$ is maximum or minimum point. *(AQA/AEB, 2000)*

8 A curve has equation $y = x(2x - 3)^5$

a) Show that $\frac{dy}{dx} = 3(2x - 3)^4(kx - 1)$ where k is a constant whose value should be stated.

b) Determine the values of x for which the curve has stationary points. *(AQA/AEB, 2001)*

9 a) Differentiate $y = x\sqrt{1 - x^2}$.

b) i) Hence show that the curve $y = x\sqrt{1 - x^2}$ has two stationary points.

ii) Show that the x-coordinates of the two stationary points are given by $x = \pm\frac{1}{\sqrt{2}}$.

10 Find the equation of the normal to the curve $y = \frac{1}{2x - 1}$ at the point where $x = 0$.

3 Trigonometric functions

This chapter will show you how to

- ✦ Recognise and use the reciprocal trigonometric functions $\sec x$, $\operatorname{cosec} x$, $\cot x$
- ✦ Solve trigonometric equations involving $\sec x$, $\operatorname{cosec} x$, $\cot x$
- ✦ Prove and use trigonometric identities involving $\sec x$, $\operatorname{cosec} x$, $\cot x$
- ✦ Differentiate the trigonometric functions $\sin x$, $\cos x$, $\tan x$

Before you start

You should know how to ...	Check in
1 Recognise the graphs of $\sin x$, $\cos x$ and $\tan x$ and their transformations.	**1** a) [graph: y against x, axis marks -2π, $-\pi$, O, π, 2π, 3π, 4π; y marks 1, -1] i) What is the equation of this graph? ii) Describe a transformation by which it may be obtained from the graph of $y = \cos x$. b) Sketch the graph of i) $y = \tan\left(x + \frac{\pi}{2}\right)$ ii) $y = \tan(x - \pi)$ for $-\pi \leqslant x \leqslant \pi$.
2 Solve trigonometric equations, including use of the identity $\cos^2 x + \sin^2 x = 1$.	**2** Solve these equations. a) $3 \sin 2x = 1$ $\quad 0 \leqslant x \leqslant 180°$ b) $2 \sin \theta + 3 \cos \theta = 0$ $\quad -\pi \leqslant \theta \leqslant \pi$ c) $5 \sin^2 \theta - 3 \cos \theta - 3 = 0$ $\quad 0 \leqslant x \leqslant 360°$ d) $\tan\left(x - \frac{\pi}{6}\right) + 2 = 0$ $\quad 0 \leqslant x \leqslant 2\pi$
3 Use the chain rule, product rule and quotient rule for differentiation.	**3** Differentiate with respect to x: a) $y = (x^3 - 2)^5$ b) $y = 2x(1 - x^2)^6$ c) $y = \dfrac{1 - x^3}{1 + x^2}$

3.1 The functions cosec θ, sec θ, cot θ

In C2 you learnt about the three main trigonometric functions: sin θ, cos θ and tan θ.

The reciprocal trigonometric functions cosecant, secant and cotangent (abbreviated to cosec, sec and cot) are defined as:

$$\operatorname{cosec}\theta = \frac{1}{\sin\theta} \qquad \sec\theta = \frac{1}{\cos\theta} \qquad \cot\theta = \frac{1}{\tan\theta}$$

Plotting values of y against θ gives the graphs for these functions.

Properties of the cosec function

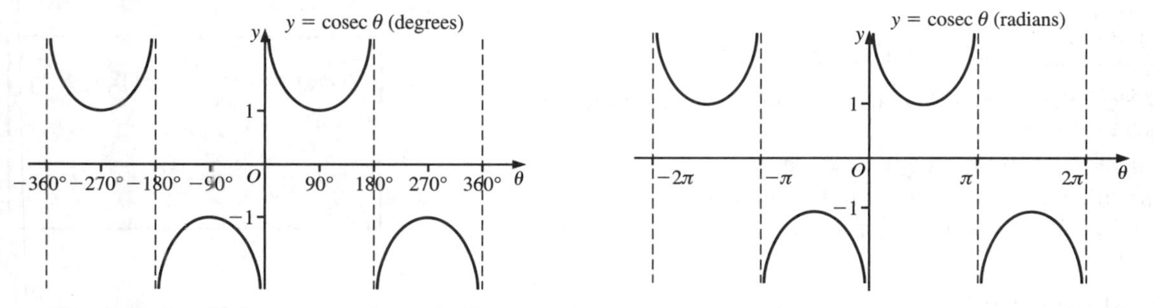

- The function $f(\theta) = \operatorname{cosec}\theta$ is periodic, of period 360° or 2π rad. That is,

  $\operatorname{cosec}(\theta + 360°) = \operatorname{cosec}\theta$ or $\operatorname{cosec}(\theta + 2\pi) = \operatorname{cosec}\theta$

- The graph of $f(\theta)$ has rotational symmetry about the origin of order 2.
- Range $f(\theta) \geqslant 1$ or $f(\theta) \leqslant -1$.
- The function $f(\theta) = \operatorname{cosec}\theta$ is not defined when $\theta = 0°, \pm 180°, \pm 360°, \dots$ (or $\pm\pi, \pm 2\pi$ rad).

> **Remember:**
> The period of a function is the smallest interval after which the function repeats the same values.

Properties of the secant function

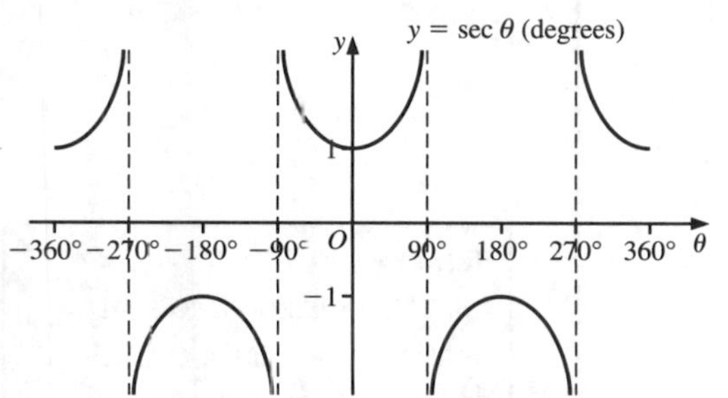

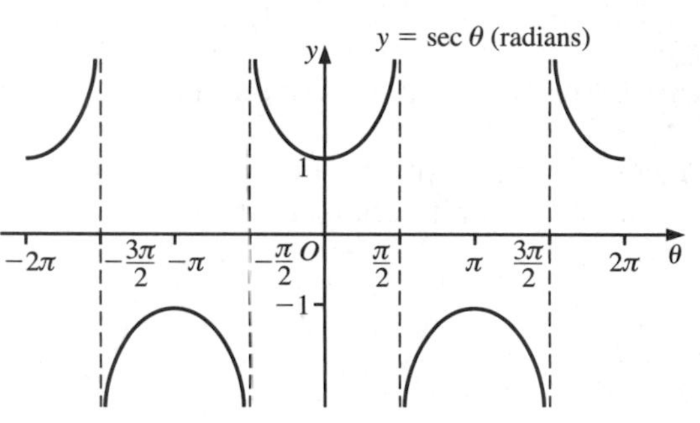

- The function $f(\theta) = \sec\theta$ is periodic, of period 360° or 2π rad. That is,

  $\sec(\theta + 360°) = \sec\theta$ or $\sec(\theta + 2\pi) = \sec\theta$

- The graph of $f(\theta)$ is symmetrical about the y-axis.
- Range $f(\theta) \geqslant 1$ or $f(\theta) \leqslant -1$.
- The function $f(\theta) = \sec\theta$ is not defined when $\theta = \pm 90°, \pm 270°, \dots$ $\left(\text{or } \pm\frac{\pi}{2}, \pm\frac{3\pi}{2} \text{ rad}\right)$.

Properties of the cotangent function

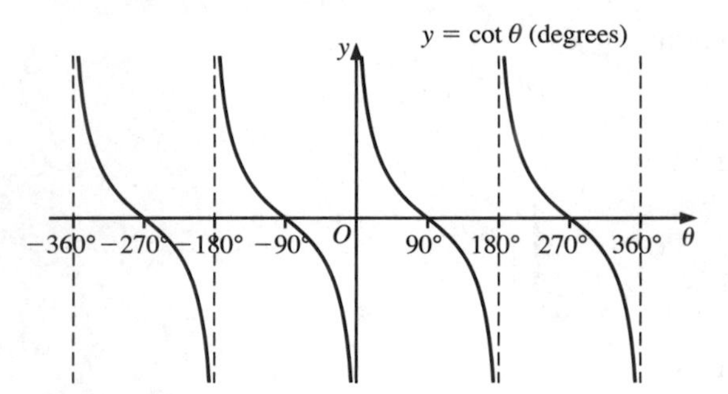

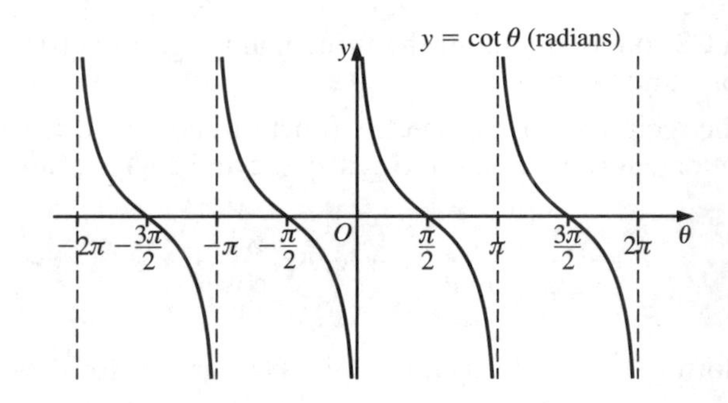

- The function $f(\theta) = \cot\theta$ is periodic, of period 180° or π rad. That is,

$$\cot(\theta + 180°) = \cot\theta \quad \text{or} \quad \cot(\theta + \pi) = \cot\pi$$

C3

- The graph of $f(\theta)$ has rotational symmetry about the origin of order 2.
- Range $f(\theta)$ can take any real value.
- The function $f(\theta) = \cot\theta$ is not defined when $\theta = 0°, \pm180°, \pm360°, \ldots$ (or $\theta = \pm\pi, \pm2\pi$ rad).

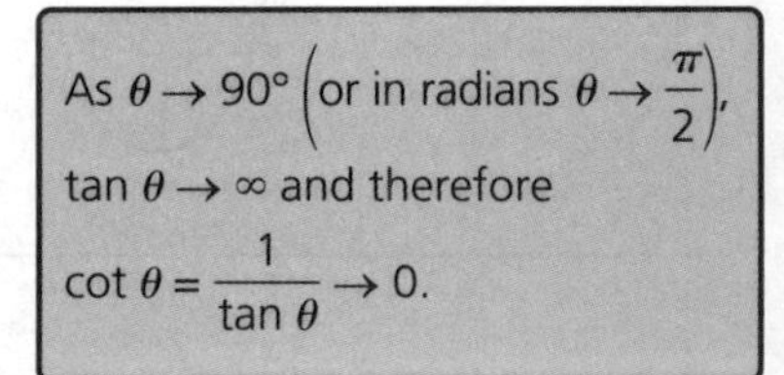
As $\theta \to 90°$ $\left(\text{or in radians } \theta \to \frac{\pi}{2}\right)$, $\tan\theta \to \infty$ and therefore $\cot\theta = \frac{1}{\tan\theta} \to 0$.

Curve sketching

This section uses the results on the transformations of graphs which you studied in C2.

Example 1

Sketch the graphs of each of these functions for $-2\pi \leqslant \theta \leqslant 2\pi$.

a) $f(\theta) = 1 + \sec\theta$ b) $f(\theta) = -3\operatorname{cosec}\theta$

c) $f(\theta) = \cot\left(\theta + \frac{\pi}{4}\right)$ d) $f(\theta) = \sec\left(\frac{1}{2}\theta\right)$

In each case state the period of the function.

a) To obtain the graph of $f(\theta) = 1 + \sec\theta$, translate the graph of $y = \sec\theta$ by 1 unit parallel to the y-axis.

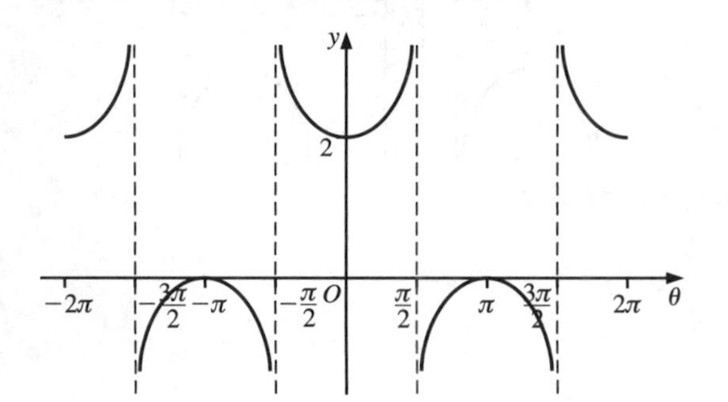

The period is 2π, which is no change from that of $\sec\theta$.

Remember:
$f(\theta) + a$ corresponds to a translation of a units parallel to the y-axis.

b) To obtain the graph of $f(\theta) = -3 \operatorname{cosec} \theta$ reflect the graph of $y = \operatorname{cosec} \theta$ in the θ-axis, then stretch parallel to the y-axis by a scale factor of 3.

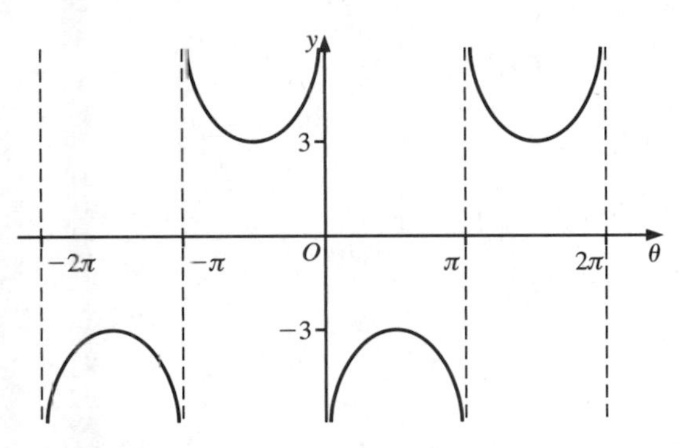

The period is 2π, which is no change from that of cosec θ.

> **Remember:**
> $af(\theta)$ corresponds to a stretch parallel to the y-axis by a scale factor of a.

c) To obtain the graph of $f(\theta) = \cot\left(\theta + \frac{\pi}{4}\right)$, translate the graph of $y = \cot \theta$ by $-\frac{\pi}{4}$ rad parallel to the θ-axis.

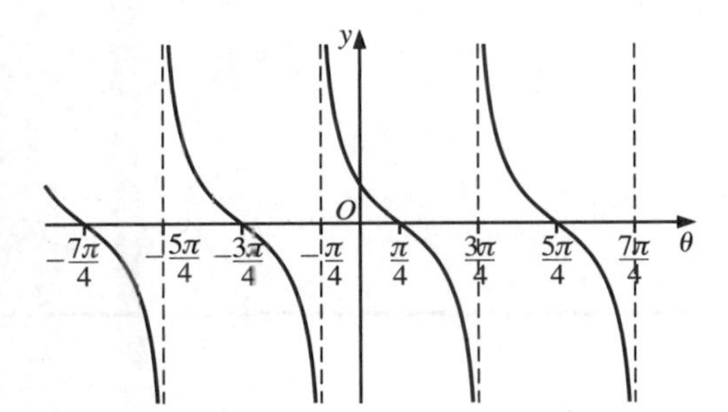

The period is π, which is no change from that of cot θ.

> **Remember:**
> $f(\theta + a)$ corresponds to a translation of $-a$ units parallel to the θ-axis.

d) To obtain the graph of $f(\theta) = \sec\left(\frac{1}{2}\theta\right)$, stretch the graph of $y = \sec \theta$ parallel to the θ-axis by a scale factor of 2.

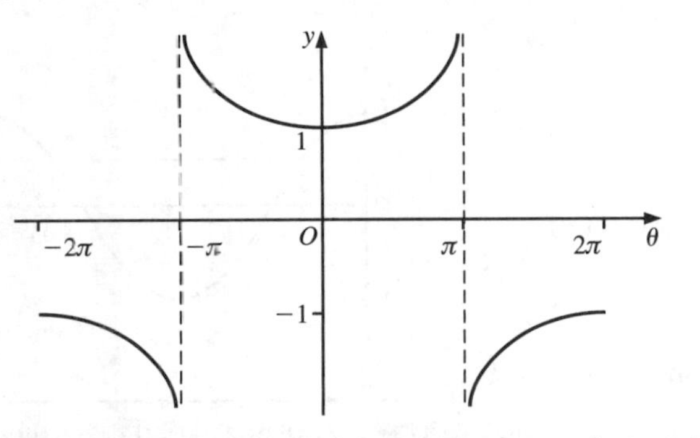

The period is 4π, which is double the period of the sec θ function.

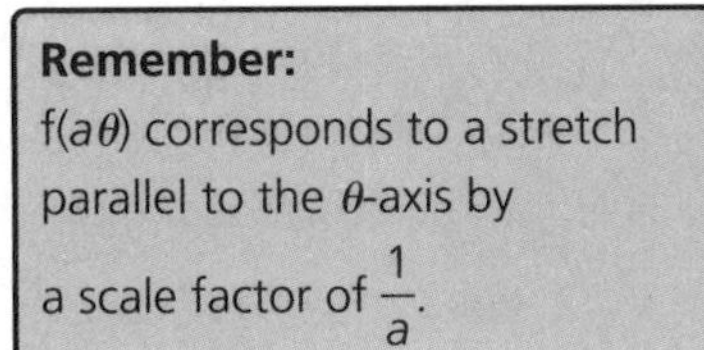

Exercise 3A

1 Sketch the graphs of each of these functions for $-360° \leqslant \theta \leqslant 360°$.

a) $y = 1 + \operatorname{cosec}\theta$ b) $y = \sec(\theta - 90°)$ c) $y = -2\operatorname{cosec}\theta$

d) $y = 1 - \sec\theta$ e) $y = \cot(\frac{1}{2}\theta)$ f) $y = \cot(\theta + 40°)$

2 Sketch the graphs of each of these functions for $0 \leqslant \theta \leqslant 2\pi$.

a) $y = \operatorname{cosec} 2\theta$ b) $y = -3\sec\theta$ c) $y = 2\cot\left(\theta - \frac{\pi}{4}\right)$

d) $y = 1 - \operatorname{cosec}\theta$ e) $y = 2\cot\theta$ f) $y = \sec(\theta + \pi)$

3 Sketch the graph of $y = 1 + \operatorname{cosec}(\theta - 60°)$, $0 \leqslant \theta \leqslant 360°$.

C3

3.2 Trigonometric equations

In C2 you studied trigonometric equations involving the three basic functions $\sin\theta$, $\cos\theta$ and $\tan\theta$. In C3 you extend those methods to $\operatorname{cosec}\theta$, $\sec\theta$ and $\cot\theta$.

As far as basic equations are concerned, the technique is very straightforward. All you need to do is transform an equation into a more recognisable form. This is illustrated in Example 2.

Example 2

Solve these equations for θ, where $0 \leqslant \theta \leqslant 2\pi$.

a) $\sec\theta = 3$ b) $5 + 2\operatorname{cosec}\theta = 0$ c) $\cot^2\theta = 4$

a) By definition $\sec\theta \equiv \dfrac{1}{\cos\theta}$. Therefore

$$\frac{1}{\cos\theta} = 3$$

Taking the reciprocal of both sides gives

$$\cos\theta = \tfrac{1}{3}$$

Drawing the graph of $y = \cos\theta$ and $y = \frac{1}{3}$ on the same set of axes for $0 \leqslant \theta \leqslant 2\pi$ gives:

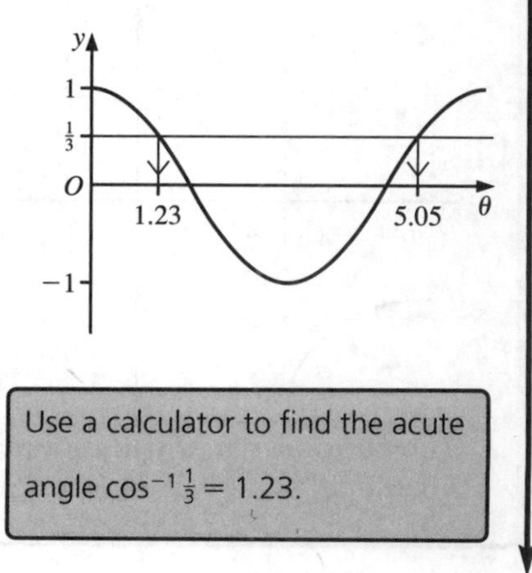

One solution is $\cos^{-1}(\frac{1}{3}) = 1.23$. The other solution in this range is $\theta = 2\pi - 1.23 = 5.05$.

Use a calculator to find the acute angle $\cos^{-1}\frac{1}{3} = 1.23$.

The solutions are $\theta = 1.23, 5.05$.

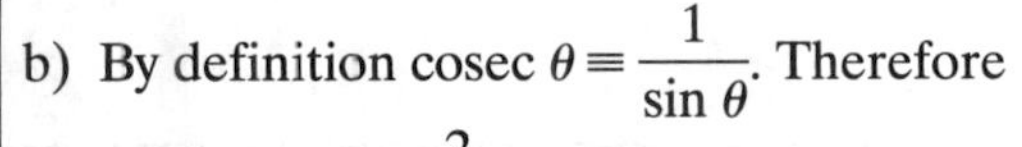

b) By definition $\operatorname{cosec}\theta \equiv \dfrac{1}{\sin\theta}$. Therefore

$$5 + \frac{2}{\sin\theta} = 0$$
$$-5 = \frac{2}{\sin\theta}$$
$$-5\sin\theta = 2$$
$$\therefore \quad \sin\theta = -\tfrac{2}{5}$$

Drawing the graph of $y = \sin\theta$ and $y = -\frac{2}{5}$ on the same set of axes for $0 \leqslant \theta \leqslant 2\pi$ gives:

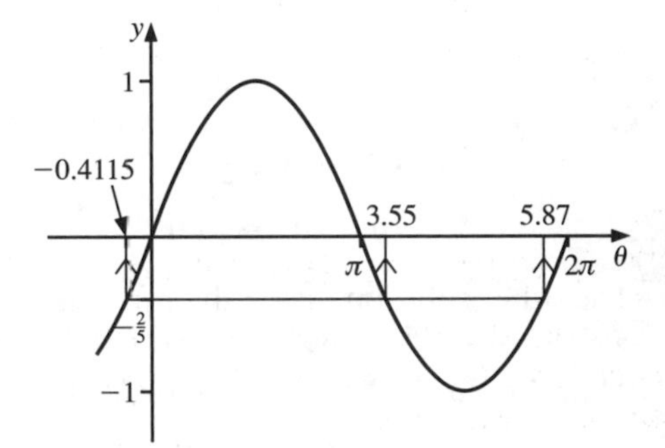

The calculator gives one solution of this equation as $\theta = -0.4115$. This solution is outside the range. The solutions in the range $0 \leqslant \theta \leqslant 2\pi$ are

$$\theta = \pi + 0.4115 = 3.55$$

and $\quad \theta = 2\pi - 0.4115 = 5.87$

The solutions are $\theta = 3.55, 5.87$ (2 d.p.)

c) By definition $\cot\theta \equiv \dfrac{1}{\tan\theta}$. Therefore

$$\frac{1}{\tan^2\theta} = 4$$
$$\tan^2\theta = \tfrac{1}{4}$$
$$\therefore \quad \tan\theta = \pm\tfrac{1}{2}$$

When $\tan\theta = \frac{1}{2}$, one solution is $\theta = 0.464$.

The other solution is $\theta = \pi + 0.464 = 3.61$.

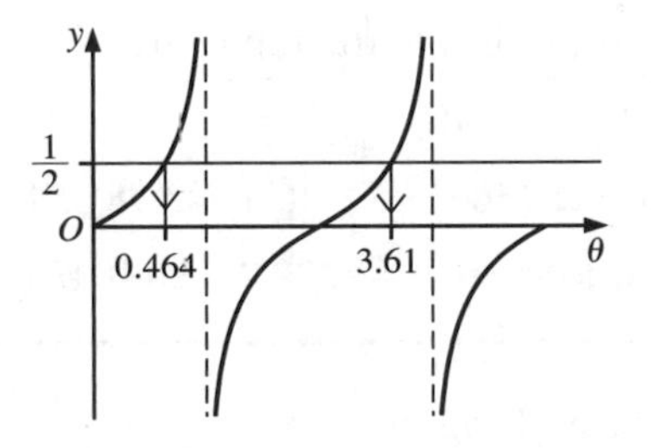

When $\tan\theta = -\frac{1}{2}$, one solution is $\theta = -0.464$.

The other solutions are $\pi - 0.464 = 2.68$
and $2\pi - 0.464 = 5.82$.

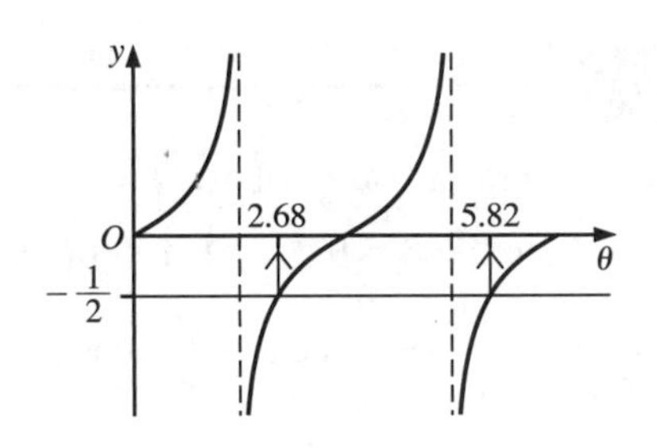

The solutions are $\theta = 0.464, 2.68, 3.61, 5.82$.

For some equations it is helpful to change the range.

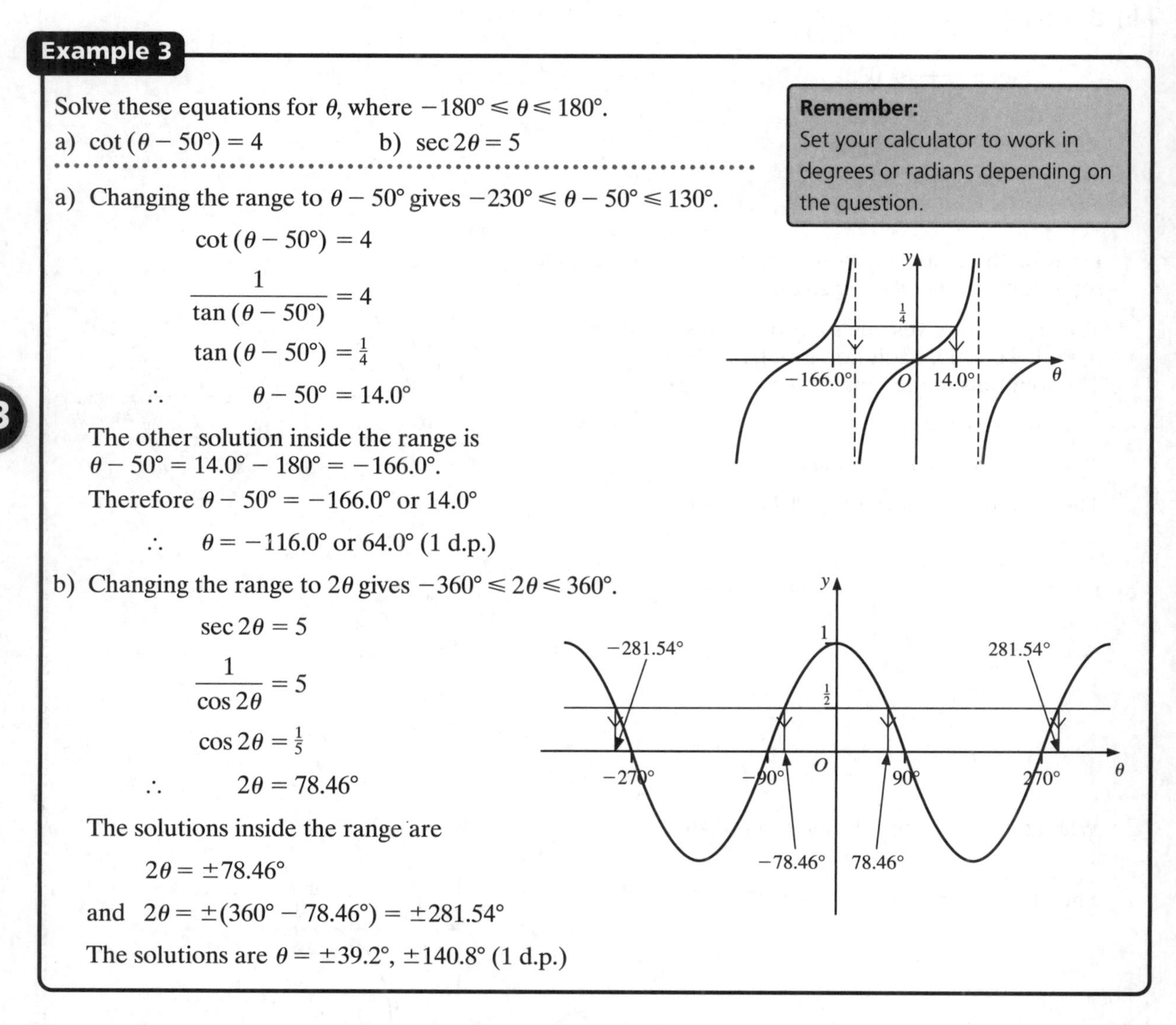

Example 3

Solve these equations for θ, where $-180° \leqslant \theta \leqslant 180°$.

a) $\cot(\theta - 50°) = 4$ b) $\sec 2\theta = 5$

a) Changing the range to $\theta - 50°$ gives $-230° \leqslant \theta - 50° \leqslant 130°$.

$$\cot(\theta - 50°) = 4$$

$$\frac{1}{\tan(\theta - 50°)} = 4$$

$$\tan(\theta - 50°) = \tfrac{1}{4}$$

$$\therefore \quad \theta - 50° = 14.0°$$

The other solution inside the range is $\theta - 50° = 14.0° - 180° = -166.0°$.

Therefore $\theta - 50° = -166.0°$ or $14.0°$

$$\therefore \quad \theta = -116.0° \text{ or } 64.0° \text{ (1 d.p.)}$$

b) Changing the range to 2θ gives $-360° \leqslant 2\theta \leqslant 360°$.

$$\sec 2\theta = 5$$

$$\frac{1}{\cos 2\theta} = 5$$

$$\cos 2\theta = \tfrac{1}{5}$$

$$\therefore \quad 2\theta = 78.46°$$

The solutions inside the range are

$$2\theta = \pm 78.46°$$

and $2\theta = \pm(360° - 78.46°) = \pm 281.54°$

The solutions are $\theta = \pm 39.2°, \pm 140.8°$ (1 d.p.)

More complicated trigonometric equations can be treated like algebraic equations.

Example 4

Solve these equations for θ, where $0 \leqslant \theta \leqslant 2\pi$.

a) $\sec^2\theta + 2\sec\theta - 8 = 0$ b) $1 + \tan\theta = 6\cot\theta$

a) Notice that $\sec^2\theta + 2\sec\theta - 8 = 0$ is a quadratic equation in $\sec\theta$. Factorising gives:

$$(\sec\theta + 4)(\sec\theta - 2) = 0$$

$$\therefore \quad \sec\theta = -4 \quad \text{or} \quad \sec\theta = 2$$

$$\therefore \quad \cos\theta = -\tfrac{1}{4} \quad \text{or} \quad \cos\theta = \tfrac{1}{2}$$

When $\cos\theta = -\frac{1}{4}$, one solution is $\theta = 1.82$ rad. In the required range the other solution is $2\pi - 1.82 = 4.46$ rad.

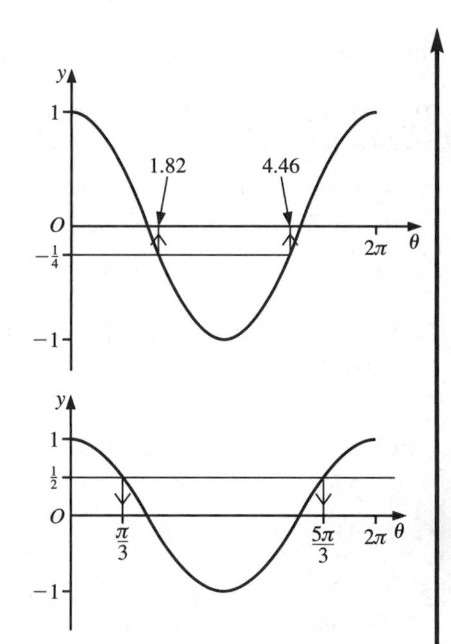

When $\cos\theta = \frac{1}{2}$, one solution is $\theta = \dfrac{\pi}{3}$. In the required range the other solution is $2\pi - \dfrac{\pi}{3} = \dfrac{5\pi}{3}$.

The solutions are $\theta = 1.82, 4.46, \dfrac{\pi}{3}$ and $\dfrac{5\pi}{3}$.

b) Rewrite the equation $1 + \tan\theta = 6\cot\theta$ in the form

$$1 + \tan\theta = \frac{6}{\tan\theta}$$

Multiply both sides by $\tan\theta$:

$$\tan\theta + \tan^2\theta = 6$$

This is a quadratic in $\tan\theta$, and rearranging and factorising gives:

$$\tan^2\theta + \tan\theta - 6 = 0$$

$$\therefore \quad (\tan\theta + 3)(\tan\theta - 2) = 0$$

$$\therefore \quad \tan\theta = -3 \quad \text{or} \quad \tan\theta = 2$$

When $\tan\theta = -3$, one solution is $\theta = -1.25$.
In the required range the solutions are $\theta = \pi - 1.25 = 1.89$ and $\theta = 2\pi - 1.25 = 5.03$.

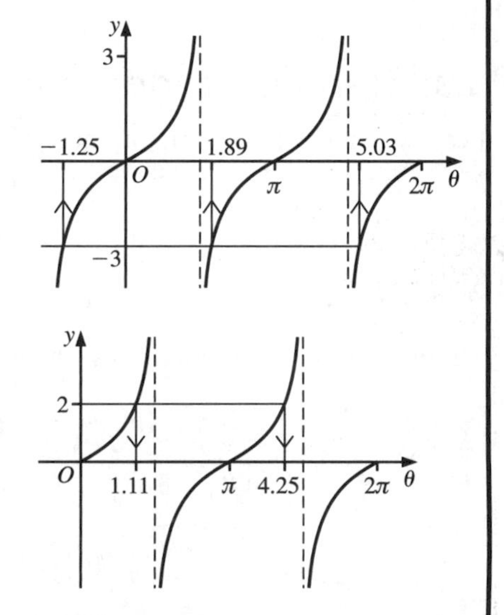

When $\tan\theta = 2$, one solution is $\theta = 1.11$.
In the required range the other solution is $\theta = \pi + 1.11 = 4.25$.

The solutions are $\theta = 1.89, 5.03, 1.11, 4.25$ (2 d.p.)

Exercise 3B

1 Solve each of these equations for $-180° \leqslant \theta \leqslant 180°$, giving your answers correct to one decimal place.

a) $\operatorname{cosec} \theta = 2$ b) $\sec \theta = 3$ c) $\cot \theta = 0.5$

d) $\cot \theta = -3$ e) $\sec \theta = 6$ f) $\operatorname{cosec} \theta = 5$

g) $\sec \theta = -1$ h) $\operatorname{cosec} \theta = -10$

2 Solve each of these equations for $0 \leqslant \theta \leqslant 2\pi$, giving your answers correct to two decimal places.

a) $\sec^2 \theta = 4$ b) $\cot^2 \theta + 3 \cot \theta = 0$

c) $\cot^2 \theta - 3 \cot \theta + 2 = 0$ d) $\sec^2 \theta + 4 \sec \theta - 5 = 0$

e) $2 \cot^2 \theta - 7 \cot \theta + 6 = 0$ f) $3 \cos \theta + 4 \sec \theta = 8$

g) $4 \sin \theta + 1 = 3 \operatorname{cosec} \theta$ h) $3 \sec \theta + 11 = 4 \cos \theta$

3 Solve each of the following equations for $0° \leqslant \theta \leqslant 180°$, giving your answers correct to one decimal place.

a) $\sec 2\theta = 3$ b) $\cot 3\theta = 4$ c) $\operatorname{cosec} 4\theta = 2$

d) $\cot 2\theta = -0.4$ e) $2 + \operatorname{cosec} 2\theta = 0$ f) $5 \cot 3\theta = 2$

g) $\sec 4\theta = -8$ h) $5 + \operatorname{cosec} 3\theta = 0$

4 Solve each of these equations for $0 \leqslant \theta \leqslant 180°$, giving your answers correct to one decimal place.

a) $\operatorname{cosec}(\theta + 20°) = 4$ b) $\sec(\theta - 50°) = -3$

c) $1 + 5 \cot(\theta - 100°) = 0$ d) $\operatorname{cosec}(\theta - 162°) = 6$

e) $\sec(\theta - 62°) = 3$ f) $\cot(\theta + 17°) = -0.4$

g) $5 \cot(\theta - 150°) = 3$ h) $\operatorname{cosec}(\theta + 210°) = 4$

3.3 Standard trigonometric identities

In C2 you met the two trigonometric identities

✦ $\tan \theta \equiv \dfrac{\sin \theta}{\cos \theta}$

✦ $\sin^2 \theta + \cos^2 \theta \equiv 1$

Note: The symbol $\equiv$ is used commonly for identities. It means that the statement is true for all values of the variable. In your examination you can use the simpler $=$ symbol.

There are three more identities which you need to know:

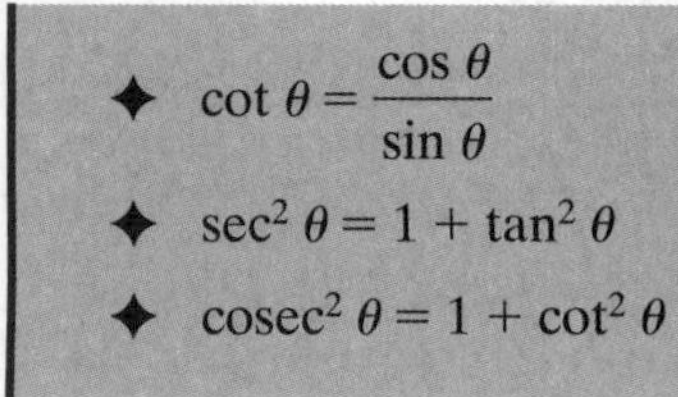

✦ $\cot \theta \equiv \dfrac{\cos \theta}{\sin \theta}$

✦ $\sec^2 \theta \equiv 1 + \tan^2 \theta$

✦ $\operatorname{cosec}^2 \theta \equiv 1 + \cot^2 \theta$

You should memorise these identities.

You can derive these three new identities from the original two.

- Since $\tan\theta = \dfrac{\sin\theta}{\cos\theta}$
  taking the reciprocal of both sides gives
  $$\cot\theta = \frac{\cos\theta}{\sin\theta}$$
- Since $\sin^2\theta + \cos^2\theta = 1$
  dividing both sides by $\cos^2\theta$ gives
  $$\frac{\sin^2\theta}{\cos^2\theta} + \frac{\cos^2\theta}{\cos^2\theta} = \frac{1}{\cos^2\theta}$$
  or $$\left(\frac{\sin\theta}{\cos\theta}\right)^2 + \left(\frac{\cos\theta}{\cos\theta}\right)^2 = \left(\frac{1}{\cos\theta}\right)^2$$
  therefore $\tan^2\theta + 1 = \sec^2\theta$
  as required.
- Again, start with the identity $\sin^2\theta + \cos^2\theta = 1$.
  Dividing both sides by $\sin^2\theta$ gives
  $$\frac{\sin^2\theta}{\sin^2\theta} + \frac{\cos^2\theta}{\sin^2\theta} = \frac{1}{\sin^2\theta}$$
  or $$\left(\frac{\sin\theta}{\sin\theta}\right)^2 + \left(\frac{\cos\theta}{\sin\theta}\right)^2 = \left(\frac{1}{\sin\theta}\right)^2$$
  therefore $1 + \cot^2\theta = \operatorname{cosec}^2\theta$
  as required.

C3

Further trigonometric equations

You are now ready to use these identities to solve some harder equations.

Example 5

Solve these equations for θ, where $0 \leqslant \theta \leqslant 360°$.

a) $3\sin\theta = 4\cos\theta$ b) $2\sec\theta = 5\operatorname{cosec}\theta$ c) $\sec\theta + \operatorname{cosec}\theta = 0$

a) $$3\sin\theta = 4\cos\theta$$

Dividing both sides by $\cos\theta$:
$$3\frac{\sin\theta}{\cos\theta} = 4\frac{\cos\theta}{\cos\theta}$$
$$\therefore \quad 3\tan\theta = 4$$
$$\therefore \quad \tan\theta = \tfrac{4}{3}$$

When $\tan\theta = \frac{4}{3}$, $\theta = 53.1°$ or $\theta = 180° + 53.1° = 233.1°$ in the required range.

The solutions are $\theta = 53.1°, 233.1°$ (1 d.p.)

It is useful to draw a sketch:

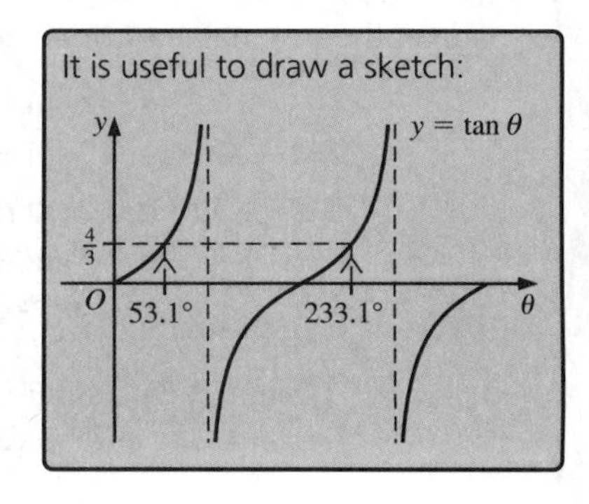

b)
$$2\sec\theta = 5\operatorname{cosec}\theta$$
$$\therefore \quad \frac{2}{\cos\theta} = \frac{5}{\sin\theta}$$
$$\frac{2\sin\theta}{\cos\theta} = 5$$
$$\therefore \quad 2\tan\theta = 5$$
$$\therefore \quad \tan\theta = \tfrac{5}{2}$$

When $\tan\theta = \frac{5}{2}$, $\theta = 68.2°$ or $\theta = 180° + 68.2° = 248.2°$ in the required range.

The solutions are $\theta = 68.2°, 248.2°$ (1 d.p.)

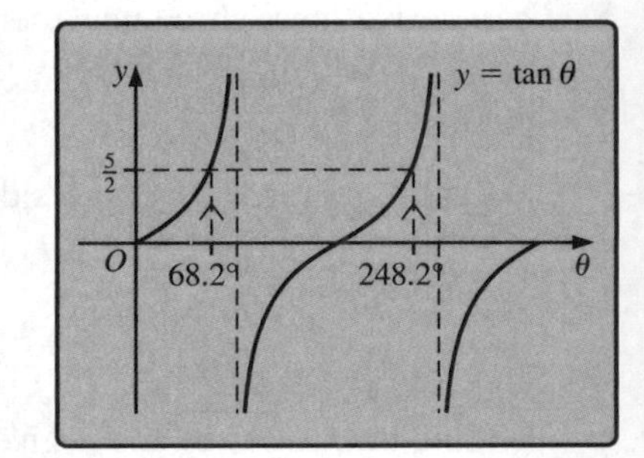

c)
$$\sec\theta + \operatorname{cosec}\theta = 0$$
$$\sec\theta = -\operatorname{cosec}\theta$$
$$\frac{1}{\cos\theta} = -\frac{1}{\sin\theta}$$
$$\frac{\sin\theta}{\cos\theta} = -1$$
$$\therefore \quad \tan\theta = -1$$

When $\tan\theta = -1$, $\theta = 135°$ or $\theta = 180° + 135° = 315°$ in the required range.

The solutions are $\theta = 135°, 315°$.

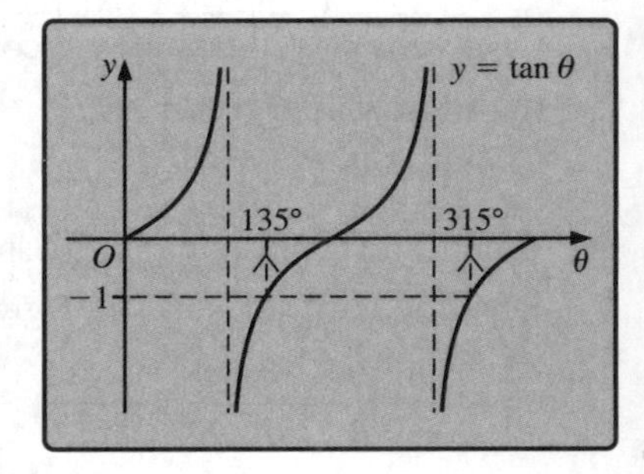

C3

Example 6

Solve these equations for θ, where $0 \leqslant \theta \leqslant 2\pi$.

a) $8\cos^2\theta - 2\sin\theta - 5 = 0$

b) $\tan^2\theta - \sec\theta = 1$

c) $10\cot\theta + 11 = 3\operatorname{cosec}^2\theta$

a) Notice that if you replace $\cos^2\theta$ with an expression in terms of $\sin^2\theta$, the original equation becomes a quadratic in $\sin\theta$.

Use the identity $\cos^2\theta = 1 - \sin^2\theta$

$$8\cos^2\theta - 2\sin\theta - 5 = 0$$
$$\therefore \quad 8(1 - \sin^2\theta) - 2\sin\theta - 5 = 0$$
$$8 - 8\sin^2\theta - 2\sin\theta - 5 = 0$$
$$8\sin^2\theta + 2\sin\theta - 3 = 0$$

Factorising and solving give:

$$(4\sin\theta + 3)(2\sin\theta - 1)$$
$$\therefore \quad \sin\theta = -\tfrac{3}{4} \quad \text{or} \quad \sin\theta = \tfrac{1}{2}$$

When $\sin\theta = -\frac{3}{4}$ one solution is $\theta = -0.85$. In the required range the solutions are $\theta = \pi + 0.85 = 3.99$ and $\theta = 2\pi - 0.85 = 5.43$.

When $\sin\theta = \frac{1}{2}$ one solution is $\theta = \dfrac{\pi}{6}$. The other solution in the required range is $\theta = \pi - \dfrac{\pi}{6} = \dfrac{5\pi}{6}$.

The solutions are $\theta = 3.99, 5.43, \dfrac{\pi}{6}, \dfrac{5\pi}{6}$.

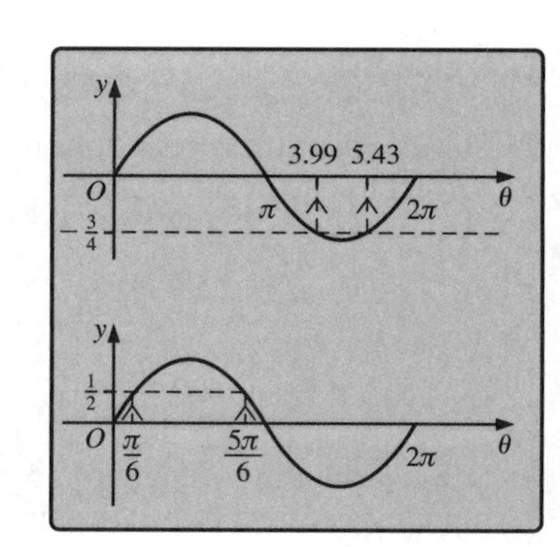

b) Notice that if you replace $\tan^2\theta$ with an expression in terms of $\sec^2\theta$, the original equation becomes a quadratic in $\sec\theta$.

$$\tan^2\theta - \sec\theta = 1$$
$$\therefore \quad (\sec^2\theta - 1) - \sec\theta = 1$$
$$\therefore \quad \sec^2\theta - \sec\theta - 2 = 0$$

Factorising and solving give:

$$(\sec\theta - 2)(\sec\theta + 1) = 0$$
$$\therefore \quad \sec\theta = 2 \quad \text{or} \quad \sec\theta = -1$$
$$\therefore \quad \cos\theta = \tfrac{1}{2} \quad \text{or} \quad \cos\theta = -1$$

When $\cos\theta = \frac{1}{2}$ one solution is $\theta = \dfrac{\pi}{3}$. The other solution in the required range is $\theta = 2\pi - \dfrac{\pi}{3} = \dfrac{5\pi}{3}$.

When $\cos\theta = -1$ the only solution in the required range is $\theta = \pi$.

The solutions are $\theta = \dfrac{\pi}{3}, \dfrac{5\pi}{3}, \pi$.

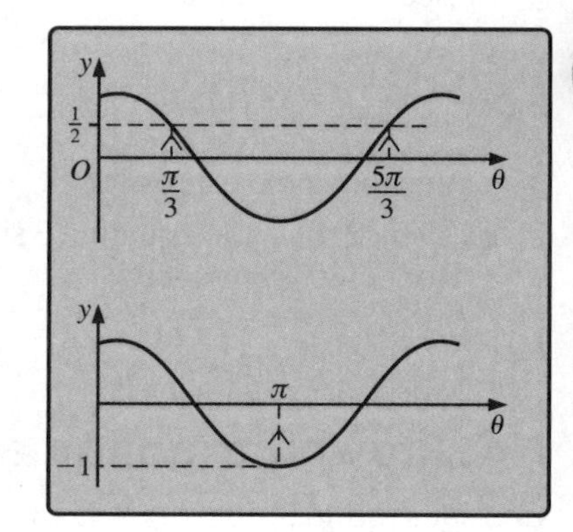

C3

c) Notice that if you replace $\text{cosec}^2\theta$ with an expression in terms of $\cot^2\theta$, the original equation becomes a quadratic in $\cot\theta$.

$$10\cot\theta + 11 = 3\,\text{cosec}^2\theta$$
$$\therefore \quad 10\cot\theta + 11 = 3(1 + \cot^2\theta)$$
$$\therefore \quad 3\cot^2\theta - 10\cot\theta - 8 = 0$$

Factorising and solving give:

$$(3\cot\theta + 2)(\cot\theta - 4)$$
$$\therefore \quad \cot\theta = -\tfrac{2}{3} \quad \text{or} \quad \cot\theta = 4$$
$$\therefore \quad \tan\theta = -\tfrac{3}{2} \quad \text{or} \quad \tan\theta = \tfrac{1}{4}$$

When $\tan\theta = -\frac{3}{2}$ one solution is $\theta = -0.98$. The solutions in the required range are $\theta = \pi - 0.98 = 2.16$ or $\theta = 2\pi - 0.98 = 5.30$.

When $\tan\theta = \frac{1}{4}$ one solution is $\theta = 0.244\,98$. The other solution in the required range is $\theta = \pi + 0.244\,98 = 3.386$.

The solutions are $\theta = 2.16, 5.30, 0.25, 3.39$ (2 d.p.).

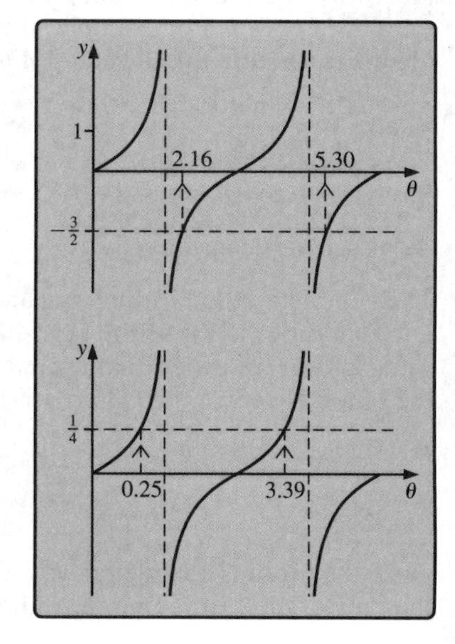

Exercise 3C

1 Solve each of these equations for $0 \leqslant \theta \leqslant 360°$, giving your answers correct to one decimal place.

a) $4 \sin \theta = 5 \cos \theta$
b) $3 \cos \theta = 8 \sin \theta$
c) $3 \sec \theta = 2 \operatorname{cosec} \theta$
d) $2 \operatorname{cosec} \theta = 5 \sec \theta$
e) $\sec \theta - \operatorname{cosec} \theta = 0$
f) $2 \operatorname{cosec} \theta + 3 \sec \theta = 0$
g) $4 \cos^2 \theta = \sin^2 \theta$
h) $3 \sec^2 \theta = 5 \operatorname{cosec}^2 \theta$

2 Solve each of these equations for $0 \leqslant \theta \leqslant 2\pi$, giving your answers in radians correct to two decimal places.

a) $3 \sin^2 \theta - 5 \cos \theta - 1 = 0$
b) $3 \cos^2 \theta = 2(1 + \sin \theta)$
c) $\sec^2 \theta = 7 + \tan \theta$
d) $11 + \tan^2 \theta = 7 \sec \theta$
e) $2 \operatorname{cosec}^2 \theta + 7 \cot \theta = 6$
f) $2 \cot^2 \theta = 13 + \operatorname{cosec} \theta$
g) $2 \sec^2 \theta = 13 (\tan \theta - 1)$
h) $6 \operatorname{cosec}^2 \theta = 5(2 - \cot \theta)$

C3

3 a) Factorise the expression $3x^3 - 10x^2 + 9x - 2$.
b) Hence solve the equation $3 \sin^3 \theta + 10 \cos^2 \theta + 9 \sin \theta - 12 = 0$ for $-180° \leqslant \theta \leqslant 180°$.

3.4 Proving trigonometric identities

Consider this identity:

$$\tan \theta + \cos \theta \equiv \sec \theta \operatorname{cosec} \theta$$

If you substitute different values of θ into the LHS and RHS you will verify this identity is true for those particular values of θ. However, this does not **prove** the identity for all values of θ. You can prove identities by using other simpler identities which you know to be true for all values of θ. For example, you know that the following identities are true for all values of θ.

- $\tan \theta \equiv \dfrac{\sin \theta}{\cos \theta}$
- $\cot \theta \equiv \dfrac{\cos \theta}{\sin \theta}$
- $\sin^2 \theta + \cos^2 \theta \equiv 1$
- $\sec^2 \theta \equiv 1 + \tan^2 \theta$
- $\operatorname{cosec}^2 \theta \equiv 1 + \cot^2 \theta$

The general method for proving an identity is to choose either the left-hand side (LHS) **or** the right-hand side (RHS) and show, by using known identities, that it can be manipulated into the form of the other. However, two alternative techniques are also possible.

In your examination, you will usually be expected to start at the LHS and work towards the RHS.

- $\text{LHS} - \text{RHS} \equiv 0$ or
- $\dfrac{\text{LHS}}{\text{RHS}} \equiv 1$

Once the proof is completed, you can write QED, which stands for the Latin 'Quod Erat Demonstrandum' – 'Which was to be proved'.

Example 7

Prove the identity $\tan\theta + \cot\theta \equiv \sec\theta\,\mathrm{cosec}\,\theta$.

$$\begin{aligned}
\text{LHS} &\equiv \tan\theta + \cot\theta \\
&\equiv \frac{\sin\theta}{\cos\theta} + \frac{\cos\theta}{\sin\theta} \\
&\equiv \frac{\sin^2\theta + \cos^2\theta}{\sin\theta\cos\theta} \\
&\equiv \frac{1}{\sin\theta\cos\theta} \quad (\text{since } \sin^2\theta + \cos^2\theta \equiv 1) \\
&\equiv \frac{1}{\sin\theta} \times \frac{1}{\cos\theta} \\
&\equiv \mathrm{cosec}\,\theta\sec\theta \equiv \text{RHS} \qquad \text{QED}
\end{aligned}$$

The $\equiv$ symbol will be used for an identity throughout the remainder of this section.

C3

Example 8

Prove the identity

$$(1 - \sin\theta + \cos\theta)^2 \equiv 2(1 - \sin\theta)(1 + \cos\theta)$$

$$\begin{aligned}
\text{LHS} &\equiv (1 - \sin\theta + \cos\theta)(1 - \sin\theta + \cos\theta) \\
&\equiv 1 - 2\sin\theta + 2\cos\theta + \sin^2\theta + \cos^2\theta - 2\sin\theta\cos\theta \\
&\equiv 2 - 2\sin\theta + 2\cos\theta - 2\sin\theta\cos\theta \quad (\text{since } \sin^2\theta + \cos^2\theta \equiv 1) \\
&\equiv 2(1 - \sin\theta + \cos\theta - \sin\theta\cos\theta) \\
&\equiv 2(1 - \sin\theta)(1 + \cos\theta) \equiv \text{RHS} \qquad \text{QED}
\end{aligned}$$

Example 9

Prove the identity

$$\frac{1 + \sin\theta}{1 - \sin\theta} \equiv (\tan\theta + \sec\theta)^2$$

$$\begin{aligned}
\text{RHS} &\equiv \left(\frac{\sin\theta}{\cos\theta} + \frac{1}{\cos\theta}\right)^2 \\
&\equiv \left(\frac{\sin\theta + 1}{\cos\theta}\right)^2 \\
&\equiv \frac{(1 + \sin\theta)^2}{\cos^2\theta} \\
&\equiv \frac{(1 + \sin\theta)^2}{1 - \sin^2\theta} \\
&\equiv \frac{(1 + \sin\theta)^2}{(1 + \sin\theta)(1 - \sin\theta)} \\
&\equiv \frac{1 + \sin\theta}{1 - \sin\theta} \equiv \text{LHS} \qquad \text{QED}
\end{aligned}$$

Exercise 3D

Prove each of these identities.

1. $\sin\theta\tan\theta + \cos\theta = \sec\theta$
2. $\operatorname{cosec}\theta + \tan\theta\sec\theta = \operatorname{cosec}\theta\sec^2\theta$
3. $\operatorname{cosec}\theta - \sin\theta = \cot\theta\cos\theta$
4. $(\sin\theta + \cos\theta)^2 - 1 = 2\sin\theta\cos\theta$
5. $(\sin\theta - \operatorname{cosec}\theta)^2 = \sin^2\theta + \cot^2\theta - 1$
6. $(\sec\theta + \tan\theta)(\sec\theta - \tan\theta) = 1$
7. $\tan^2\theta + \sin^2\theta = (\sec\theta + \cos\theta)(\sec\theta - \cos\theta)$
8. $\sec^2\theta + \cot^2\theta = \operatorname{cosec}^2\theta + \tan^2\theta$
9. $(\sin\theta + \cos\theta)(1 - \sin\theta\cos\theta) = \sin^3\theta + \cos^3\theta$
10. $\tan^4\theta + \tan^2\theta = \sec^4\theta - \sec^2\theta$
11. $\cos^4\theta - \sin^4\theta = \cos^2\theta - \sin^2\theta$
12. $\sin\theta + \cos\theta = \dfrac{1 - 2\cos^2\theta}{\sin\theta - \cos\theta}$
13. $\dfrac{\sin\theta}{1 + \cos\theta} + \dfrac{1 + \cos\theta}{\sin\theta} = 2\operatorname{cosec}\theta$
14. $\dfrac{\operatorname{cosec}\theta}{\cot\theta + \tan\theta} = \cos\theta$
15. $\dfrac{1}{1 + \tan^2\theta} + \dfrac{1}{1 + \cot^2\theta} = 1$
16. $\dfrac{1 - \sin\theta}{\cos\theta} = \dfrac{1}{\sec\theta + \tan\theta}$

C3

3.5 Differentiation of sin *x*, cos *x* and tan *x*

If x is measured in radians then:

- $\dfrac{d}{dx}(\sin x) = \cos x$
- $\dfrac{d}{dx}(\cos x) = -\sin x$
- $\dfrac{d}{dx}(\tan x) = \sec^2 x$

You should memorise these derivatives.

The proofs of these results require theorems on compound angles which you will not meet until C4. So, for the time being, you just need to be able to apply them in different situations.

Example 10

Find $\frac{dy}{dx}$ for each of these functions.

a) $y = \sin 3x$ b) $y = \cos(4x - 1)$ c) $y = \tan(x^3)$

a) If you let $u = 3x$, then $y = \sin u$.

Differentiating each expression gives

$$\frac{dy}{du} = \cos u \quad \text{and} \quad \frac{du}{dx} = 3$$

By the chain rule,

See section 2.1.

$$\frac{dy}{dx} = \frac{dy}{du}\frac{du}{dx}$$
$$= (\cos u)(3)$$
$$= 3\cos u$$
$$\therefore \quad \frac{dy}{dx} = 3\cos 3x$$

b) By the chain rule, when $y = \cos(4x - 1)$

$$\frac{dy}{dx} = -\sin(4x - 1) \times \frac{d}{dx}(4x - 1)$$
$$= -\sin(4x - 1) \times (4)$$
$$= -4\sin(4x - 1)$$
$$\therefore \quad \frac{dy}{dx} = -4\sin(4x - 1)$$

c) By the chain rule, when $y = \tan(x^3)$

$$\frac{dy}{dx} = \sec^2(x^3) \times \frac{d}{dx}(x^3)$$
$$= \sec^2(x^3) \times (3x^2)$$
$$= 3x^2\sec^2(x^3)$$
$$\therefore \quad \frac{dy}{dx} = 3x^2\sec^2(x^3)$$

You can use the inverse function of a function rule to integrate trigonometric functions.

See section 2.2.

Example 11

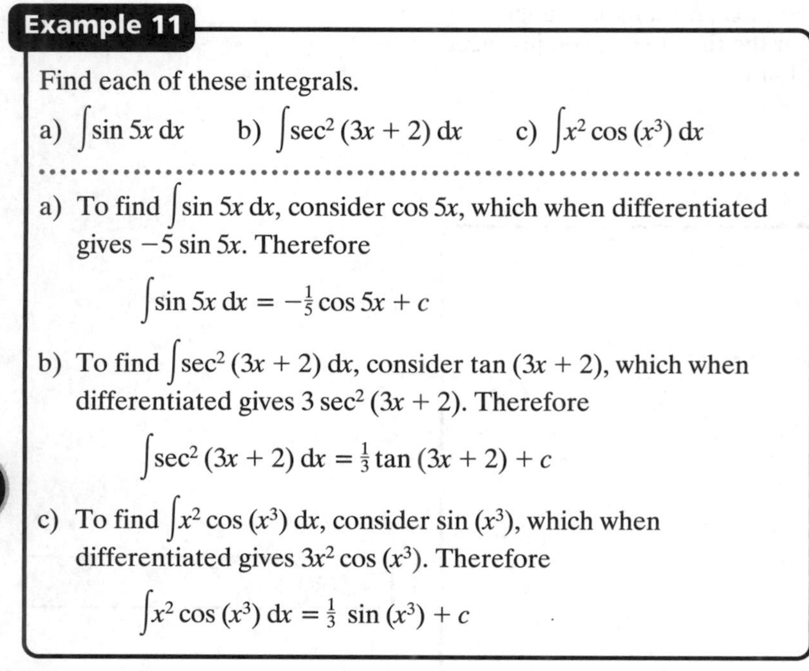

Find each of these integrals.

a) $\int \sin 5x \, dx$ b) $\int \sec^2 (3x + 2) \, dx$ c) $\int x^2 \cos (x^3) \, dx$

a) To find $\int \sin 5x \, dx$, consider $\cos 5x$, which when differentiated gives $-5 \sin 5x$. Therefore

$$\int \sin 5x \, dx = -\tfrac{1}{5} \cos 5x + c$$

b) To find $\int \sec^2 (3x + 2) \, dx$, consider $\tan (3x + 2)$, which when differentiated gives $3 \sec^2 (3x + 2)$. Therefore

$$\int \sec^2 (3x + 2) \, dx = \tfrac{1}{3} \tan (3x + 2) + c$$

c) To find $\int x^2 \cos (x^3) \, dx$, consider $\sin (x^3)$, which when differentiated gives $3x^2 \cos (x^3)$. Therefore

$$\int x^2 \cos (x^3) \, dx = \tfrac{1}{3} \sin (x^3) + c$$

In general,

- ✦ $\int \cos kx \, dx = \dfrac{1}{k} \sin kx + c$
- ✦ $\int \sin kx \, dx = -\dfrac{1}{k} \cos kx + c$

You should memorise these formulae.

You can also differentiate powers of $\sin x$, $\cos x$ and $\tan x$.

Example 12

Differentiate each of these with respect to x.

a) $y = \sin^3 x$ b) $y = 3 \tan^5 x$

a) It helps to write $\sin^3 x$ as $(\sin x)^3$

Now, by the chain rule

$$\frac{dy}{dx} = 3 (\sin x)^2 \times \frac{d}{dx}(\sin x)$$
$$= 3 (\sin x)^2 \times \cos x$$
$$\therefore \quad \frac{dy}{dx} = 3 \cos x \sin^2 x$$

b) This time write $\tan^5 x$ as $(\tan x)^5$

By the chain rule,

$$\frac{dy}{dx} = 3 \times 5 (\tan x)^4 \times \frac{d}{dx}(\tan x)$$
$$= 15 (\tan x)^4 \times (\sec^2 x)$$
$$\therefore \quad \frac{dy}{dx} = 15 \sec^2 x \tan^4 x$$

Exercise 3E

1 Find $\frac{dy}{dx}$ for each of these.

a) $y = \sin 3x$
b) $y = \cos 2x$
c) $y = \tan 5x$
d) $y = -\sin 6x$
e) $y = 2\cos 7x$
f) $y = -6\cos 5x$
g) $y = 8\sin \frac{1}{2}x$
h) $y = \tan (x + 3)$
i) $y = \sin (x - 4)$
j) $y = 3\sin \left(x + \frac{\pi}{4}\right)$
k) $y = -2\tan (4x - 7)$
l) $y = 8\sin \left(\frac{3x - \pi}{2}\right)$

2 Differentiate each of these with respect to x.

a) $\sin (x^2)$
b) $\tan (x^3)$
c) $2\cos (x^2 - 1)$
d) $3\sin (2x^3 + 3)$
e) $-4\sin (1 - x^2)$
f) $6\cos (4 - 3x^4)$
g) $-\cos (x^2 - 2)$
h) $\tan (x^3 - 3)$
i) $\frac{1}{2}\cos (6x^2 + 1)$
j) $-7\tan (2 - x^4)$
k) $6\sin \sqrt{x}$
l) $\cos \left(\frac{1}{x}\right)$

3 Find each of these integrals.

a) $\int 2\cos 2x \, dx$
b) $\int \sin 4x \, dx$
c) $\int \sec^2 4x \, dx$
d) $\int \sec^2 (2x - 1) \, dx$
e) $\int -6\sin (3x + 2) \, dx$
f) $\int \sin \left(\frac{5x - \pi}{4}\right) dx$
g) $\int x\cos (x^2) \, dx$
h) $\int 8x^3 \sec^2 (x^4) \, dx$
i) $\int 3x\cos (x^2 - 7) \, dx$
j) $\int 12x\cos (x^2 - 4) \, dx$
k) $\int x^2 \sin (3 - x^3) \, dx$
l) $\int \frac{\sin \sqrt{x}}{\sqrt{x}} \, dx$

4 Find $f'(x)$ for each of these.

a) $f(x) = \sin^2 x$
b) $f(x) = \cos^3 x$
c) $f(x) = \sqrt{\cos x}$
d) $f(x) = \tan^6 x$
e) $f(x) = 2\sin^7 x$
f) $f(x) = -3\cos^6 x$
g) $f(x) = \sin^4 5x$
h) $f(x) = \tan^4 3x$
i) $f(x) = 2\sqrt{\cos 4x}$

3.6 Products and quotients

You can also apply the product and quotient rules of differentiation to trigonometric functions.

Remember:
$$\frac{d}{dx}(uv) = u\frac{dv}{dx} + v\frac{du}{dx}$$

Example 13

Use the product rule to differentiate each of these functions.

a) $x^2 \sin x$ b) $\cos 3x \sin 2x$

a) When $y = x^2 \sin x$, using the product rule gives:
$$\frac{dy}{dx} = x^2 \cos x + 2x \sin x$$

$u = x^2$ $v = \sin x$

$\frac{du}{dx} = 2x$ $\frac{dv}{dx} = \cos x$

b) When $y = \cos 3x \sin 2x$, using the product rule gives:
$$\frac{dy}{dx} = \cos 3x \times 2\cos 2x + \sin 2x \times (-3\sin 3x)$$
$$= 2\cos 3x \cos 2x - 3\sin 2x \sin 3x$$

$u = \cos 3x$ $v = \sin 2x$

$\frac{du}{dx} = -3\sin 3x$ $\frac{dv}{dx} = 2\cos 2x$

C3

Example 14

Use the quotient rule to differentiate each of these functions.

a) $\dfrac{x^3}{\tan x}$ b) $\dfrac{\sin x}{\cos x}$ c) $\dfrac{1+\sin x}{1+\cos x}$

Remember:
$$\frac{d}{dx}\left(\frac{u}{v}\right) = \frac{v\frac{du}{dx} - u\frac{dv}{dx}}{v^2}$$

a) When $y = \dfrac{x^3}{\tan x}$, using the quotient rule gives:
$$\frac{dy}{dx} = \frac{\tan x \times 3x^2 - x^3 \sec^2 x}{\tan^2 x}$$
$$= \frac{3x^2 \tan x - x^3 \sec^2 x}{\tan^2 x}$$

$u = x^3$ $v = \tan x$

$\frac{du}{dx} = 3x^2$ $\frac{dv}{dx} = \sec^2 x$

b) When $y = \dfrac{\sin x}{\cos x}$, using the quotient rule gives:
$$\frac{dy}{dx} = \frac{\cos x \cos x - \sin x(-\sin x)}{\cos^2 x}$$
$$= \frac{\cos^2 x + \sin^2 x}{\cos^2 x} = \frac{1}{\cos^2 x}$$
$$= \sec^2 x$$

$u = \sin x$ $v = \cos x$

$\frac{du}{dx} = \cos x$ $\frac{dv}{dx} = -\sin x$

Since $\tan x = \dfrac{\sin x}{\cos x}$, $\dfrac{d}{dx}(\tan x) = \sec^2 x$.

c) When $y = \dfrac{1+\sin x}{1+\cos x}$, using the quotient rule gives:
$$\frac{dy}{dx} = \frac{(1+\cos x)\cos x - (1+\sin x)(-\sin x)}{(1+\cos x)^2}$$
$$= \frac{\cos x + \cos^2 x + \sin x + \sin^2 x}{(1+\cos x)^2}$$
$$= \frac{1+\cos x+\sin x}{(1+\cos x)^2}$$

$u = 1 + \sin x$ $v = 1 + \cos x$

$\frac{du}{dx} = \cos x$ $\frac{dv}{dx} = -\sin x$

Exercise 3F

1 Use the product rule to differentiate each of these expressions.

a) $x \tan x$ b) $x^3 \cos x$

c) $x^2 \sin 2x$ d) $x^4 \tan 3x$

e) $\sin 2x \sin 3x$ f) $\sin x \cos 5x$

g) $\tan 4x \tan 6x$ h) $3x \sin x$

i) $5x \tan 2x$ j) $2x^2 \cos 3x$

k) $(1 + \sin x)(1 + \cos x)$ l) $x^4(1 + \sin 2x)$

2 Use the quotient rule to differentiate each of these expressions.

a) $\dfrac{x}{\sin x}$ b) $\dfrac{x^2}{\tan x}$

c) $\dfrac{\tan x}{x}$ d) $\dfrac{x}{\cos 2x}$

e) $\dfrac{\sin 2x}{x^3}$ f) $\dfrac{\cos x}{\sin x}$

g) $\dfrac{\sin 3x}{\sin x}$ h) $\dfrac{\cos 2x}{\tan 3x}$

i) $\dfrac{x}{1 + \sin x}$ j) $\dfrac{1 + \tan x}{x^2}$

k) $\dfrac{1 + \cos x}{1 - \cos x}$ l) $\dfrac{\sin x + \cos x}{\sin x - \cos x}$

3 Differentiate each of these expressions with respect to x.

a) $x \sin x$ b) $x^2 \cos x$

c) $x \cos 3x$ d) $x^3 \tan 6x$

e) $x \sin^5 x$ f) $3x^2 \cos^4 x$

g) $\dfrac{x}{\tan x}$ h) $\dfrac{\cos 2x}{x + 1}$

i) $\dfrac{1}{1 + \tan x}$ j) $\dfrac{1 + \sin 2x}{\cos 2x}$

k) $\dfrac{x}{1 + \cos x}$ l) $\dfrac{1 + \sin x}{1 + \cos x}$

3.7 Applications

You can use differentiation to find the equations of tangents to trigonometric curves.

Example 15

Find the equation of the tangent to the curve $y = x + \tan x$ at the point where $x = \frac{\pi}{4}$.

You need $\frac{dy}{dx}$ when $x = \frac{\pi}{4}$. Since $y = x + \tan x$,

$$\frac{dy}{dx} = 1 + \sec^2 x$$

When $x = \frac{\pi}{4}$: $\left.\frac{dy}{dx}\right|_{x=\frac{\pi}{4}} = 1 + \sec^2\left(\frac{\pi}{4}\right)$

$$= 1 + \frac{1}{\cos^2\left(\frac{\pi}{4}\right)} = 1 + \frac{1}{0.5} = 3$$

The gradient of the tangent line is 3. Therefore, the equation of the tangent is of the form $y = 3x + c$.

When $x = \frac{\pi}{4}$, $y = \left(\frac{\pi}{4} + 1\right)$. Therefore, the tangent passes through the point $\left(\frac{\pi}{4}, \frac{\pi}{4} + 1\right)$. So,

$$\frac{\pi}{4} + 1 = 3\left(\frac{\pi}{4}\right) + c \quad \text{giving} \quad c = 1 - \frac{\pi}{2}$$

The equation of the tangent is

$$y = 3x + \left(1 - \frac{\pi}{2}\right) \quad \text{or} \quad 2y - 6x = 2 - \pi$$

Remember:
To find the gradient of a tangent to a curve at a point you need to find $\frac{dy}{dx}$ at that point.

You can also use differentiation to find the stationary points on the graph of a trigonometric equation.

Example 16

Given that $y = \frac{\sin x}{2 + \cos x}$

a) show that $\frac{dy}{dx} = \frac{2\cos x + 1}{(2 + \cos x)^2}$

b) Hence find the values of x at the two points on the curve $y = \frac{\sin x}{2 + \cos x}$, in the range $0 \leqslant x \leqslant 2\pi$, where the gradient is zero.

a) Using the quotient rule gives

$$\frac{d}{dx}\left(\frac{\sin x}{2+\cos x}\right) = \frac{(2+\cos x)\cos x - \sin x(-\sin x)}{(2+\cos x)^2}$$

$$= \frac{2\cos x + \cos^2 x + \sin^2 x}{(2+\cos x)^2}$$

$$\therefore \quad \frac{dy}{dx} = \frac{2\cos x + 1}{(2+\cos x)^2}$$

$u = \sin x$ $\quad v = 2 + \cos x$

$\frac{du}{dx} = \cos x$ $\quad \frac{dv}{dx} = -\sin x$

This is the gradient of the curve.

b) $$\frac{2\cos x + 1}{(2+\cos x)^2} = 0$$

$2\cos x + 1 = 0$ so $\cos x = -\frac{1}{2}$

$x = 2.09$ or 4.19

$\left(\text{or } \frac{2\pi}{3} \text{ or } \frac{4\pi}{3}\right)$.

C3

You can use integration to find the area under a curve.

Example 17

Find the area of the region A between the curve $y = \cos 2x$, the x-axis and the y-axis.

The graph shows the required area, A.

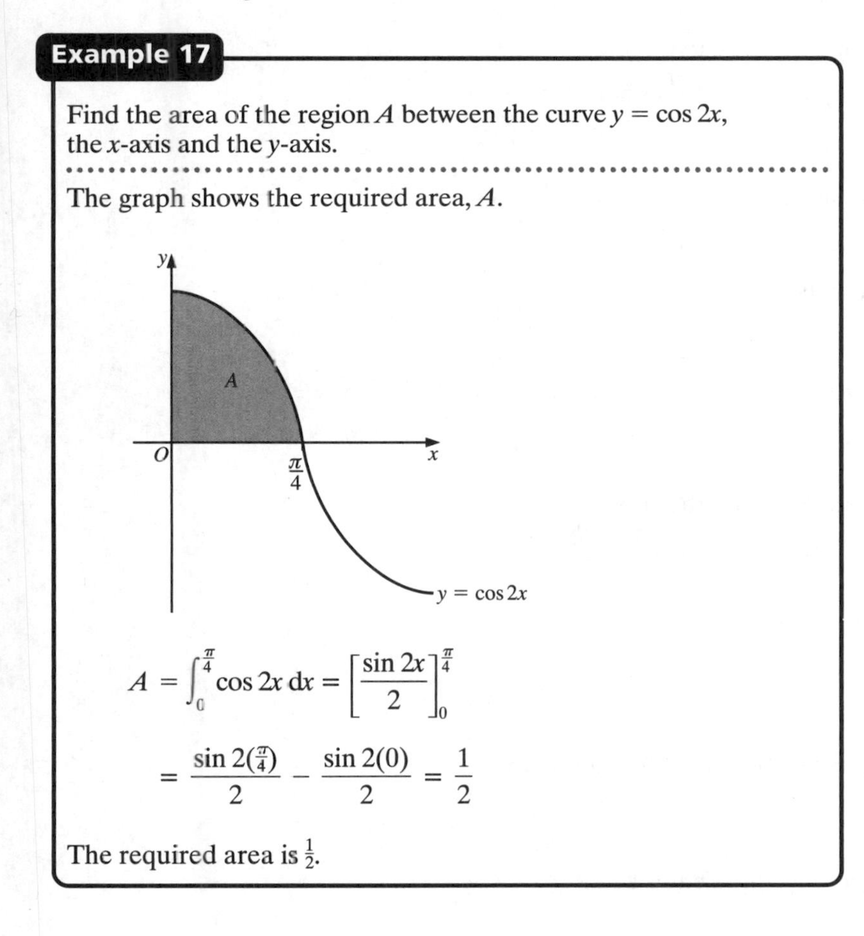

$$A = \int_0^{\frac{\pi}{4}} \cos 2x \, dx = \left[\frac{\sin 2x}{2}\right]_0^{\frac{\pi}{4}}$$

$$= \frac{\sin 2(\frac{\pi}{4})}{2} - \frac{\sin 2(0)}{2} = \frac{1}{2}$$

The required area is $\frac{1}{2}$.

Exercise 3G

This exercise revises the calculus techniques which were developed in earlier modules and applies these techniques to trigonometric functions.

Tangents and normals

1 Find the equation of the tangent to the curve $y = x + \sin x$ at the point where $x = \frac{\pi}{2}$.

2 Find the equation of the tangent and normal to the curve $y = x \cos x$ at the point where $x = \pi$.

3 a) Find the equations of normals to the curve $y = \cos 2x$ at the points $A\left(\frac{\pi}{4}, 0\right)$ and $B\left(\frac{3\pi}{4}, 0\right)$.

b) These normals meet at the point C. Find the coordinates of the point C.

c) Calculate the area of the triangle ABC.

4 Find the equation of the tangent and the normal to the curve $y = \dfrac{1}{1 + 2\sin x}$ at the point where $x = \pi$.

5 Find the x values of the two points on the curve $y = \sin x\,(2\cos x + 1)$, in the range $-\frac{\pi}{2} \leqslant x \leqslant \frac{\pi}{2}$, where the gradient is $-\frac{1}{2}$.

6 a) Show that there are two points on the curve $y = \dfrac{\sin x}{1 + \cos x}$, in the range $0 \leqslant x \leqslant 2\pi$, where the gradient is $\frac{2}{3}$.

b) Find the x values of these points.

Stationary points

7 a) Given that $y = \sin x\,(1 - \cos x)$, show that

$$\frac{dy}{dx} = (1 + 2\cos x)(1 - \cos x)$$

b) Hence find the x values of the points on the curve $y = \sin x\,(1 - \cos x)$, in the range $0 \leqslant x \leqslant \pi$, where the gradient is zero.

8 a) Given that $y = \dfrac{x - \sin x}{1 + \cos x}$, show that

$$\frac{dy}{dx} = \frac{x \sin x}{(1 + \cos x)^2}$$

b) Hence find the coordinates of the two points on the curve

$$y = \frac{x - \sin x}{1 + \cos x}$$

in the range $0 \leqslant x \leqslant 2\pi$, where the gradient is zero.

9 Find the stationary values on each of these curves in the range $0 \leqslant x \leqslant 2\pi$.

a) $y = x + \cos x$ b) $y = \dfrac{\sin x}{2 - \sin x}$

10 Evaluate these definite integrals.

a) $\int_0^{\frac{\pi}{2}} (1 - \cos x)\, dx$ b) $\int_0^{\frac{\pi}{6}} \sin 3x\, dx$

c) $\int_0^{\frac{\pi}{4}} 1 + \sin 2x\, dx$ d) $\int_{-\frac{\pi}{4}}^{\frac{\pi}{4}} \sec^2 x\, dx$

11 Find the area between the curve $y = \sin x$ and the x-axis from $x = 0$ to $x = \pi$.

12 Find the area between the curve $y = 3\cos x + 2\sin x$ and the x-axis from $x = 0$ to $x = \dfrac{\pi}{2}$.

C3

13 In the interval $0 \leqslant x < \pi$, the line $y = \frac{1}{2}$ meets the curve $y = \sin x$ at the points A and B.

a) Find the coordinates of A and B.

b) Calculate the area enclosed between the curve and the line between A and B.

Summary

You should know how to ...	Check out
1 Sketch the graphs of reciprocal trigonometric functions.	**1** On the same axes, sketch the graphs of $y = \sin x$ and $y = \operatorname{cosec} x$. Use a domain of $-2\pi < x < 2\pi$.
2 Solve trigonometric equations involving $\sec x$. $\cot x$, and $\operatorname{cosec} x$	**2** Solve these equations for all solutions in the range given. a) $\sec x = 2$ $\quad 0 \leqslant x \leqslant 360°$ b) $\operatorname{cosec} x = -3$ $\quad -\pi < x < \pi$ c) $\operatorname{cosec}^2 x = -2\cot x$ $\quad -\pi \leqslant x \leqslant \pi$
3 Prove trigonometric identities: $\sin x^2 + \cos^2 x = 1$ $\sec^2 x = 1 + \tan^2 x$ $\operatorname{cosec}^2 x = 1 + \cot^2 x$	**3** Prove the identities: a) $\dfrac{\sin\theta}{\sec^2\theta - 1} = \cos\theta\cot\theta$ b) $\tan\theta + \cot\theta = \sec\theta\operatorname{cosec}\theta$
4 Differentiate and integrate trigonometric functions.	**4** a) Differentiate $y = x^2 \sin 3x$ b) Integrate $\int \cos 2x\, dx$

Revision exercise 3

1 A curve has equation $y = \dfrac{2x}{\sin x}$, $0 < x < \pi$.

a) Find $\dfrac{dy}{dx}$.

b) The point P on the curve has coordinates $\left(\dfrac{\pi}{2}, \pi\right)$.

i) Show that the equation of the tangent to the curve at P is $y = 2x$.

ii) Find the equation of the normal to the curve at P, giving your answer in the form $y = mx + c$. *(AQA, 2003)*

2 a) Show that the equation $\tan^2\theta + \sec\theta = 11$ can be written as $x^2 + x - 12 = 0$ where $x = \sec\theta$.

b) Hence solve the equation $\tan^2\theta + \sec\theta = 11$, giving all solutions to the nearest 0.1° in the interval $0° < \theta < 360°$. *(AQA, 2003)*

3 A curve is defined for $0 \leqslant x \leqslant \pi$ by the equation $y = 2x - 1 + \sin 2x$.

a) Find $\dfrac{dy}{dx}$.

b) The region bounded by the curve, the x-axis and the lines $x = \dfrac{\pi}{2}$ and $x = \pi$ is R.

Given that R is above the x-axis show that the area of R is $\dfrac{3\pi^2}{4} - \dfrac{\pi}{2} - 1$. *(AQA, 2003)*

4 a) Find $\dfrac{dy}{dx}$ when i) $y = x\tan 3x$ ii) $y = \dfrac{\sin x}{x}$

b) Show that $\displaystyle\int_0^{\frac{\pi}{8}} x\sin 2x\,dx = \frac{4-\pi}{16\sqrt{2}}$ Hint: look ahead to p.100 *(AQA, 2002)*

5 A curve has equation $y = (x^2 + 5x + 4)\cos 3x$.

a) Find $\dfrac{dy}{dx}$.

b) Find the equation of the tangent to the curve at the point where $x = 0$. *(AQA, 2002)*

6 Find the equation of the tangent to the curve $y = \dfrac{2+x}{\cos x}$ at the point on the curve where $x = 0$. *(AQA, 2002)*

7 a) Find $\int(6\tan\theta - \sec^2\theta)\,d\theta$. Hint: look ahead to p.80

b) Find the solution of $6\tan\theta - \sec^2\theta = 7$ in the interval $0 \leqslant \theta \leqslant 2\pi$ giving each answer in radians to one decimal place. *(AQA, 2002)*

8 Solve the equation $\cos\theta + \sec\theta = 3$ giving all solutions in degrees to the nearest 0.1° in the interval $0° < \theta < 360°$. *(AQA/AEB, 2000)*

4 Exponentials and logarithms

This chapter will show you how to

- ✦ Reorganise and use the exponential function e^x.
- ✦ Differentiate and integrate with exponential functions
- ✦ Reorganise and use the natural logarithm function $\ln x$.
- ✦ Differentiate and integrate with natural logarithms

Before you start

You should know how to ...	Check in
1 Recognise the graph of the exponential function $y = a^x$.	**1** Sketch the graph of: a) $y = 2^x$ b) $y = 3^{-x}$
2 Use the laws of logarithms to simplify logarithmic expressions.	**2** Express as a single logarithm: a) $\log_a 8 + \log_a 3 - 2\log_a 2$ b) $\frac{1}{2}\log_2 36 + \frac{1}{3}\log_2 27 - 2\log_2 9$
3 Solve equations of the form $a^x = b$. You should also be able to use the techniques of differentiation described in Chapter 2.	**3** Solve these equations for x. a) $4^x = 32$ b) $9 \times 5^x = 3$

C3

4.1 Differentiating and integrating exponential functions

In C2 you met equations which contained expressions such as 2^x, 3^x, that is expressions of the form a^x. These are examples of **exponential** functions. You can use a numerical method to explore the derivatives of these functions. Begin by considering $y = 2^x$.

In the diagram, $P(x, 2^x)$ and $Q(x + 0.1, 2^{x+0.1})$ are nearby points on the curve $y = 2^x$.

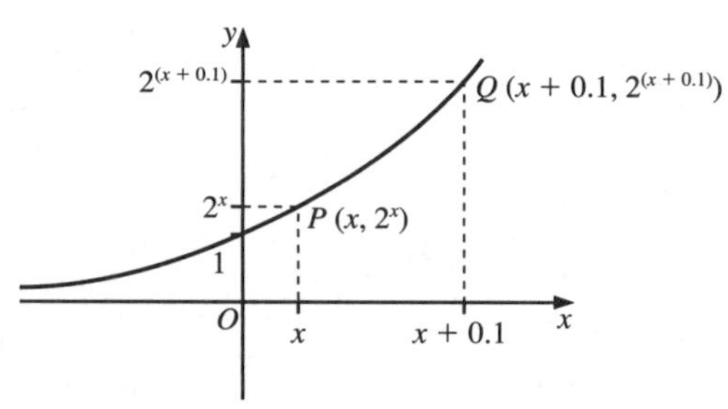

The gradient of the chord PQ is given by

$$m_{PQ} = \frac{2^{(x+0.1)} - 2^x}{0.1} = \frac{2^x(2^{0.1} - 1)}{0.1} = 0.718 \times 2^x$$

Now consider the gradient of PQ as Q moves closer to P.

When Q is $Q(x + 0.01, 2^{(x+0.01)})$ the gradient is

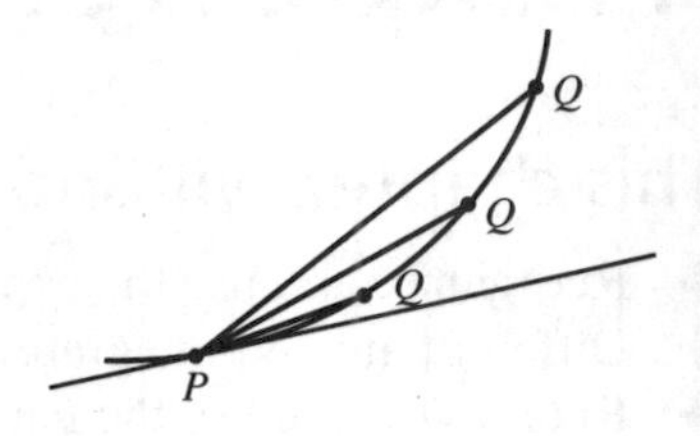

$$m_{PQ} = \frac{2^{(x+0.01)} - 2^x}{0.01} = \frac{2^x(2^{0.01} - 1)}{0.01} = 0.696 \times 2^x$$

When Q is $Q((x + 0.001), 2^{(x+0.001)})$ the gradient is

$$m_{PQ} = \frac{2^{(x+0.001)} - 2^x}{0.001} = \frac{2^x(2^{0.001} - 1)}{0.001} = 0.693 \times 2^x$$

When Q is $Q(x + 0.0001, 2^{(x+0.0001)})$ the gradient is

$$m_{PQ} = \frac{2^{(x+0.0001)} - 2^x}{0.0001} = \frac{2^x(2^{0.0001} - 1)}{0.0001} = 0.693 \times 2^x$$

As Q moves closer to P along the curve, the chord approximates to the tangent to the curve at P. The gradient of the chord appears to be tending to 0.693×2^x.

A similar analysis of the curve $y = 3^x$ goes like this:

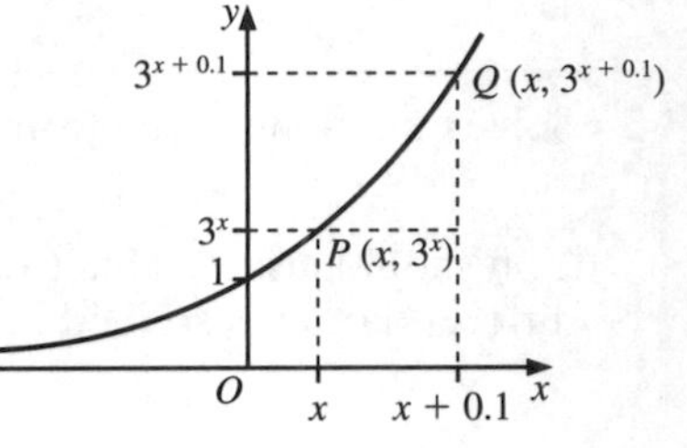

When Q is $Q(x + 0.1, 3^{(x+0.1)})$ the gradient is

$$m_{PQ} = \frac{3^{(x+0.1)} - 3^x}{0.1} = \frac{3^x(3^{0.1} - 1)}{0.1} = 1.161 \times 3^x$$

When Q is $Q(x + 0.01, 3^{(x+0.01)})$ the gradient is

$$m_{PQ} = \frac{3^{(x+0.01)} - 3^x}{0.01} = \frac{3^x(3^{0.01} - 1)}{0.01} = 1.105 \times 3^x$$

When Q is $Q(x + 0.001, 3^{(x+0.001)})$ the gradient is

$$m_{PQ} = \frac{3^{(x+0.001)} - 3^x}{0.001} = \frac{3^x(3^{0.001} - 1)}{0.001} = 1.099 \times 3^x$$

When Q is $Q(x + 0.0001, 3^{(x+0.0001)})$ the gradient is

$$m_{PQ} = \frac{3^{(x+0.0001)} - 3^x}{0.0001} = \frac{3^x(3^{0.0001} - 1)}{0.0001} = 1.099 \times 3^x$$

As Q moves closer to P along the curve, the chord approximates to the tangent to the curve at P. The gradient of the chord appears to be tending to 1.099×3^x.

Notice that in the case of $y = 2^x$ the gradient of the curve is less than $y = 2^x$, whereas in the case of $y = 3^x$ the gradient of the curve is greater than $y = 3^x$. In fact there exists an exponential function $y = a^x$ such that the gradient of the function equals $1 \times a^x$.

From this investigation you see that $2 < a < 3$. In fact, $a = 2.718\,28$ to five decimal places. The value of a is denoted by the symbol e.

The function $y = e^x$ is called the exponential function and $\frac{dy}{dx} = e^x$.

$y = e^x$ is the only function that is its own derivative.

The graph of $y = e^x$ is shown in the diagram.

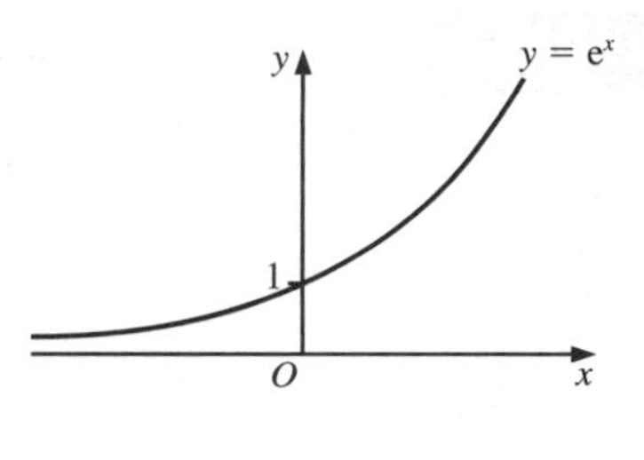

e is called a transcendental number, because it is not the root of an algebraic equation with rational coefficients. It was discovered by the brilliant Swiss mathematician Leonard Euler (1707–1783). Euler also introduced the use of π (another transcendental number), i (the imaginary square root of -1) and the summation symbol, Σ.

C3

Example 1

Find $\frac{dy}{dx}$ for each of these functions.

a) $y = e^{3x}$ b) $y = e^{5x+2}$ c) $y = e^{x^2}$

a) By the chain rule, when $y = e^{3x}$,

$$\frac{dy}{dx} = e^{3x} \times \frac{d}{dx}(3x)$$
$$= e^{3x}(3)$$
$$= 3e^{3x}$$
$$\therefore \quad \frac{dy}{dx} = 3e^{3x}$$

b) When $y = e^{5x+2}$,

$$\frac{dy}{dx} = e^{5x+2} \times \frac{d}{dx}(5x+2)$$
$$= e^{5x+2}(5)$$
$$= 5e^{5x+2}$$
$$\therefore \quad \frac{dy}{dx} = 5e^{5x+2}$$

c) When $y = e^{x^2}$,

$$\frac{dy}{dx} = e^{x^2} \times \frac{d}{dx}(x^2)$$
$$= e^{x^2}(2x)$$
$$= 2xe^{x^2}$$
$$\therefore \quad \frac{dy}{dx} = 2xe^{x^2}$$

In general,

When $y = e^{kx}$, $\frac{dy}{dx} = ke^{kx}$

You must learn this result.

You can now use the inverse function of a function rule to integrate exponential functions.

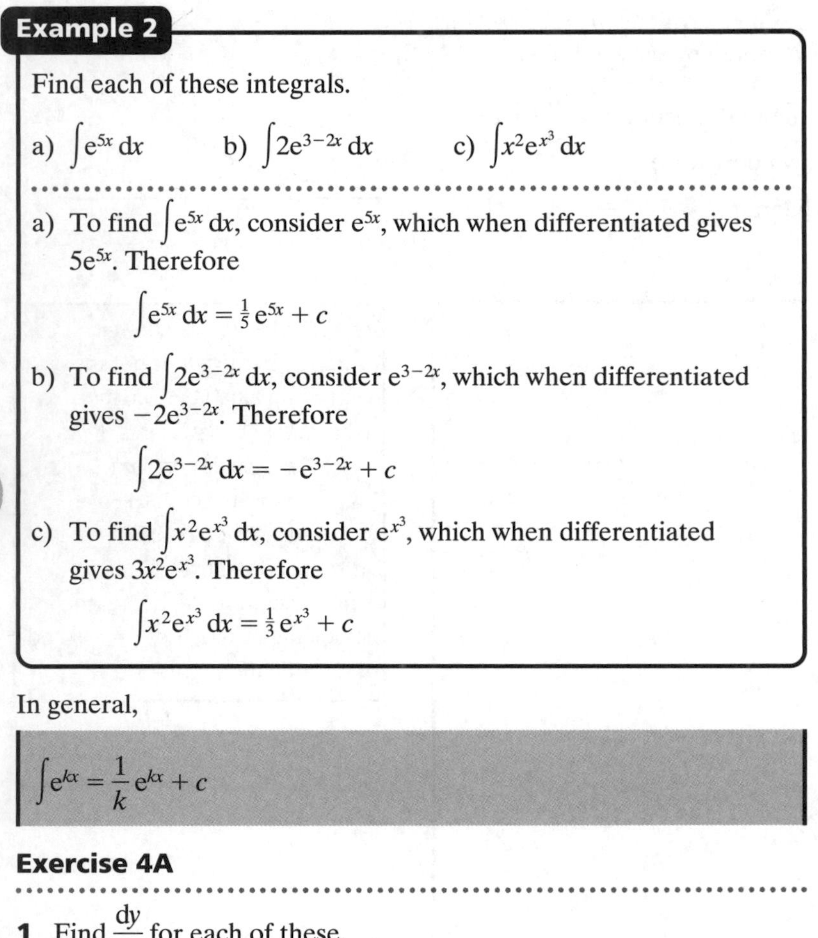

Example 2

Find each of these integrals.

a) $\int e^{5x}\,dx$ b) $\int 2e^{3-2x}\,dx$ c) $\int x^2 e^{x^3}\,dx$

a) To find $\int e^{5x}\,dx$, consider e^{5x}, which when differentiated gives $5e^{5x}$. Therefore

$$\int e^{5x}\,dx = \tfrac{1}{5}e^{5x} + c$$

b) To find $\int 2e^{3-2x}\,dx$, consider e^{3-2x}, which when differentiated gives $-2e^{3-2x}$. Therefore

$$\int 2e^{3-2x}\,dx = -e^{3-2x} + c$$

c) To find $\int x^2 e^{x^3}\,dx$, consider e^{x^3}, which when differentiated gives $3x^2e^{x^3}$. Therefore

$$\int x^2 e^{x^3}\,dx = \tfrac{1}{3}e^{x^3} + c$$

In general,

$$\int e^{kx} = \frac{1}{k}e^{kx} + c$$

Exercise 4A

1 Find $\dfrac{dy}{dx}$ for each of these.

a) e^{2x} b) e^{6x} c) e^{-x}
d) $2e^{3x}$ e) $7e^{-3x}$ f) e^{2x+5}
g) e^{3x-1} h) $2e^{4x+3}$ i) $5e^{3x-8}$
j) $4e^{2-x}$ k) $6e^{1-2x}$ l) $5e^{2-7x}$

2 Differentiate each of these with respect to x.

a) e^{x^3} b) e^{x^6} c) $5e^{x^2}$
d) $3e^{-x^2}$ e) e^{x^2+1} f) e^{2x^3-3}
g) $-e^{2x^4}$ h) $4e^{6-x^3}$ i) $2e^{5-x^4}$
j) $e^{\frac{1}{x}}$ k) $e^{\sqrt{x}}$ l) $-5e^{\frac{1}{x^2}}$

3 Find each of these integrals.

a) $\int e^{2x}\,dx$ b) $\int e^{5x}\,dx$ c) $\int e^{-x}\,dx$
d) $\int 8e^{4x}\,dx$ e) $\int 2e^{3x}\,dx$ f) $\int e^{5-x}\,dx$
g) $\int e^{4x+3}\,dx$ h) $\int 6e^{3x+1}\,dx$ i) $\int xe^{x^2}\,dx$
j) $\int x^4e^{x^5}\,dx$ k) $\int 2x^2e^{x^3}\,dx$ l) $\int xe^{-x^2}\,dx$

4.2 Natural logarithms

Logarithms to the base e are called **natural logarithms**. The notation $\ln x$ is used as the standard abbreviation for $\log_e x$. We will use $\ln x$ from this point onwards.

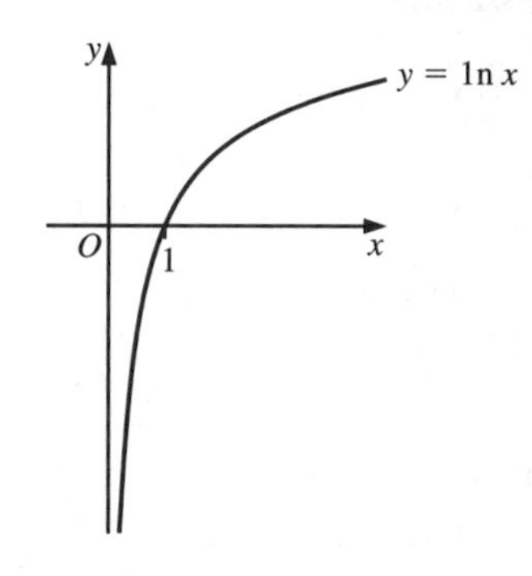

The function $\ln x$ is the inverse function of e^x. Notice that

$$\ln e^x = x \ln e = x \log_e e = x \times 1$$

Therefore,

$$\ln(e^x) = x \quad \text{and} \quad e^{\ln x} = x$$

The graph of $y = \ln x$ is shown.

Since $\ln x$ is the inverse function of e^x, the graph of $y = \ln x$ is a reflection, in the line $y = x$, of the graph of e^x.

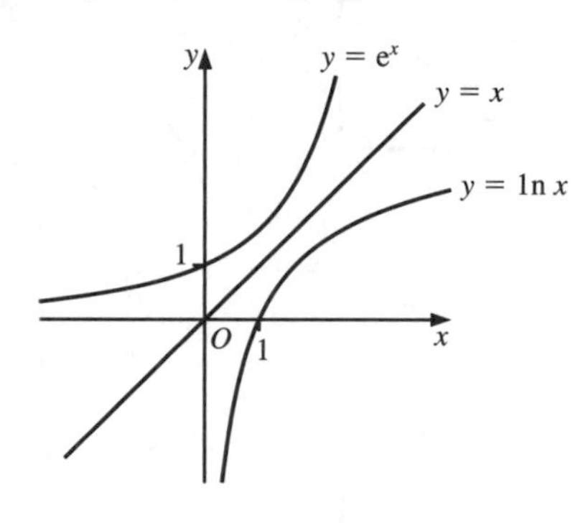

C3

Differentiating natural logarithms

If $y = \ln x$, then by definition $e^y = x$.

Differentiating e^y with respect to y gives

$$e^y = \frac{dx}{dy}$$

$$\frac{dy}{dx} = \frac{1}{e^y}$$

$$\therefore \quad \frac{dy}{dx} = \frac{1}{x}$$

$$\frac{dy}{dx} = \frac{1}{\frac{dx}{dy}}$$

If $y = \ln x$, then

$$\frac{dy}{dx} = \frac{1}{x}$$

You should learn these results.

It follows from this result that

$$\int \frac{1}{x}\,dx = \ln x + c$$

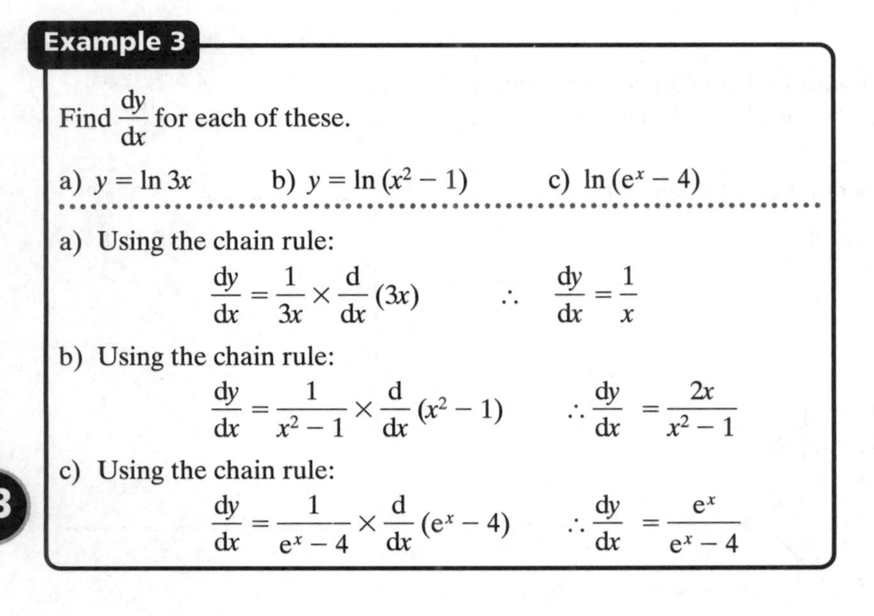

Example 3

Find $\frac{dy}{dx}$ for each of these.

a) $y = \ln 3x$ b) $y = \ln (x^2 - 1)$ c) $\ln (e^x - 4)$

a) Using the chain rule:

$$\frac{dy}{dx} = \frac{1}{3x} \times \frac{d}{dx}(3x) \qquad \therefore \quad \frac{dy}{dx} = \frac{1}{x}$$

b) Using the chain rule:

$$\frac{dy}{dx} = \frac{1}{x^2 - 1} \times \frac{d}{dx}(x^2 - 1) \qquad \therefore \quad \frac{dy}{dx} = \frac{2x}{x^2 - 1}$$

c) Using the chain rule:

$$\frac{dy}{dx} = \frac{1}{e^x - 4} \times \frac{d}{dx}(e^x - 4) \qquad \therefore \quad \frac{dy}{dx} = \frac{e^x}{e^x - 4}$$

If $y = \ln (f(x))$, then

$$\frac{dy}{dx} = \frac{1}{f(x)} \times f'(x) = \frac{f'(x)}{f(x)}$$

So:

$$\int \frac{f'(x)}{f(x)}\, dx = \ln (f(x)) + c, \text{ provided } f(x) > 0$$

Example 4

Find each of these integrals.

a) $\int \frac{1}{4x + 1}\, dx$ b) $\int \frac{x^2}{x^3 + 3}\, dx$ c) $\int \frac{2e^x}{e^x + 3}\, dx$

a) To find $\int \frac{1}{4x + 1}\, dx$, consider $\ln (4x + 1)$ which when differentiated gives $\frac{4}{4x + 1}$.

Therefore, $\int \frac{1}{4x + 1}\, dx = \frac{1}{4} \ln (4x + 1) + c$

b) To find $\int \frac{x^2}{x^3 + 3}\, dx$, consider $\ln (x^3 + 3)$ which when differentiated gives $\frac{3x^2}{x^3 + 3}$.

Therefore, $\int \frac{x^2}{x^3 + 3}\, dx = \frac{1}{3} \ln (x^3 + 3) + c$

c) To find $\int \frac{2e^x}{e^x + 3}\,dx$, consider $\ln(e^x + 3)$ which when differentiated gives $\frac{e^x}{e^x + 3}$.

Therefore, $\int \frac{2e^x}{e^x + 3}\,dx = 2\ln(e^x + 3) + c$

Definite integrals involving logarithms

You know that

$$\int \frac{1}{x}\,dx = \ln x + c \qquad [1]$$

You can use this result to find areas under curves.

Example 5

a) Sketch the curve $y = \frac{2}{3x + 2}$.

b) Find the area under the curve $y = \frac{2}{3x + 2}$ between $x = 2$ and $x = 6$, giving your answer in the form $a \ln b$.

a) The curve is a transformation of the graph $y = \frac{1}{x}$, with an asymptote at $x = -\frac{2}{3}$.

b) The area is given by $\int_2^6 \frac{2}{3x + 2}\,dx$.

$$\int_2^6 \frac{2}{3x + 2}\,dx = \left[\tfrac{2}{3}\ln(3x + 2)\right]_2^6$$

$$= \tfrac{2}{3}\ln 20 - \tfrac{2}{3}\ln 8$$

$$= \tfrac{2}{3}\ln\left(\tfrac{20}{8}\right) = \tfrac{2}{3}\ln\left(\tfrac{5}{2}\right)$$

However, the function $\ln x$ is only valid provided $x > 0$, as you can see from its graph (on page 79). Therefore, the integral [1] is only valid when $x > 0$.

We therefore have a problem, since $\frac{1}{x}$ exists for negative values of x.

Consider the area between the curve $y = \frac{1}{x}$, the x-axis and the lines $x = -1$ and $x = -2$.

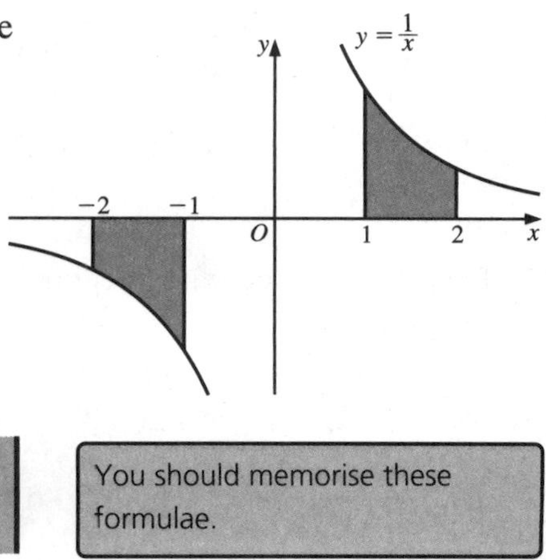

Clearly, this area exists and is identical to the area between the curve, the x-axis and the lines $x = 1$ and $x = 2$. Therefore, for both positive and negative values of x,

$$\int \frac{1}{x}\,dx = \ln|x| + c \quad \text{and} \quad \int \frac{f'(x)}{f(x)}\,dx = \ln|f(x)| + c$$

You should memorise these formulae.

It is usual practice to write the modulus sign only in the case of definite integrals.

Since the function $y = \dfrac{1}{x}$ is not defined when $x = 0$, the definite integral

$$\int_a^b \frac{1}{x}\,dx$$

is not defined if the interval $[a, b]$ includes $x = 0$. In other words a and b must both have the same sign for the integral to be valid.

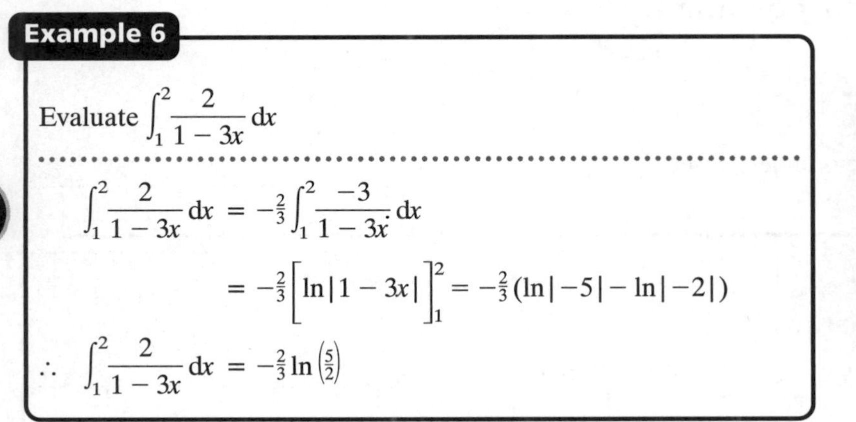

Example 6

Evaluate $\displaystyle\int_1^2 \frac{2}{1-3x}\,dx$

$$\int_1^2 \frac{2}{1-3x}\,dx = -\tfrac{2}{3}\int_1^2 \frac{-3}{1-3x}\,dx$$

$$= -\tfrac{2}{3}\Big[\ln|1-3x|\Big]_1^2 = -\tfrac{2}{3}(\ln|-5| - \ln|-2|)$$

$$\therefore \quad \int_1^2 \frac{2}{1-3x}\,dx = -\tfrac{2}{3}\ln\left(\tfrac{5}{2}\right)$$

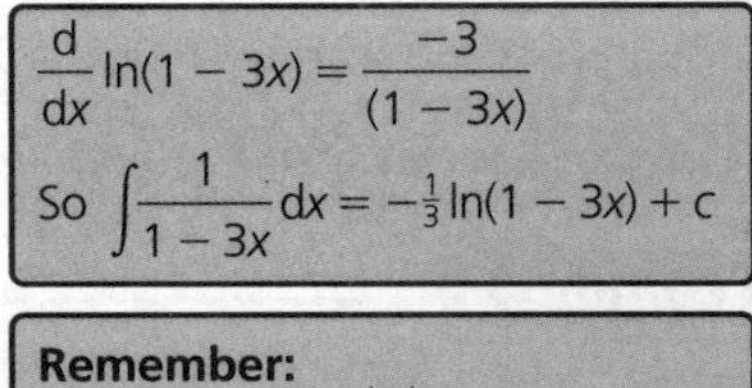

$\dfrac{d}{dx}\ln(1-3x) = \dfrac{-3}{(1-3x)}$

So $\displaystyle\int \frac{1}{1-3x}\,dx = -\tfrac{1}{3}\ln(1-3x) + c$

Remember:

$\ln a - \ln b = \ln\left(\dfrac{a}{b}\right)$

C3

Exercise 4B

1 Find $\dfrac{dy}{dx}$ for each of these.

a) $y = \ln(1 + 2x)$ b) $y = \ln(1 - 4x)$ c) $y = \ln(1 + x^2)$

d) $y = \ln(x^3 - 2)$ e) $y = \ln(x^3 - 3x)$ f) $y = \ln(e^x + 4)$

g) $y = \ln(1 + e^{6x})$ h) $y = \ln(\sqrt{x})$

2 Integrate each of these with respect to x.

a) $\dfrac{1}{1+x}$ b) $\dfrac{2}{x}$ c) $\dfrac{4}{2x-1}$ d) $\dfrac{3}{5+6x}$

e) $\dfrac{2x}{x^2+1}$ f) $\dfrac{4x}{2-x^2}$ g) $\dfrac{2x-1}{x^2-x}$ h) $\dfrac{e^x}{1+e^x}$

3 Evaluate these definite integrals, leaving each of your answers in the form $a \ln b$.

a) $\displaystyle\int_1^3 \frac{1}{(1+x)}\,dx$ b) $\displaystyle\int_0^2 \frac{1}{(6-x)}\,dx$

c) $\displaystyle\int_3^6 \frac{1}{(3+2x)}\,dx$ d) $\displaystyle\int_0^2 \frac{1}{(9-4x)}\,dx$

4.3 Products and quotients

You can also apply the product and quotient rules to expressions involving e^x and $\ln x$.

Example 7

Use the product rule to differentiate each of these functions.

a) $x^3 \ln x$ b) $(2 - x^2)e^x$

a)
$$\frac{dy}{dx} = x^3 \frac{1}{x} + \ln x \times (3x^2)$$
$$= x^2 + 3x^2 \ln x$$
$$\therefore \quad \frac{dy}{dx} = x^2(1 + 3\ln x)$$

$u = x^3$ $v = \ln x$

$\frac{du}{dx} = 3x^2$ $\frac{dv}{dx} = \frac{1}{x}$

b)
$$\frac{dy}{dx} = (2 - x^2)(e^x) + e^x \times (-2x)$$
$$\therefore \quad \frac{dy}{dx} = e^x(2 - 2x - x^2)$$

$u = 2 - x^2$ $v = e^x$

$\frac{du}{dx} = -2x$ $\frac{dv}{dx} = e^x$

C3

Example 8

Use the quotient rule to differentiate each of these functions.

a) $\frac{e^{2x}}{1 + e^x}$ b) $\frac{\ln(1 + x)}{x}$

a)
$$\frac{d}{dx}\left(\frac{e^{2x}}{1 + e^x}\right) = \frac{(1 + e^x)2e^{2x} - e^{2x}(e^x)}{(1 + e^x)^2}$$
$$= \frac{2e^{2x} + 2e^{3x} - e^{3x}}{(1 + e^x)^2}$$
$$\therefore \quad \frac{dy}{dx} = \frac{e^{2x}(2 + e^x)}{(1 + e^x)^2}$$

$u = e^{2x}$ $v = 1 + e^x$

$\frac{du}{dx} = 2e^{2x}$ $\frac{dv}{dx} = e^x$

b)
$$\frac{d}{dx}\left(\frac{\ln(1 + x)}{x}\right) = \frac{\frac{x}{1 + x} - \ln(1 + x)}{x^2}$$
$$= \frac{x - (1 + x)\ln(1 + x)}{x^2(1 + x)}$$
$$\therefore \quad \frac{dy}{dx} = \frac{x - (1 + x)\ln(1 + x)}{x^2(1 + x)}$$

$u = \ln(1 + x)$ $v = x$

$\frac{du}{dx} = \frac{1}{1 + x}$ $\frac{dv}{dx} = 1$

The next examples combine work on logarithms and exponentials with work on trigonometric functions.

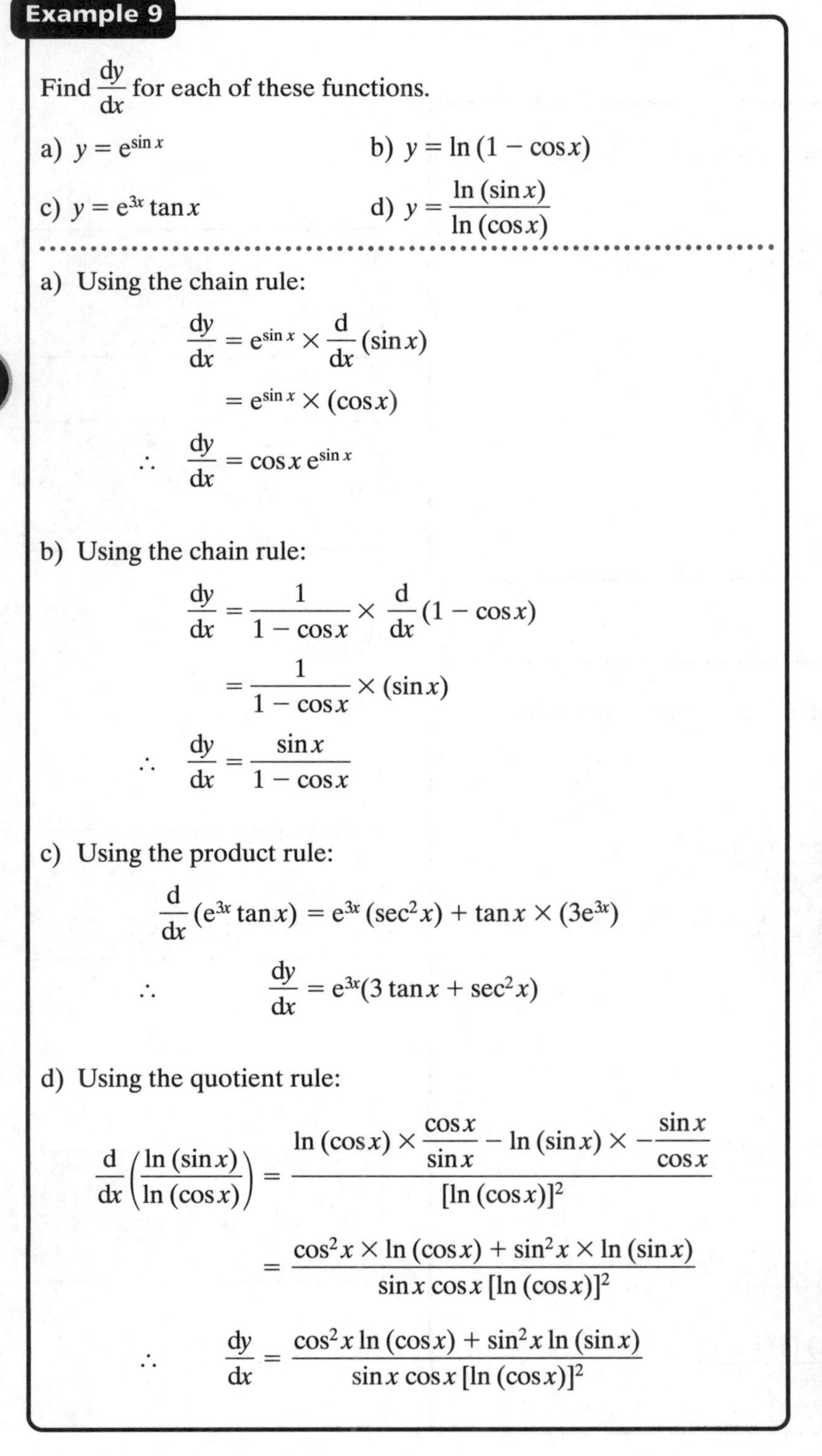

Example 9

Find $\frac{dy}{dx}$ for each of these functions.

a) $y = e^{\sin x}$

b) $y = \ln(1 - \cos x)$

c) $y = e^{3x}\tan x$

d) $y = \dfrac{\ln(\sin x)}{\ln(\cos x)}$

a) Using the chain rule:

$$\frac{dy}{dx} = e^{\sin x} \times \frac{d}{dx}(\sin x)$$

$$= e^{\sin x} \times (\cos x)$$

$$\therefore \quad \frac{dy}{dx} = \cos x \, e^{\sin x}$$

b) Using the chain rule:

$$\frac{dy}{dx} = \frac{1}{1 - \cos x} \times \frac{d}{dx}(1 - \cos x)$$

$$= \frac{1}{1 - \cos x} \times (\sin x)$$

$$\therefore \quad \frac{dy}{dx} = \frac{\sin x}{1 - \cos x}$$

c) Using the product rule:

$$\frac{d}{dx}(e^{3x}\tan x) = e^{3x}(\sec^2 x) + \tan x \times (3e^{3x})$$

$$\therefore \quad \frac{dy}{dx} = e^{3x}(3\tan x + \sec^2 x)$$

$u = e^{3x}$ $\quad v = \tan x$

$\frac{du}{dx} = 3e^{3x}$ $\quad \frac{dv}{dx} = \sec^2 x$

d) Using the quotient rule:

$$\frac{d}{dx}\left(\frac{\ln(\sin x)}{\ln(\cos x)}\right) = \frac{\ln(\cos x) \times \frac{\cos x}{\sin x} - \ln(\sin x) \times -\frac{\sin x}{\cos x}}{[\ln(\cos x)]^2}$$

$$= \frac{\cos^2 x \times \ln(\cos x) + \sin^2 x \times \ln(\sin x)}{\sin x \cos x \, [\ln(\cos x)]^2}$$

$$\therefore \quad \frac{dy}{dx} = \frac{\cos^2 x \ln(\cos x) + \sin^2 x \ln(\sin x)}{\sin x \cos x \, [\ln(\cos x)]^2}$$

$u = \ln(\sin x)$ $\quad v = \ln(\cos x)$

$\frac{du}{dx} = \frac{\cos x}{\sin x}$ $\quad \frac{dv}{dx} = \frac{-\sin x}{\cos x}$

Example 10

Find each of these integrals.

a) $\int \sec^2 x\, e^{\tan x}\, dx$ b) $\int \frac{2\cos x}{1+\sin x}\, dx$

a) To find $\int \sec^2 x\, e^{\tan x}\, dx$, consider $e^{\tan x}$ which when differentiated gives $\sec^2 x\, e^{\tan x}$.

Therefore

$$\int \sec^2 x\, e^{\tan x}\, dx = e^{\tan x} + c$$

b) To find $\int \frac{2\cos x}{1+\sin x}\, dx$, consider $\ln(1+\sin x)$ which when differentiated gives $\frac{\cos x}{1+\sin x}$.

Therefore

$$\int \frac{2\cos x}{1+\sin x}\, dx = 2\ln(1+\sin x) + c$$

C3

Exercise 4C

1 Use the product rule to differentiate each of these expressions.

a) $x\ln x$ b) x^2e^x c) x^3e^{3x}

d) $x\ln(1+x)$ e) $e^x\ln x$ f) $(1-x^3)e^{2x}$

g) $\ln(1+2x)\ln(1+3x)$ h) $e^{5x}\ln(2+x^2)$

2 Use the quotient rule to differentiate each of these expressions.

a) $\frac{x}{e^x}$ b) $\frac{x^2}{\ln x}$ c) $\frac{\ln x}{x}$

d) $\frac{x}{e^{3x}}$ e) $\frac{e^{-4x}}{x^3}$ f) $\frac{\ln x}{e^{2x}}$

g) $\frac{1+e^x}{1-e^x}$ h) $\frac{2+\ln x}{3-\ln x}$

3 Find $\frac{dy}{dx}$ for each of these expressions.

a) $y = e^{\cos x}$ b) $y = e^x\sin x$ c) $y = (1-e^{2x})^3$

d) $y = \ln(1+\tan x)$ e) $y = \frac{1+\sin x}{e^{3x}}$ f) $y = \sin x\ln(1-2x)$

g) $y = e^{\cos 2x}$ h) $y = \tan x\ln x$ i) $y = e^{-3x}\sin 2x$

j) $y = \frac{\ln x}{1+\tan x}$ k) $y = \cos(1+\ln x)$ l) $y = e^{-5x}\cos 3x$

4 Integrate each of these expressions with respect to x.

a) $\sin x\, e^{\cos x}$ b) $e^x \sin(e^x)$ c) $\dfrac{\sin x}{1+\cos x}$

d) $\sec^2 2x\, e^{\tan 2x}$ e) $\dfrac{1+\sec^2 x}{x+\tan x}$ f) $\dfrac{\sin 2x - \cos 2x}{\sin 2x + \cos 2x}$

4.4 Applications

You can apply these techniques to finding equations of tangents and normals to logarithmic and exponential curves.

Example 11

C3

Find the equation of the tangent and the normal to the curve $y = \ln\left(\dfrac{x-1}{x+1}\right)$ at the point P where $x = 3$.

Differentiating $y = \ln\left(\dfrac{x-1}{x+1}\right)$ gives

Use the quotient rule.

$$\frac{dy}{dx} = \frac{1}{\left(\dfrac{x-1}{x+1}\right)}\left[\frac{(x+1)-(x-1)}{(x+1)^2}\right]$$

$$= \left(\frac{x+1}{x-1}\right)\left[\frac{2}{(x+1)^2}\right]$$

$$\therefore \quad \frac{dy}{dx} = \frac{2}{x^2-1}$$

When $x = 3$: $\quad y = \ln\left(\dfrac{3-1}{3+1}\right) = \ln\left(\dfrac{1}{2}\right) = -\ln 2$

At $P(3, -\ln 2)$: $\quad \dfrac{dy}{dx} = \dfrac{2}{3^2-1} = \dfrac{1}{4}$

The tangent is of the form $y = \frac{1}{4}x + c$. Using $P(3, -\ln 2)$,

$$-\ln 2 = \tfrac{3}{4} + c$$

$$\therefore \quad c = -\ln 2 - \tfrac{3}{4}$$

The equation of the tangent line is

$$y = \tfrac{1}{4}x - \ln 2 - \tfrac{3}{4}$$

or alternatively,

$$x - 4y = 4\ln 2 + 3$$

When the gradient of the tangent at P is $\frac{1}{4}$, the gradient of the normal at P is -4. Therefore, the normal is of the form $y = -4x + c$. Using $P(3, -\ln 2)$ gives

$$-\ln 2 = -4(3) + c$$

$$\therefore \quad c = 12 - \ln 2$$

The equation of the normal is $y = -4x + 12 - \ln 2$.

Example 12

Find and classify the stationary points on the curve $y = x^2e^x$.
Hence sketch the curve.

At stationary points, $\frac{dy}{dx} = 0$. Using the product rule,

$$\frac{dy}{dx} = x^2e^x + e^x \times 2x$$

When $\frac{dy}{dx} = 0$,

$$x^2e^x + 2xe^x = 0$$

$$\therefore \quad xe^x(x + 2) = 0$$

Solving gives $x = 0$ and $x = -2$.

When $x = 0, y = 0$. When $x = -2, y = 4e^{-2}$.

The points $(0, 0)$ and $(-2, 4e^{-2})$ are stationary points.

To determine their nature, consider $\frac{d^2y}{dx^2}$.

Now

$$\frac{d^2y}{dx^2} = x^2e^x + 2xe^x + 2xe^x + 2e^x$$

$$= e^x(x^2 + 4x + 2)$$

When $x = 0$: $\left.\frac{d^2y}{dx^2}\right|_{x=0} = 2 > 0$

Therefore, $(0, 0)$ is a minimum.

When $x = -2$: $\left.\frac{d^2y}{dx^2}\right|_{x=-2} = -2e^{-2} < 0$

Therefore, $(-2, 4e^{-2})$ is a maximum.

The sketch of $y = x^2e^x$ is shown.

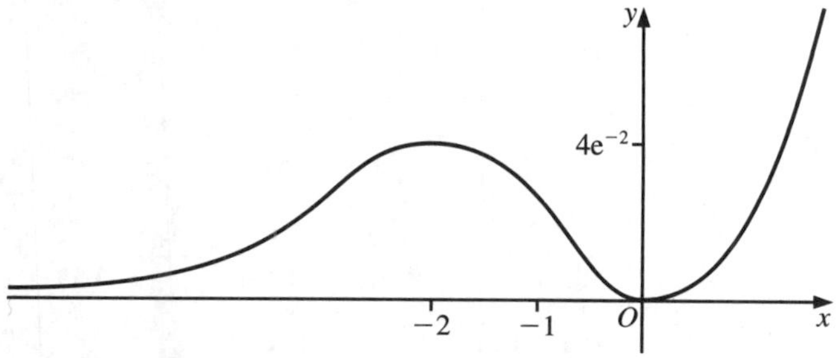

C3

Example 13

The line $2x + 3y = 14$ meets the curve $y = \dfrac{12}{x+2}$ at the points P and Q.

a) Find the coordinates of the points P and Q.

b) Find the area of the region bounded by the curve and the line.

a) Substituting $y = \dfrac{12}{x+2}$ into the equation $2x + 3y = 14$ gives

$$2x + 3 \times \frac{12}{x+2} = 14$$

$$2x + \frac{36}{x+2} = 14$$

$$2x(x+2) + 36 = 14(x+2)$$

$$2x^2 + 4x + 36 = 14x + 28$$

$$2x^2 - 10x + 8 = 0$$

$$x^2 - 5x + 4 = 0$$

$$\therefore \quad (x-1)(x-4) = 0$$

So $x = 1$ or $x = 4$.

When $x = 1$, $y = \dfrac{12}{1+2} = 4$, and when $x = 4$, $y = \dfrac{12}{4+2} = 2$.

Points P and Q are given by $P(1, 4)$ and $Q(4, 2)$.

b) The chord PQ defines a trapezium with the x-axis, and the area of the trapezium is given by

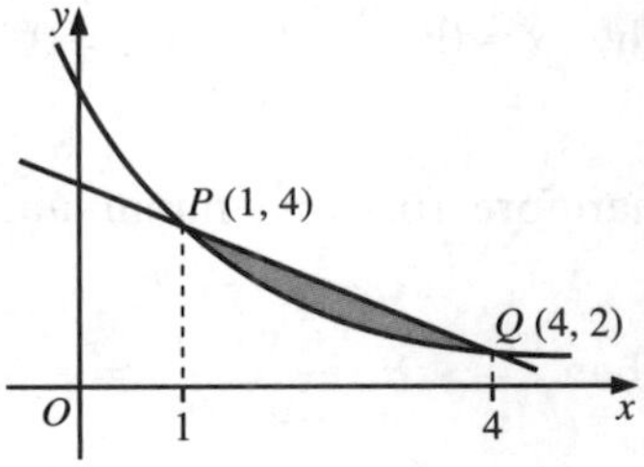

$$\text{Area}_{\text{trapezium}} = \frac{(4+2)}{2} \times 3 = 9$$

The area under the curve $y = \dfrac{12}{x+2}$ between P and Q is given by

$$\text{Area}_{\text{curve}} = \int_1^4 \frac{12}{x+2}\,dx$$

$$= \Big[12 \ln(x+2)\Big]_1^4$$

$$= 12 \ln(6) - 12 \ln(3)$$

$$= 12 \ln\left(\frac{6}{3}\right)$$

$$= 12 \ln 2$$

The required area is given by

$\text{Area}_{\text{trapezium}} - \text{Area}_{\text{curve}} = 9 - 12 \ln 2.$

Exercise 4D

This exercise revises the calculus techniques which were developed in earlier modules, and applies the techniques to the exponential and logarithmic functions.

Tangents and normals

1 Find the equation of the tangent to the curve $y = x + e^{2x}$ at the point where $x = 0$.

2 Find the equation of the tangent and the normal to the curve $y = \ln(1 + x)$ at the point where $x = 2$.

3 Find the equation of the tangent and the normal to the curve $y = xe^x$ at the point where $x = 1$.

4 The tangent to the curve $y = x \ln x$ at the point (e, e) meets the x-axis at A and the y-axis at B.

a) Find the equation of this tangent.

b) Hence find the distance AB.

5 Find the equation of the tangent and the normal to the curve $y = e^x \ln x$ at the point where $x = 1$.

6 Find the coordinates of the points on the curve $y = x^2 + \ln x$ where the gradient is 3.

7 Show that there are two points on the curve $y = \ln(1 + x^2)$ where the gradient is $\frac{5}{13}$. Find the coordinates of these points.

8 Find the coordinates of the point on the curve $y = \ln(e^x + e^{-x})$ where the gradient is $\frac{3}{5}$.

Stationary points

9 Given that $y = x^2e^{-x}$, show that $\frac{dy}{dx} = x(2 - x)e^{-x}$. Hence find the coordinates of the two points on the curve $y = x^2e^{-x}$ where the gradient is zero.

10 Given that $y = \frac{\ln x}{x}$ for $x > 0$, show that $\frac{dy}{dx} = \frac{1 - \ln x}{x^2}$. Hence find the coordinates of the point on the curve $y = \frac{\ln x}{x}$ where the gradient is zero.

11 Given that $y = \frac{e^x}{x^2 - 3}$, show that $\frac{dy}{dx} = \frac{e^x(x + 1)(x - 3)}{(x^2 - 3)^2}$. Hence find the coordinates of the two points on the curve $y = \frac{e^x}{x^2 - 3}$ where the gradient is zero.

12 Find and classify the stationary values on each of these curves.

a) $y = 2\ln(1 + x) - \ln x \quad (x > 0)$

b) $y = \frac{e^x}{x^3}$

c) $y = x(3 - \ln x) \quad (x > 0)$

d) $y = e^x(x - 1)^2$

Areas

13 Find the area between the curve $y = e^{2x}$ and the x-axis from $x = 0$ to $x = 3$.

14 Find the area between the curve $y = \frac{2}{x+3}$ and the x-axis from $x = 2$ to $x = 7$.

15 Find the area between the curve $y = 4e^{2x} - 3e^{x}$ and the x-axis from $x = 1$ to $x = 2$.

16 The line $y = \frac{1}{3}$ meets the curve $y = \frac{1}{x+1}$ at the point P.

a) Find the coordinates of P.

b) Calculate the area bounded by the line, the curve and the y-axis.

17 The line $y = x + 1$ meets the curve $y = \frac{8}{5-x}$ at the points P and Q.

a) Find the coordinates of P and Q.

b) Show that the area enclosed between the curve and the line between P and Q is $6 - 8 \ln 2$.

18 The region R is bounded by the curve $y = 3 + \frac{2}{x+1}$, the x-axis, the y-axis and the line $x = 4$. Show that the area of R is $12 + 2 \ln 5$.

C3

Summary

You should know how to ...	Check out
1 Differentiate exponential functions.	**1** Differentiate: a) $y = 2e^{3x}$ b) $y = e^{-x}$ c) $y = \frac{3}{e^{4x}}$
2 Integrate exponential functions.	**2** Evaluate: a) $\int e^{3x}\,dx$ b) $\int e^{-x}\,dx$ c) $\int \frac{1}{e^{4x}}\,dx$
3 Differentiate and integrate natural logarithms.	**3** a) Differentiate: i) $y = 3 \ln x$ ii) $y = 4 \ln 2x$ b) Evaluate: i) $\int \frac{1}{3x}\,dx$ ii) $\int \frac{2x}{x^2 - 4}\,dx$
4 Sketch the graphs of exponential and logarithmic functions	**4** a) Sketch the graphs of: i) $y = 2e^{-x}$ ii) $y = 5e^{3x}$ b) Sketch the graphs of: i) $y = \ln x$ ii) $y = \ln 3x$ c) i) Find the inverse function $f^{-1}(x)$ of $f(x) = e^{2x}$. ii) Sketch the graphs of $f(x)$ and $f^{-1}(x)$ on the same axes.

Revision exercise 4

1 a) i) Find $\int (e^{2x} + 1)\,dx$.

ii) Hence show that $\int_0^{\ln 2} (e^{2x} + 1)\,dx = \frac{3}{2} + \ln 2$.

b) The diagram shows the graph of $y = e^{2x} + 1$.
Find the y-coordinate of the point where the graph intersects

i) the y-axis

ii) the line $x = \ln 2$.

c) The function f is defined on the restricted domain $0 \leqslant x \leqslant \ln 2$ by $f(x) = e^{2x} + 1$.

i) Find the range of the function f.

ii) On one pair of axes sketch the graphs of $y = f(x)$ and $y = f^{-1}(x)$.

iii) Find an expression for $f^{-1}(x)$. *(AQA, 2004)*

C3

2 a) i) Find $\frac{dy}{dx}$ when $y = e^x \sin 2x$.

ii) Hence find the equation of the tangent to the curve $y = e^x \sin 2x$ at the origin.

b) Show that the equation of the normal to the curve $y = e^x \sin 2x$ at the point where $x = \pi$ is $2e^{\pi}y + x = \pi$. *(AQA, 2004)*

3 a) i) Verify that $\frac{1}{4}x < \ln x$ when $x = 2$.

ii) Verify that $\frac{1}{4}x > \ln x$ when $x = 10$.

iii) Draw on the same diagram sketches of the graphs with equations $y = \frac{1}{4}x$ and $y = \ln x$ for $x > 0$.

iv) Hence state the number of roots of the equation $\frac{1}{4}x = \ln x, x > 0$.

b) The curve C, with equation $y = \ln x - \frac{1}{4}x, x > 0$ has only one stationary point.

i) Find $\frac{dy}{dx}$ ii) Find $\frac{d^2y}{dx^2}$

iii) Find the coordinates of the stationary point.

iv) Determine whether the stationary point is a maximum or a minimum. *(AQA, 2004)*

4 The diagram shows a sketch of the curve with equation $y = 8 - e^{3x}$ which crosses the y-axis at the point A and the x-axis at the point B.

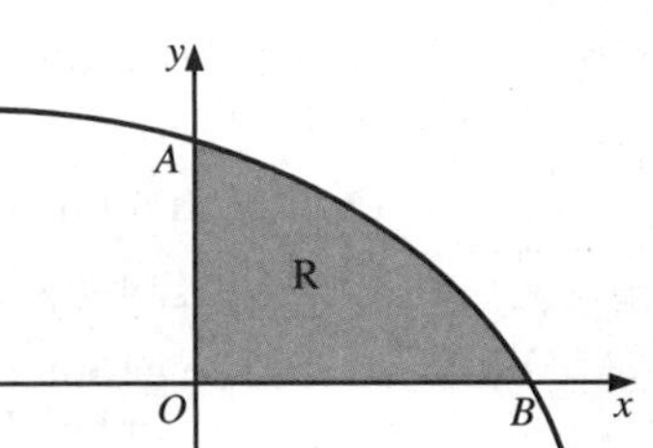

a) Find the y-coordinate of A.

b) Find the **exact** value of the x-coordinate of B.

c) Show that the gradient of the curve at B is -24.

d) i) Find $\int (8 - e^{3x})\,dx$.

ii) Hence show that the area of the shaded region R bounded by the curve $y = 8 - e^{3x}$ and the coordinate axes is $8 \ln 2 - \frac{7}{3}$.

e) i) Sketch the graph of the curve $y = |8 - e^{3x}|$.

ii) Solve the equation $|8 - e^{3x}| = 19$ giving your answer in an **exact** form. *(AQA, 2004)*

5 The diagram shows a sketch of the graph of $y = e^{2x} + 2x^{-1}$ for $x > 0$.

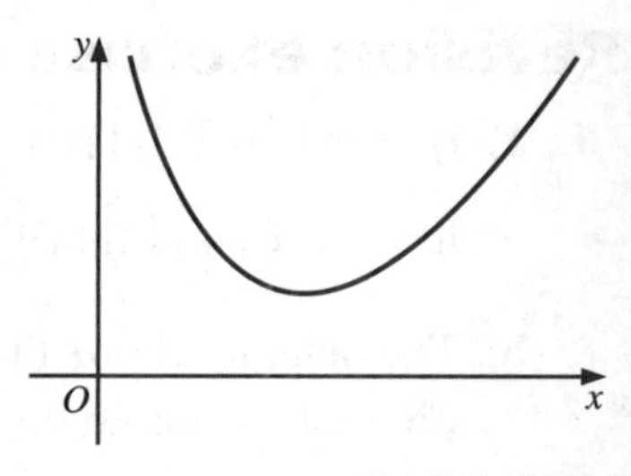

a) Find $\dfrac{dy}{dx}$.

b) Show that, at the stationary point on the graph, $x^2e^{2x} = 1$.

c) Deduce that at the stationary point $xe^x = 1$ and hence $\ln x + x = 0$.

d) Show that the equation $\ln x + x = 0$ has a root between 0.5 and 0.6.

e) Find $\int (e^{2x} + 2x^{-1})\,dx$.

(AQA, 2004)

6 On one particular day the volume, V m^3, of liquid in a tank changes with time, t hours after midnight, according to the formula $V = 8 + 6e^{-\frac{1}{12}t}$.

a) Find the volume of liquid in the tank when $t = 0$.

b) Find the rate of change, in m^3 per hour, of the volume of liquid in the tank when $t = 12$, interpreting the sign of your answer.

c) Find, to the nearest minute, the time when the volume of liquid in the tank is 11 m^3.

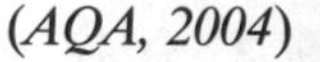
(AQA, 2004)

7 A curve has equation $y = \frac{1}{4}x^3 - 6\ln x + 1, x > 0$.

a) i) Find $\dfrac{dy}{dx}$.

ii) Hence show that the gradient of the curve at the point where $x = \frac{2}{3}$ is $-8\frac{2}{3}$.

b) i) Given that the curve has just one stationary point, find the x-coordinate of this stationary point.

ii) Find $\dfrac{d^2y}{dx^2}$.

iii) Show that $\dfrac{d^2y}{dx^2} = 4.5$ at the stationary point and hence state the nature of the stationary point.

c) P and Q are two points on the curve $y = \frac{1}{4}x^3 - 6\ln x + 1$.
The x-coordinate of P is 4 and the x-coordinate of Q is 8.
Find the gradient of the chord PQ in the form $a + b\ln 2$ where a and b are constants to be found.

(AQA, 2004)

8 The diagram shows the graph of $y = f(x)$ where f is defined for all $x > 0$ by $f(x) = 2 + \ln x$.

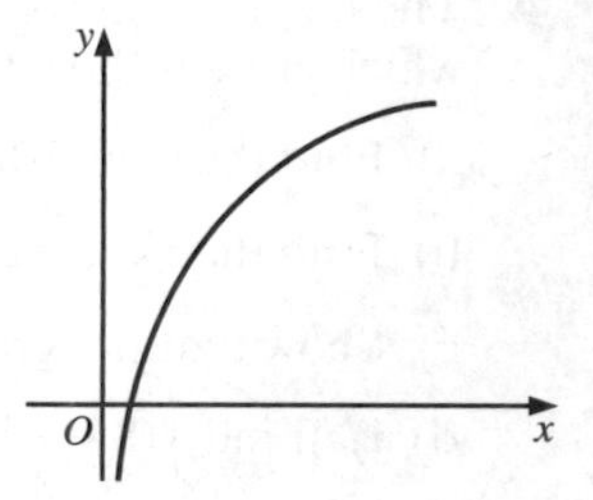

a) i) Differentiate $f(x)$ to find $f'(x)$.

ii) Find the gradient of the curve at the point where $x = e$.

b) Describe the geometrical transformation by which the graph of $y = 2 + \ln x$ can be obtained from the graph of $y = \ln x$.

c) i) State the range of the function f.

ii) State the domain and range of the inverse function f^{-1}.

iii) Find an expression for $f^{-1}(x)$.

d) The function g is defined for all x by $g(x) = ex^3$. Show that

i) $fg(x) = 3(1 + \ln x)$ ii) $fg(x) = 9 \Rightarrow x = e^2$

(AQA, 2003)

9 The function f is defined for $x > 0$ by $f(x) = e^{-2x} + \frac{3}{x} + 3$.

a) i) Differentiate $f(x)$ with respect to x to find $f'(x)$.
ii) Hence prove that f is a decreasing function.

b) Find the range of f.

c) Show that the area of the region bounded by the curve $y = e^{-2x} + \frac{3}{x} + 3$,
the x-axis and the lines $x = 1$ and $x = 2$ is $\frac{e^2 - 1}{2e^4} + 3(\ln 2 + 1)$. *(AQA, 2003)*

10 It is given that $f(x) = 2x^3 + 3x^2 + 7$.

a) Find the derivative $f'(x)$, factorising your answer.

b) Hence, show that, $\int_0^2 \frac{x(x+1)}{f(x)}\,dx = k \ln 5$, stating the value of the constant k. *(AQA, 2003)*

11 a) The diagram shows the graph of $y = f(x)$ where the function f is defined for all values of x by $f(x) = 5e^{-x}$.

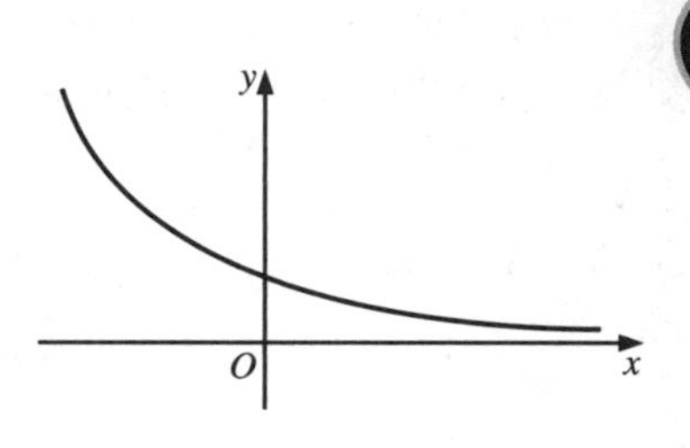

i) Write down the coordinates of the point where the graph intersects the y-axis.
ii) State the range of the function f.
iii) Find the value of $f(\ln 6)$, giving your answer as a fraction.

b) The function g is defined for all values of x by $g(x) = x + 10$.
i) Show that $gf(x) = 5(e^{-x} + 2)$.
ii) State the range of the function gf.
iii) Sketch the graph of $y = gf(x)$.
iv) Show that $gf(x) = 11 \Rightarrow x = \ln 5$.

c) A dish of water is left to cool in a room where the temperature is 10 °C. At time t minutes, where $t \geqslant 0$, the temperature of the water in degrees Celsius is $5(e^{-t} + 2)$.
i) State the temperature of the water at time $t = 0$.
ii) Calculate the time at which the temperature of the water reaches 11 °C. Give your answer to the nearest tenth of a minute.

(AQA, 2003)

12 The diagram shows a sketch of the curve with equation $y = 4 - e^{2x}$ which crosses the y-axis at the point A and the x-axis at the point B $(\ln 2, 0)$.

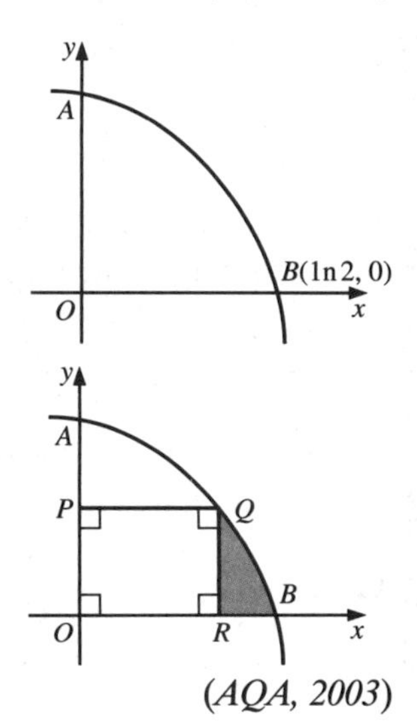

a) Find the coordinates of A.

b) i) For the curve $y = 4 - e^{2x}$ find $\frac{dy}{dx}$.
ii) Find the gradient of the curve at B.

c) The line $y = 2$ cuts the y-axis at the point P and the curve at the point Q. $OPQR$ is a rectangle.
i) Find the area of the rectangle $OPQR$. ii) Find $\int (4 - e^{2x})\,dx$.
iii) Find the area of the shaded region bounded by the curve, the x-axis and the line QR. Give your answer in the form $p \ln 2 + q$ where p and q are constants to be determined.
iv) Show that the area of the region above PQ, bounded by the curve, the y-axis and the line PQ, is half the area of the shaded region.

(AQA, 2003)

13 A curve has equation $y = x^2 - 3x + \ln x + 2, x > 0$.

a) i) Find $\dfrac{dy}{dx}$.

ii) Hence show that the gradient of the curve at the point where $x = 2$ is $\frac{3}{2}$.

b) i) Show that the x-coordinates of the stationary points of the curve satisfy the equation $2x^2 - 3x + 1 = 0$.

ii) Hence find the x-coordinates of each of the stationary points.

iii) Find $\dfrac{d^2y}{dx^2}$.

iv) Find the value of $\dfrac{d^2y}{dx^2}$ at each of the stationary points.

v) Hence show that the y-coordinate of the maximum point is $\frac{3}{4} - \ln 2$. *(AQA, 2002)*

14 The diagram shows a sketch of the curve $y = e^{2x} - 2$.

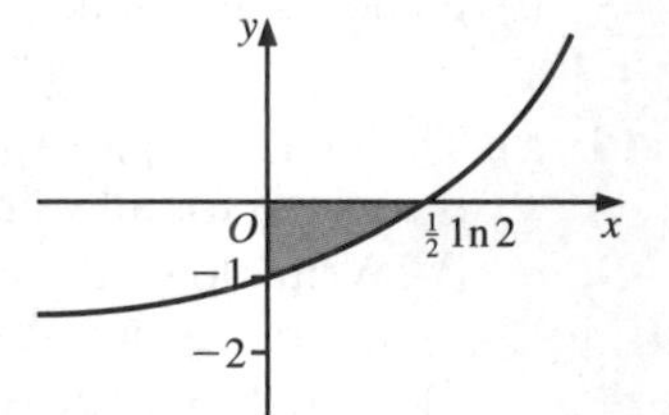

a) On separate diagrams, sketch the graphs of the following curves, showing the coordinates of the points where the graph intersects the coordinate axes.

i) $y = |e^{2x} - 2|$ ii) $y = e^{2x} - 5$

b) i) Find $\int (e^{2x} - 2)\, dx$.

ii) Hence show that the area of the shaded region bounded by the curve $y = e^{2x} - 2$ and the coordinate axes is $\ln 2 - \frac{1}{2}$.

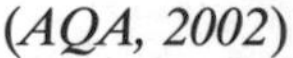
(AQA, 2002)

15 The diagram shows a sketch of the curve $y = e^{2x} - 3$ which crosses the y-axis at the point A and the x-axis at the point B.

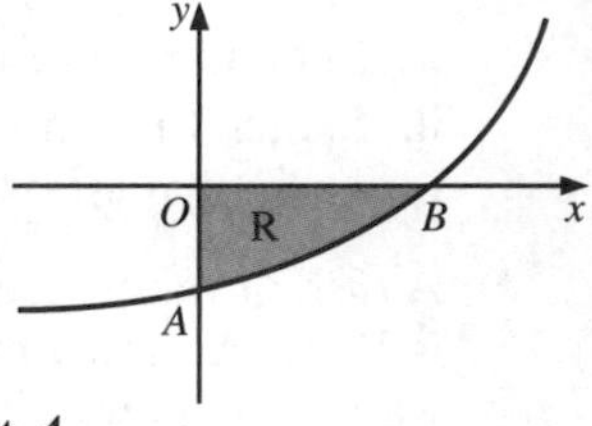

a) Find the y-coordinate of A.

b) Find the exact value of the x-coordinate of B.

c) i) Find $\dfrac{dy}{dx}$.

ii) Show that the gradient of the curve at B is three times its gradient at A.

d) Find the area of the shaded region R bounded by the curve and the coordinate axes. Give your answer in the form $p + q \ln r$ where p, q and r are constants to be determined. *(AQA, 2002)*

16 Find $\dfrac{dy}{dx}$ for each of the following cases:

a) $y = e^{2x} \sin 3x$ b) $y = (2x^2 + 1)^5$ *(AQA, 2001)*

17 The volume of oil, V m³, in a tank changes with time t hours $(1 \leqslant t \leqslant T)$ according to the formula $V = 32 - 10 \ln t$, where T represents the time when the tank is empty of oil.

a) State the volume of oil in the tank when $t = 1$.

b) Find the rate of change, in m³ per hour, of the volume of oil in the tank at the time when $t = 8$, interpreting the sign of your answer.

c) Determine the value of T. *(AQA, 2001)*

18 Describe in each of the following cases, a single transformation which maps the graph of $y = e^x$ onto the graph of the function given.

a) $y = e^{3x}$ b) $y = e^{x-3}$ c) $y = \ln x$ *(AQA, 2001)*

5 Integration

This chapter will show you how to

- ✦ Integrate by substitution
- ✦ Integrate by parts
- ✦ Recognise and use standard integrals
- ✦ Calculate volumes of revolution

Before you start

You should know how to ...	Check in
1 Find an indefinite integral.	**1** Integrate these functions. a) $y = 2x^3 - 3x$ b) $y = \frac{1}{x^2\sqrt{x}}$ c) $y = \frac{2}{3x^2} - \frac{1}{x}$ d) $y = \sin 3x$ e) $y = e^{4x}$
2 Use definite integration to find the area under a curve. You should also be able to use the techniques of differentiation described in Chapters 2, 3 and 4.	**2** Find the area between the curve $y = e^{-2x}$, the x-axis, and the lines $x = 1$ and $x = -1$. y, $y = e^{-2x}$, 1, -1, O, 1, x

C3

5.1 Integration by substitution

Suppose you want to find $\int (3x - 2)^4 \, dx$. In Chapter 2 you learnt how to treat this as an inverse function of a function. However, some questions require a more formal technique called **integration by substitution.**

This method is the reverse of the chain rule for differentiation:

$$\frac{dy}{dx} = \frac{dy}{du}\frac{du}{dx}$$

To find $\int (3x - 2)^4 \, dx$, let $u = 3x - 2$.

Then $\int (3x - 2)^4 \, dx = \int u^4 \, dx$.

By the chain rule $\int u^4 \, dx = \int u^4 \frac{dx}{du} \, du$, and this allows you to change the variable u entirely.

You cannot integrate u^4 with respect to x. You can only integrate u^4 with respect to u.

Since $u = 3x - 2$, $\frac{du}{dx} = 3$. Inverting this gives $\frac{dx}{du} = \frac{1}{3}$.

So $\int (3x-2)^4\,dx = \int u^4\,dx = \int u^4 \frac{dx}{du}\,du = \int u^4 \times \frac{1}{3}\,du = \int \frac{u^4}{3}\,du = \frac{u^5}{15} + c$

Since $u = 3x - 2$, the answer is

$$\int (3x-2)^4\,dx = \frac{(3x-2)^5}{15} + c$$

The final answer must be given in terms of x.

You will often be given the appropriate substitution to use.

Example 1

Use the substitution $u = x^2 + 1$ to find $\int x(x^2+1)^3\,dx$.

Given $u = x^2 + 1$, then $\frac{du}{dx} = 2x$. Substituting gives

$$\int x(x^2+1)^3\,dx = \int x(x^2+1)^3 \frac{dx}{du}\,du$$
$$= \int x(u)^3 \frac{1}{2x}\,du$$
$$= \int \frac{u^3}{2}\,du$$
$$= \frac{u^4}{8} + c$$

Since $u = x^2 + 1$,

$$\int x(x^2+1)^3\,dx = \frac{(x^2+1)^4}{8} + c$$

C3

Example 2

Use the substitution $u = 2x + 1$ to find $\int x(2x+1)^3\,dx$.

Given $u = 2x + 1$, then $\frac{du}{dx} = 2$ and $x = \frac{u-1}{2}$. Substituting gives

$$\int x(2x+1)^3\,dx = \int x(2x+1)^3 \frac{dx}{du}\,du$$
$$= \int \frac{u-1}{2} u^3 \frac{1}{2}\,du = \int \frac{u^4}{4} - \frac{u^3}{4}\,du$$
$$= \frac{u^5}{20} - \frac{u^4}{16} + c$$
$$= \frac{u^4}{80}(4u - 5) + c$$

Since $u = 2x + 1$,

$$\int x(2x+1)^3\,dx = \frac{(2x+1)^4}{80}[4(2x+1) - 5] + c$$
$$= \frac{(2x+1)^4}{80}(8x - 1) + c$$

Remember:
You can integrate polynomials term by term.
$\int \frac{u^4 - u^3}{4}\,du = \int \frac{u^4}{4}\,du - \int \frac{u^3}{4}\,du$

The same method can be applied to more complex integrals.

Example 3

Find each of these integrals by using the substitution suggested.

a) $\int (x + 1)(3x - 2)^5 \, dx, \quad u = 3x - 2$

b) $\int x\sqrt{x - 1} \, dx, \quad u = x - 1$

c) $\int \frac{3x - 4}{\sqrt{2x + 1}} \, dx, \quad u = \sqrt{2x + 1}$

a) Given $u = 3x - 2$, then $\frac{du}{dx} = 3$ and $x = \frac{u + 2}{3}$. Substituting gives

$$I = \int (x + 1)(3x - 2)^5 \, dx = \int (x + 1)(3x - 2)^5 \frac{dx}{du} \, du$$

$$= \int \left(\frac{u + 2}{3} + 1\right) u^5 \frac{1}{3} \, du$$

$$= \int \left(\frac{u^6}{9} + \frac{5u^5}{9}\right) du$$

$$\therefore \quad I = \frac{u^7}{63} + \frac{5u^6}{54} + c = \frac{u^6}{378}(6u + 35) + c$$

Since $u = 3x - 2$,

$$I = \frac{(3x - 2)^6}{378}[6(3x - 2) + 35] + c$$

$$\therefore \quad I = \frac{(3x - 2)^6}{378}(18x + 23) + c$$

I represents the integral to be found.

b) Given $u = x - 1$, then $\frac{du}{dx} = 1$ and $x = u + 1$. Substituting gives

$$I = \int x\sqrt{x - 1} \, dx = \int x\sqrt{x - 1} \frac{dx}{du} \, du$$

$$= \int (u + 1)u^{\frac{1}{2}} \, du$$

$$= \int u^{\frac{3}{2}} + u^{\frac{1}{2}} \, du$$

$$\therefore \quad I = \frac{2u^{\frac{5}{2}}}{5} + \frac{2u^{\frac{3}{2}}}{3} + c = \frac{2u^{\frac{3}{2}}}{15}(3u + 5) + c$$

Since $u = x - 1$,

$$I = \frac{2(x - 1)^{\frac{3}{2}}}{15}[3(x - 1) + 5] + c$$

$$\therefore \quad I = \frac{2(x - 1)^{\frac{3}{2}}}{15}(3x + 2) + c$$

c) Given $u = \sqrt{2x + 1}$, then

$$\frac{du}{dx} = \frac{1}{2}(2x + 1)^{-\frac{1}{2}} \times 2 = \frac{1}{(2x + 1)^{\frac{1}{2}}} = \frac{1}{u}$$

Also,

$$u^2 = 2x + 1 \quad \therefore \quad x = \frac{u^2 - 1}{2}$$

Therefore,

$$I = \int \frac{3x - 4}{\sqrt{2x + 1}} \frac{dx}{du} du$$

$$= \int \frac{3x - 4}{u} u \, du$$

$$= \int \left[3\left(\frac{u^2 - 1}{2}\right) - 4\right] du$$

$$= \int \left(\frac{3u^2}{2} - \frac{11}{2}\right) du$$

$$\therefore \quad I = \frac{u^3}{2} - \frac{11u}{2} + c = \frac{u}{2}(u^2 - 11) + c$$

Since $u = (2x + 1)^{\frac{1}{2}}$,

$$I = \frac{(2x + 1)^{\frac{1}{2}}}{2}[(2x + 1) - 11] + c$$

$$= \frac{(2x + 1)^{\frac{1}{2}}}{2}(2x - 10) + c$$

$$\therefore \quad I = (x - 5)\sqrt{2x + 1} + c$$

$$\frac{dx}{du} = \frac{1}{\frac{du}{dx}} = u$$

Exercise 5A

1 Find these integrals, in each case using the suggested substitution.

a) $\int (2x - 1)^3 \, dx$, $u = 2x - 1$

b) $\int (3x + 5)^4 \, dx$, $u = 3x + 5$

c) $\int (4x - 3)^5 \, dx$, $u = 4x - 3$

d) $\int (2 - 5x)^2 \, dx$, $u = 2 - 5x$

e) $\int x(x^2 - 1)^3 \, dx$, $u = x^2 - 1$

f) $\int x(1 - 2x^2)^3 \, dx$, $u = 1 - 2x^2$

g) $\int 2x(3x^2 + 5)^2 \, dx$, $u = 3x^2 + 5$

h) $\int x^2(x^3 + 1)^2 \, dx$, $u = x^3 + 1$

2 Find these integrals, in each case using the suggested substitution.

a) $\int x(x - 3)^2 \, dx$, $u = x - 3$

b) $\int x(x + 4)^3 \, dx$, $u = x + 4$

c) $\int (x - 4)(x - 1)^3 \, dx$, $u = x - 1$

d) $\int x(2x - 3)^2 \, dx$, $u = 2x - 3$

e) $\int (3x + 1)(2x - 5)^2 \, dx$, $u = 2x - 5$

f) $\int \frac{x}{x + 3} \, dx$, $u = x + 3$

g) $\int \frac{x}{(x + 1)^2} \, dx$, $u = x + 1$

h) $\int \frac{x + 2}{(2x - 3)^3} \, dx$, $u = 2x - 3$

3 Find these integrals, in each case using the suggested substitution.

a) $\int x\sqrt{x+1}\,dx$, $u = x + 1$
b) $\int x\sqrt{x-1}\,dx$, $u = \sqrt{x-1}$
c) $\int (x-4)\sqrt{x+5}\,dx$, $u = x + 5$
d) $\int (3x-2)\sqrt{1-2x}\,dx$, $u = \sqrt{1-2x}$
e) $\int \frac{x}{\sqrt{x+1}}\,dx$, $u = x + 1$
f) $\int \frac{x}{\sqrt{x-3}}\,dx$, $u = \sqrt{x-3}$
g) $\int \frac{x-2}{\sqrt{x-4}}\,dx$, $u = x - 4$
h) $\int \frac{x+3}{\sqrt{5-x}}\,dx$, $u = \sqrt{5-x}$

4 Integrate each of these with respect to x.

a) $x(x+3)^3$
b) $\frac{x+2}{x-1}$
c) $x\sqrt{5-x}$
d) $\frac{x-3}{(x+2)^2}$
e) $\frac{x}{\sqrt{2x+1}}$
f) $(x-3)(5-2x)^4$
g) $\frac{2x-1}{x+7}$
h) $\frac{x}{\sqrt{(x+1)^3}}$
i) $\frac{x+2}{\sqrt{4-x}}$
j) $\frac{x+3}{(3-x)^2}$
k) $x^2(x-1)^4$
l) $x\sqrt{(1-x)^3}$

In these questions you must choose a suitable substitution.

C3

5.2 Definite integrals

For a definite integral you can substitute for the limits of the integral. This is illustrated in the following two examples.

Example 4

Use the substitution $u = 2x - 3$ to find $\int_4^9 \frac{6}{2x-3}\,dx$.

Given $u = 2x - 3$, then $\frac{du}{dx} = 2$.

The limits must also be changed from x limits to u limits, by calculating the value of u when $x = 4$ and when $x = 9$.

When $x = 4$, $u = 2(4) - 3 = 5$. When $x = 9$, $u = 2(9) - 3 = 15$.

$$\text{Therefore} \quad \int_4^9 \frac{5}{2x-3}\,dx = \int_5^{15} \frac{6}{2x-3}\frac{dx}{du}\,du$$

$$= \int_5^{15} \frac{6}{u} \times \frac{1}{2}\,du$$

$$= \int_5^{15} \frac{3}{u}\,du$$

$$= \Big[3\ln u\Big]_5^{15}$$

$$= 3\ln 15 - 3\ln 5$$

$$= 3\ln\left(\frac{15}{5}\right)$$

$$= 3\ln 3$$

Example 5

Evaluate $\int_3^4 \frac{3x}{\sqrt{x-2}}\,dx$ using the substitution $u = \sqrt{x-2}$.

Given $u = (x-2)^{\frac{1}{2}}$, then

$$\frac{du}{dx} = \frac{1}{2}(x-2)^{-\frac{1}{2}} = \frac{1}{2(x-2)^{\frac{1}{2}}} = \frac{1}{2u}.$$

When $x = 3$, $u = \sqrt{3-2} = 1$. When $x = 4$, $u = \sqrt{4-2} = \sqrt{2}$.

Also,

$$u^2 = x - 2 \quad \therefore \quad x = u^2 + 2$$

Therefore,

$$\int_3^4 \frac{3x}{\sqrt{x-2}}\,dx = \int_1^{\sqrt{2}} \frac{3(u^2+2)}{u} 2u\,du = \int_1^{\sqrt{2}} (6u^2 + 12)\,du$$

$$= \left[2u^3 + 12u\right]_1^{\sqrt{2}} = 16\sqrt{2} - 14$$

C3

Exercise 5B

1 Use an appropriate substitution to evaluate each of these integrals.

a) $\int_0^2 (4x+1)^2\,dx$ b) $\int_2^3 (2x-3)^2\,dx$ c) $\int_{-1}^1 x(x^2+3)^2\,dx$

d) $\int_2^6 \frac{4}{x+2}\,dx$ e) $\int_2^5 \frac{8}{2x-1}\,dx$ f) $\int_0^2 \frac{8}{8-3x}\,dx$

2 Evaluate each of these.

a) $\int_3^4 \frac{x}{x-2}\,dx$ b) $\int_3^5 x(x-3)^2\,dx$ c) $\int_1^6 x\sqrt{x+3}\,dx$

d) $\int_1^3 \frac{x^2}{2x-1}\,dx$ e) $\int_4^7 \frac{5-x}{\sqrt{x-3}}\,dx$ f) $\int_1^3 (3x+1)(2-x)^4\,dx$

5.3 Integration by parts

This method is used to integrate the product of two functions. You already know that

$$\frac{d(uv)}{dx} = u\frac{dv}{dx} + v\frac{du}{dx}$$

Integrating both sides with respect to x gives

$$uv = \int u\frac{dv}{dx}\,dx + \int v\frac{du}{dx}\,dx$$

$$\int u\frac{dv}{dx}\,dx = uv - \int v\frac{du}{dx}\,dx$$

The product to be integrated has two parts:

- ✦ The function u, which is differentiated
- ✦ the function $\frac{dv}{dx}$, which is integrated.

This method is the reverse of the product rule for differentiation (section 2.3).

But how do you decide which part to differentiate and which part to integrate? Choose the function that becomes *simpler* when you differentiate to be u.

Example 6

Find $\int x(x+3)^3\,dx$.

The two parts of the integral are x and $(x+3)^3$.

The function x becomes simpler when differentiated.

So let $u = x$ and $\frac{dv}{dx} = (x+3)^3$.

When $u = x$, then $\frac{du}{dx} = 1$.

When $\frac{dv}{dx} = (x+3)^3$, then $v = \frac{1}{4}(x+3)^4$.

Therefore,

$$\int x(x+3)^3\,dx = \frac{x(x+3)^4}{4} - \int \frac{(x+3)^4}{4} \times 1\,dx$$
$$= \frac{x(x+3)^4}{4} - \frac{(x+3)^5}{20} + c$$
$$= \frac{(x+3)^4}{20}[5x - (x+3)] + c$$
$$= \tfrac{1}{20}(x+3)^4)(4x-3) + c$$

$u = x \quad v = \frac{1}{4}(x+3)^4$

$\frac{du}{dx} = 1 \quad \frac{dv}{dx} = (x+3)^3$

C3

In the next example, integration by parts is applied to an exponential function.

Example 7

Find $\int xe^x\,dx$.

In this case, both x and e^x can be easily integrated but x becomes simpler when differentiated. So let $u = x$ and $\frac{dv}{dx} = e^x$.

Therefore,

$$\int xe^x\,dx = xe^x - \int e^x \times 1\,dx$$
$$= xe^x - e^x + c$$

$u = x \quad v = e^x$

$\frac{du}{dx} = 1 \quad \frac{dv}{dx} = e^x$

Sometimes you may be unable to integrate one part of the function. Then you must let that part be u.

Example 8

Find $\int x^3 \ln x \, dx$.

In this case, you do not know how to integrate $\ln x$.

Therefore, let $u = \ln x$ and $\frac{dv}{dx} = x^3$.

Therefore,

$$\int x^3 \ln x \, dx = (\ln x)\frac{x^4}{4} - \int \frac{x^4}{4}\left(\frac{1}{x}\right) dx$$
$$= \frac{x^4}{4}\ln x - \int \frac{x^3}{4} dx$$
$$= \frac{x^4}{4}\ln x - \frac{x^4}{16} + c$$
$$= \frac{x^4}{16}(4\ln x - 1) + c$$

$u = \ln x \qquad v = \frac{1}{4}x^4$

$\frac{du}{dx} = \frac{1}{x} \qquad \frac{dv}{dx} = x^3$

In some cases, you may need to perform the process of integration by parts more than once. The next example illustrates such a case.

Example 9

Find $\int x^2 \cos x \, dx$.

In this case, let $u = x^2$ and $\frac{dv}{dx} = \cos x$. This is because the term x^2 becomes a constant, namely 2, when differentiated twice.

Therefore,

$$I = \int x^2 \cos x \, dx = x^2 \sin x - \int \sin x \, (2x) \, dx$$

$u = x^2 \qquad v = \sin x$

$\frac{du}{dx} = 2x \qquad \frac{dv}{dx} = \cos x$

At this point, you can see that the process of integrating by parts needs to be applied to $\int 2x \sin x \, dx$.

Let $u = 2x$ and $\frac{dv}{dx} = \sin x$, since $2x$ becomes simpler when differentiated.

Therefore,

$$\int 2x \sin x \, dx = 2x(-\cos x) - \int (-\cos x) \times 2 \, dx$$
$$= -2x \cos x + \int 2 \cos x \, dx$$
$$= -2x \cos x + 2 \sin x \, dx + c$$

$u = 2x \qquad v = -\cos x$

$\frac{du}{dx} = 2 \qquad \frac{dv}{dx} = \sin x$

Therefore,

$$I = x^2 \sin x - (-2x \cos x + 2 \sin x) + c$$
$$= x^2 \sin x + 2x \cos x - 2 \sin x + c$$
$$= (x^2 - 2) \sin x + 2x \cos x + c$$

So far, we have not found a way of integrating $\ln x$. However, you can find $\int \ln x \, dx$ using integration by parts.

Example 10

Find $\int \ln x \, dx$.

Write $\int \ln x \, dx$ as $\int \ln x \times 1 \, dx$ and then let $u = \ln x$ and $\frac{dv}{dx} = 1$.

Therefore,

$$\int \ln x \, dx = (\ln x) \times x - \int x\left(\frac{1}{x}\right) dx$$
$$= x \ln x - \int 1 \, dx$$
$$= x \ln x - x + c$$

$u = \ln x$ $\quad v = x$

$\frac{du}{dx} = \frac{1}{x}$ $\quad \frac{dv}{dx} = 1$

C3

Exercise 5C

All questions in this exercise should be tackled by integration by parts.

1 Integrate each of these functions.

a) $x(x-1)^2$ b) $x(x+1)^3$ c) $x(4-x)^3$
d) $x(2x+3)^5$ e) $(x-1)(x+2)^2$ f) $(x+3)(x-4)^5$
g) $(3x-1)(2x+3)^2$ h) $(2-5x)(4-x)^4$

2 Find each of these integrals.

a) $\int \frac{x}{(x-1)^2} \, dx$ b) $\int \frac{x}{(x+1)^2} \, dx$ c) $\int \frac{x-2}{(2x-3)^2} \, dx$
d) $\int \frac{3x-4}{(x+2)^4} \, dx$ e) $\int \frac{x}{\sqrt{2x-3}} \, dx$ f) $\int \frac{x+4}{\sqrt{3x-2}} \, dx$
g) $\int \frac{3x+1}{\sqrt{1-2x}} \, dx$ h) $\int x\sqrt{4-x} \, dx$ i) $\int (2-5x)\sqrt{3-2x} \, dx$

3 Integrate each of these with respect to x.

a) $x \cos x$ b) $x \sin 2x$ c) xe^{3x} d) xe^{-x}
e) $(6x-1)\cos 3x$ f) $x \ln x$ g) $x^2 \ln x$ h) $\sqrt{x} \ln x$

4 Evaluate each of these integrals.

a) $\int_0^2 (x-3)^2 \, dx$ b) $\int_0^{\frac{\pi}{2}} x \sin 2x \, dx$ c) $\int_{-1}^1 (x+1)e^x \, dx$
d) $\int_3^6 \frac{x}{\sqrt{x-2}} \, dx$ e) $\int_2^4 x^3 \ln x \, dx$ f) $\int_{\frac{\pi}{6}}^{\frac{\pi}{3}} x \cos 3x \, dx$

5 Integrate each of these with respect to x.

a) $x^2 \sin x$ b) $x^2(x+3)^3$ c) x^2e^x
d) $x^2 \cos 2x$ e) x^2e^{-2x} f) $(x+1)^2 \sin x$

6 Evaluate each of these integrals.

a) $\int_0^1 x^2 e^{2x}\,dx$ b) $\int_0^{\pi} x^2 \sin x\,dx$ c) $\int_{-1}^{1} x^2(x+3)^3\,dx$

d) $\int_0^{\frac{\pi}{4}} x^2 \cos 2x\,dx$ e) $\int_3^6 \frac{x^2}{\sqrt{x-2}}\,dx$ f) $\int_0^{\pi} (x-\pi)^2 \sin x\,dx$

5.4 Standard integrals

You should be able to use these two standard integrals:

i) $\int \frac{dx}{a^2+x^2} = \frac{1}{a}\tan^{-1}\left(\frac{x}{a}\right) + c$

ii) $\int \frac{dx}{\sqrt{a^2-x^2}} = \sin^{-1}\left(\frac{x}{a}\right) + c$

C3

The formula book gives this information:

f(x)	f′(x)
$\sin^{-1} x$	$\frac{1}{\sqrt{1-x^2}}$
$\tan^{-1} x$	$\frac{1}{1+x^2}$

Proof of (i)

Let $x = a\tan u$, then $\frac{dx}{du} = a\sec^2 u$. Therefore,

Using integration by substitution.

$$\int \frac{dx}{a^2+x^2} = \int \frac{1}{a^2+(a\tan u)^2}\frac{dx}{du}\,du = \int \frac{1}{a^2(1+\tan^2 u)} a\sec^2 u\,du$$

$$\therefore \quad I = \int \frac{du}{a} \quad (\text{since } 1+\tan^2 u = \sec^2 u)$$

$$= \frac{u}{a} + c$$

Since $x = a\tan u$, $\tan u = \frac{x}{a}$, and $u = \tan^{-1}\left(\frac{x}{a}\right)$. Therefore,

$$I = \frac{1}{a}\tan^{-1}\left(\frac{x}{a}\right) + c$$

as required.

Proof of (ii)

Let $x = a\sin u$, then $\frac{dx}{du} = a\cos u$. Therefore,

$$\int \frac{dx}{\sqrt{a^2-x^2}} = \int \frac{1}{\sqrt{a^2-a^2\sin^2 u}}\frac{dx}{du}\,du = \int \frac{1}{\sqrt{a^2(1-\sin^2 u)}} a\cos u\,du$$

$$\therefore \quad I = \int \frac{\cos u}{\sqrt{\cos^2 u}}\,du \quad (\text{since } 1-\sin^2 u = \cos^2 u)$$

$$= u + c$$

Since $x = a\sin u$, $\sin u = \frac{x}{a}$, and $u = \sin^{-1}\left(\frac{x}{a}\right)$. Therefore,

$$I = \sin^{-1}\left(\frac{x}{a}\right) + c$$

as required.

The examples which follow show how you can use these standard integrals.

Example 11

Find each of these integrals.

a) $\int \frac{1}{\sqrt{16 - x^2}}\,dx$ b) $\int \frac{1}{\sqrt{36 - 4x^2}}\,dx$ c) $\int \frac{1}{1 + 25x^2}\,dx$

a) $\int \frac{1}{\sqrt{16 - x^2}}\,dx = \int \frac{1}{\sqrt{4^2 - x^2}}\,dx$

$$= \sin^{-1}\left(\frac{x}{4}\right) + c$$

b) $\int \frac{1}{\sqrt{36 - 4x^2}}\,dx = \int \frac{1}{\sqrt{4(9 - x^2)}}\,dx = \frac{1}{2}\int \frac{1}{\sqrt{3^2 - x^2}}\,dx$

$$= \frac{1}{2}\sin^{-1}\left(\frac{x}{3}\right) + c$$

c) $\int \frac{1}{1 + 25x^2}\,dx = \int \frac{1}{25(\frac{1}{25} + x^2)}\,dx = \frac{1}{25}\int \frac{1}{(\frac{1}{5})^2 + x^2}\,dx$

$$= \frac{1}{25} \times \frac{1}{(\frac{1}{5})}\tan^{-1}\left[\frac{x}{(\frac{1}{5})}\right] + c$$

$$= \frac{1}{5}\tan^{-1}5x + c$$

Simplify the expression so that the coefficient of x^2 is 1.

C3

Example 12

Evaluate the integral

$$\int_0^2 \frac{1}{\sqrt{4 - x^2}}\,dx$$

$$\int_0^2 \frac{1}{\sqrt{4 - x^2}}\,dx = \left[\sin^{-1}\left(\frac{x}{2}\right)\right]_0^2$$

$$= \sin^{-1}(1) - \sin^{-1}(0)$$

$$= \frac{\pi}{2} - 0$$

$$= \frac{\pi}{2}$$

The next two examples combine two methods of integration.

Example 13

Evaluate $\int_1^2 \frac{x-4}{x^2+9}\,dx$

$$\int_1^2 \frac{x-4}{x^2+9}\,dx = \int_1^2 \frac{x}{x^2+9}\,dx - \int_1^2 \frac{4}{x^2+9}\,dx$$

$$= \frac{1}{2}\int_1^2 \frac{2x}{x^2+9}\,dx - \int_1^2 \frac{4}{x^2+9}\,dx$$

$$= \frac{1}{2}\Big[\ln(x^2+9)\Big]_1^2 - 4\left[\frac{1}{3}\tan^{-1}\left(\frac{x}{3}\right)\right]_1^2$$

$$= \frac{1}{2}(\ln 13 - \ln 10) - 4\left[\frac{1}{3}\tan^{-1}\left(\frac{2}{3}\right) - \frac{1}{3}\tan^{-1}\left(\frac{1}{3}\right)\right]$$

$$= \frac{1}{2}\ln\left(\frac{13}{10}\right) - \frac{4}{3}\left[\tan^{-1}\left(\frac{2}{3}\right) - \tan^{-1}\left(\frac{1}{3}\right)\right]$$

$\frac{d}{dx}\ln(x^2+9) = \frac{2x}{x^2+9}$

So $\int \frac{x}{x^2+9}\,dx = \frac{1}{2}\ln(x^2+9) + c$

C3

Example 14 uses integration by substitution and a standard integral.

Example 14

Use the substitution $u = e^x$ to find

$$\int \frac{1}{e^x + e^{-x}}\,dx$$

Let $u = e^x$, then $\frac{du}{dx} = e^x = u$. Therefore,

$$\int \frac{1}{e^x+e^{-x}}\,dx = \int \frac{1}{u+\frac{1}{u}}\,\frac{dx}{du}\,du$$

$$= \int \frac{1}{u+\frac{1}{u}}\,\frac{1}{u}\,du$$

$$= \int \frac{1}{u^2+1}\,du$$

$$= \tan^{-1} u + c$$

$$= \tan^{-1}(e^x) + c$$

Exercise 5D

1 Find each of these integrals.

a) $\int \frac{1}{\sqrt{4-x^2}}\,dx$ b) $\int \frac{1}{\sqrt{25-x^2}}\,dx$ c) $\int \frac{1}{\sqrt{1-9x^2}}\,dx$

d) $\int \frac{6}{\sqrt{1-36x^2}}\,dx$ e) $\int \frac{1}{\sqrt{4-9x^2}}\,dx$ f) $\int \frac{1}{\sqrt{25-16x^2}}\,dx$

g) $\int \frac{4}{4+x^2}\,dx$ h) $\int \frac{1}{100+x^2}\,dx$ i) $\int \frac{1}{1+9x^2}\,dx$

j) $\int \frac{1}{1+25x^2}\,dx$ k) $\int \frac{1}{9+4x^2}\,dx$ l) $\int \frac{14}{49+16x^2}\,dx$

2 Evaluate each of these.

a) $\int_0^1 \frac{1}{\sqrt{1-x^2}}\,dx$ b) $\int_0^1 \frac{1}{1+x^2}\,dx$ c) $\int_0^3 \frac{1}{9+x^2}\,dx$

d) $\int_{-4}^4 \frac{3}{\sqrt{16-x^2}}\,dx$ e) $\int_{-2}^0 \frac{1}{\sqrt{4-x^2}}\,dx$ f) $\int_{-6}^6 \frac{24}{36+x^2}\,dx$

3 a) By using the substitution $u = e^x$, show that

$$\int \frac{e^{2x}}{1+e^x}\,dx = \int \frac{u}{1+u}\,du.$$

b) Deduce that $\int \frac{e^{2x}}{1+e^x}\,dx = e^x - \ln(1+e^x) + c$.

4 a) By using the substitution $u = e^x$, show that

$$\int \frac{e^x}{1+e^{2x}}\,dx = \int \frac{1}{1+u^2}\,du.$$

b) Deduce that $\int \frac{e^x}{1+e^{2x}}\,dx = \tan^{-1}(e^x) + c$.

5.5 Applications

In C2 you learnt how to integrate in order to find the area under a curve, or the area between two curves. You are now ready to integrate in order to calculate a volume of revolution.

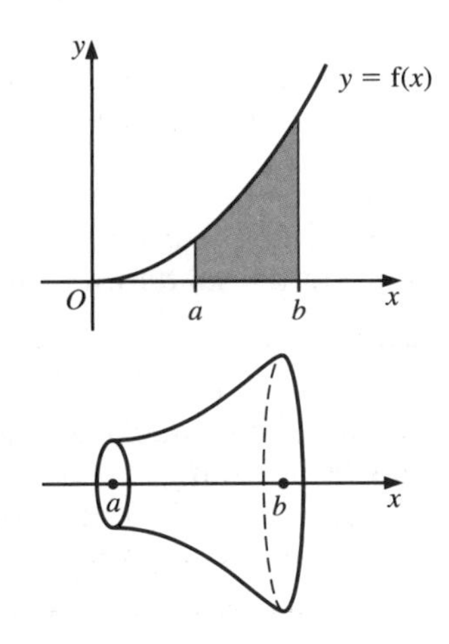

Volumes of revolution about the x-axis

Consider the area under a curve $y = f(x)$ between $x = a$ and $x = b$, as shown.

Now consider the solid formed when this area is rotated through 360° (or 2π radians) about the x-axis.

The volume of this solid is called a **volume of revolution**.

The volume of this solid can be calculated by imagining that the solid is made up of lots of thin discs.

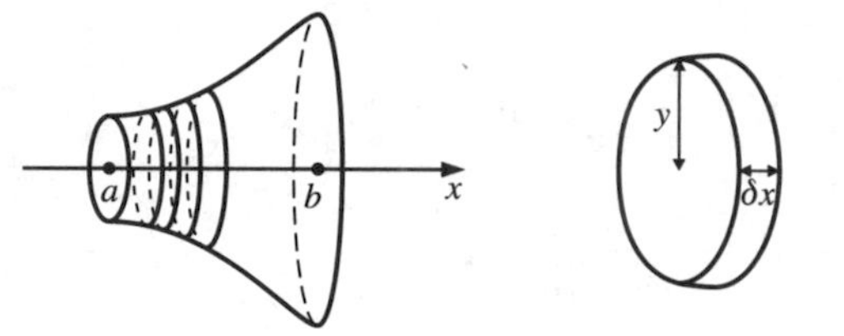

A typical disc has thickness δx and radius y. So its volume is given by $\pi y^2\ \delta x$. To find the volume of the whole solid you can use calculus to sum up the volumes of all of these discs between $x = a$ and $x = b$.

> The volume of the solid formed when the area under the curve $y = \mathrm{f}(x)$, between $x = a$ and $x = b$, is rotated through 360° about the x-axis is given by $\pi\int_a^b y^2\,\mathrm{d}x$.

You should memorise this formula.

Example 15

Find the volume of the solid formed when the area between the curve $y = x^2 + 2$ and the x-axis from $x = 1$ to $x = 3$ is rotated through 2π radians about the x-axis.

The volume V is given by

$$V = \pi\int_1^3 y^2\,\mathrm{d}x$$

Now $y^2 = (x^2 + 2)^2 = x^4 + 4x^2 + 4$. Therefore,

$$V = \pi\int_1^3 (x^4 + 4x^2 + 4)\,\mathrm{d}x$$

$$= \pi\left[\frac{x^5}{5} + \frac{4x^3}{3} + 4x\right]_1^3$$

$$= \pi\left(\frac{483}{5} - \frac{83}{15}\right)$$

$$\therefore\quad V = \frac{1366\pi}{15}$$

The volume of the solid formed is $\dfrac{1366\pi}{15}$.

You learnt in C1 that you could find the area between a curve and a straight line by integrating each function and subtracting one from the other. The same technique is used to find the volume generated when the area between a curve and a straight line is rotated about the x-axis.

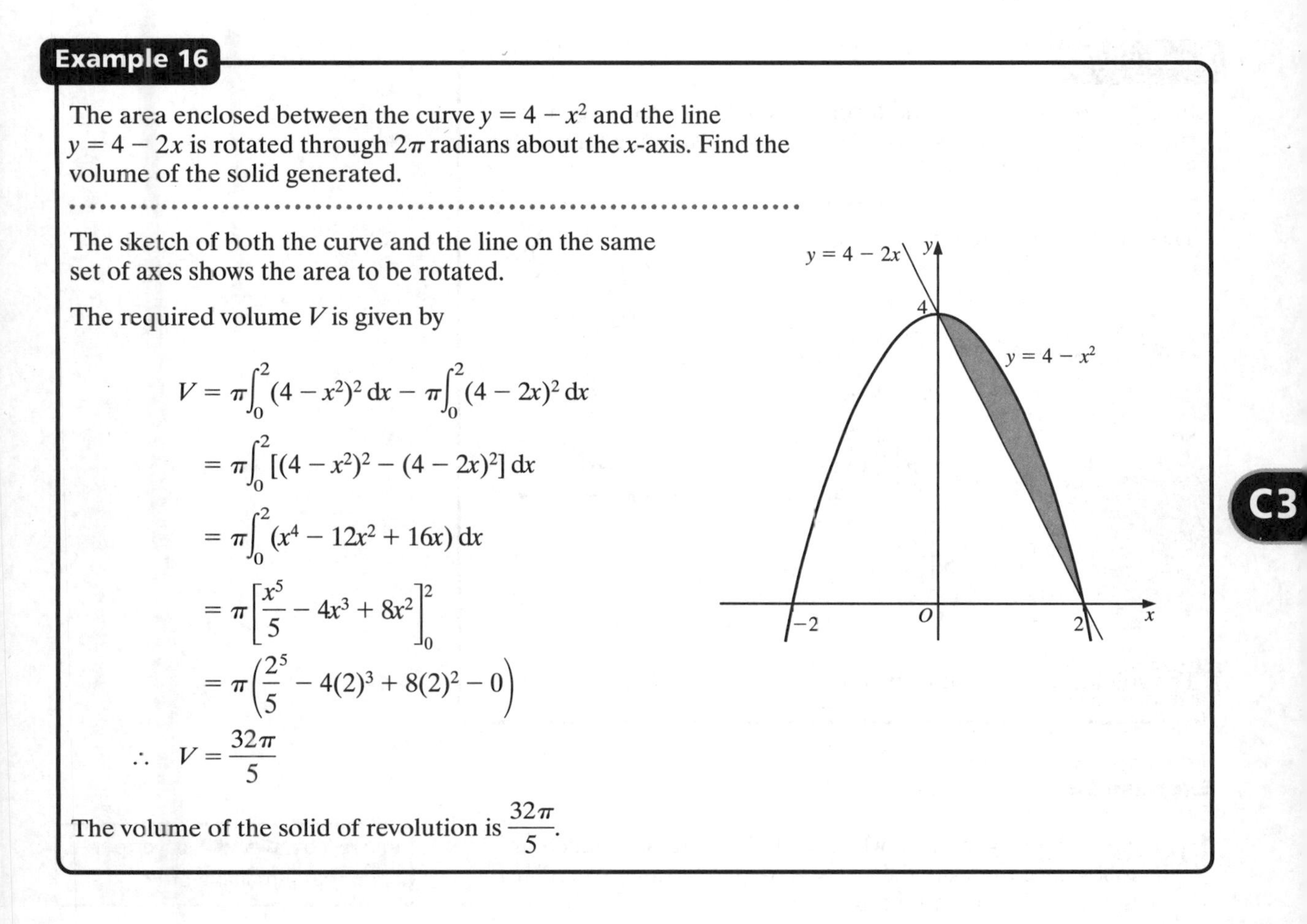

Example 16

The area enclosed between the curve $y = 4 - x^2$ and the line $y = 4 - 2x$ is rotated through 2π radians about the x-axis. Find the volume of the solid generated.

The sketch of both the curve and the line on the same set of axes shows the area to be rotated.

The required volume V is given by

$$V = \pi\int_0^2 (4 - x^2)^2\,dx - \pi\int_0^2 (4 - 2x)^2\,dx$$

$$= \pi\int_0^2 [(4 - x^2)^2 - (4 - 2x)^2]\,dx$$

$$= \pi\int_0^2 (x^4 - 12x^2 + 16x)\,dx$$

$$= \pi\left[\frac{x^5}{5} - 4x^3 + 8x^2\right]_0^2$$

$$= \pi\left(\frac{2^5}{5} - 4(2)^3 + 8(2)^2 - 0\right)$$

$$\therefore \quad V = \frac{32\pi}{5}$$

The volume of the solid of revolution is $\frac{32\pi}{5}$.

C3

Volumes of revolution about the y-axis

The volume of the solid of revolution formed by rotating an area through 2π radians about the y-axis can be found in a way similar to that about the x-axis.

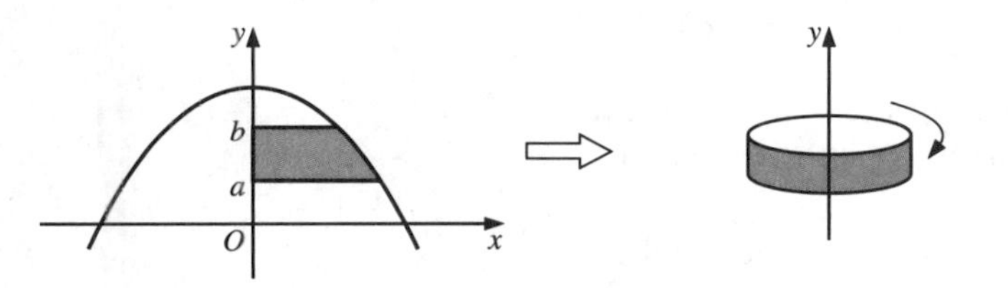

The volume of such a solid of revolution is given by

$$V = \pi\int_a^b x^2\,dy$$

Remember that dy implies that the limits a and b are y limits. You should memorise this formula.

Example 17

Find the volume of the solid formed when the area between the curve $y = x^3$ and the y-axis from $y = 1$ to $y = 8$ is rotated through 2π radians about the y-axis.

The required volume V is given by

$$V = \pi \int_1^8 x^2 \, dy$$

Now $y = x^3$. Therefore

$$y^{\frac{2}{3}} = (x^3)^{\frac{2}{3}} \qquad \therefore \quad y^{\frac{2}{3}} = x^2$$

So

$$V = \pi \int_1^8 y^{\frac{2}{3}} \, dy = \pi \left[\frac{3}{5} y^{\frac{5}{3}} \right]_1^8$$

$$= \pi \left[\frac{96}{5} - \frac{3}{5} \right]$$

$$\therefore \quad V = \frac{93\pi}{5}$$

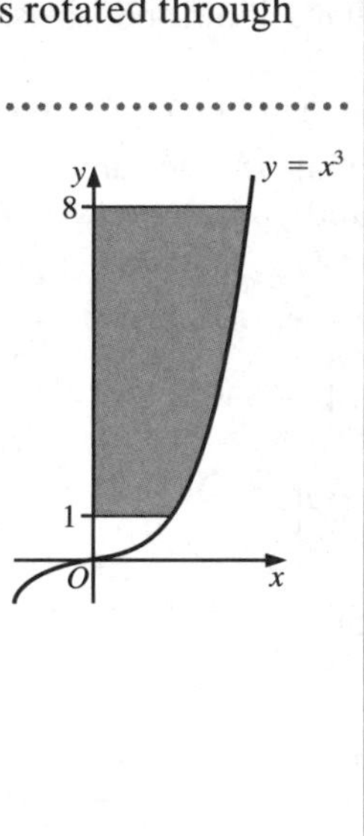

The volume of the solid formed is $\dfrac{93\pi}{5}$.

C3

Exercise 5E

In questions 1 and 2 leave your answer as a multiple of π.

1 Find the volume generated when each of the areas, bounded by the following curves and the x-axis, is rotated through 2π radians about the x-axis between the given lines.

a) $y = x; x = 0$ and $x = 6$
b) $y = x^2; x = 0$ and $x = 5$
c) $y = \sqrt{x}; x = 0$ and $x = 4$
d) $y = \dfrac{1}{x^2}; x = 1$ and $x = 2$
e) $y = 3\sqrt{x}; x = 2$ and $x = 4$
f) $y = 2x + 1; x = 1$ and $x = 3$
g) $y = 5 - x; x = 2$ and $x = 5$
h) $y = x^2 + 1; x = 0$ and $x = 3$
i) $y = \sqrt{x^2 + 3x}; x = 2$ and $x = 6$
j) $y = \sqrt{3x^2 + 8}; x = 1$ and $x = 3$
k) $y = \dfrac{x^2 - 2}{x^2}; x = \frac{1}{4}$ and $x = \frac{1}{2}$
l) $y = \sqrt{x} - 3; x = 4$ and $x = 9$

2 Find the volume generated when each of the areas in the positive quadrant, bounded by the following curves and lines, is rotated through 2π radians about the y-axis.

a) $y = \frac{1}{2}x; y = 0$ and $y = 6$
b) $y = x^2; y = 0$ and $y = 9$
c) $y = x^3; y = 0$ and $y = 8$
d) $y = \sqrt{x}; y = 0$ and $y = 3$
e) $y = 3x; y = 3$ and $y = 6$
f) $y = x^4; y = 1$ and $y = 4$
g) $y = \dfrac{1}{x}; y = 2$ and $y = 4$
h) $y = x - 1; y = 2$ and $y = 5$
i) $y = \frac{1}{2}x + 3; y = 4$ and $y = 6$
j) $y = x^2; y = 2$ and $y = 6$
k) $y = \sqrt{x^2 + 1}; y = 1$ and $y = 3$
l) $y = \sqrt{2x^2 - 1}; y = 1$ and $y = 6$

Remember:
It helps to draw a sketch.

3 The curve $y = x^2$ meets the line $y = 4$ at the points P and Q.

a) Find the coordinates of P and Q.

b) Calculate the volume generated when the region bounded by the curve and the line is rotated through 360° about the x-axis.

4 The curve $y = x^2 + 1$ meets the line $y = 2$ at the points A and B.

a) Find the coordinates of A and B.

The region bounded by the curve and the line is rotated through 2π radians about the x-axis.

b) Calculate the volume of the solid generated.

5 The region bounded by the lines $y = x + 1$, $y = 3$ and the y-axis is rotated through 360° about the x-axis. Calculate the volume of the solid generated.

6 Calculate the volume generated when the region bounded by the curve $y = \frac{4}{x}$ and the lines $x = 1$ and $y = 1$ is rotated 360° about the x-axis.

7 The region bounded by the curve $y = \sqrt{x}$, the x-axis and the line $x = 4$ is rotated 2π radians about the y-axis. Calculate the volume of the solid generated.

8 The region R is bounded by the curve $y = x^2 + 2$, the line $x = 1$, and the x- and y-axes. Calculate the volume of the solid generated when R is rotated through 360° about the y-axis.

9 Calculate the volume generated when the region bounded by the curve $y = 9 - x^2$, the line $x = 2$, and the x-axis is rotated through 360° about the y-axis.

10 The line $y = 3x$ meets the curve $y = x^2$ at the points O and P.

a) Calculate the coordinates of P.

b) Find the volume of the solid generated when the area enclosed by the line and the curve is rotated through 2π radians about i) the x-axis, ii) the y-axis.

11 a) On one set of axes sketch the graphs of the curves $y = x(1 - x)$ and $y = 2x(1 - x)$.

b) Calculate the volume generated when the finite region bounded by the two curves is rotated through 360° about the x-axis.

In questions 12–14 give your answer to 3 significant figures.

12 Find the volume generated when the region bounded by the curve $y = \sqrt{x}$ and the line $y = \frac{1}{5}x$ is rotated through 360° about the y-axis.

13 The curve $y = x^2$ meets the curve $y = 8 - x^2$ at the points P and Q.

a) Find the coordinates of P and Q.

b) Calculate the volume generated when the region bounded by the two curves is rotated through π radians about i) the x-axis, ii) the y-axis.

14 Find the volume generated when the region bounded by the curves $y = 2x^2$ and line $y = 3 - x^2$ is rotated through 180° about a) the x-axis, b) the y-axis.

5.6 Mixed problems

You now know how to calculate areas under and between curves, and how to calculate volumes of revolution. You have also learnt how to integrate using the techniques of substitution and parts. So you are now ready to throw the whole lot together and tackle some more advanced questions using these new techniques.

Example 18

The diagram shows part of the curve with equation $y = \frac{6x}{1+x}$, together with the line $y = 2x$. The curve and the line intersect at O and P.

a) Find the coordinates of the point P.

b) Calculate the area of the region bounded by the curve and the straight line between O and P.

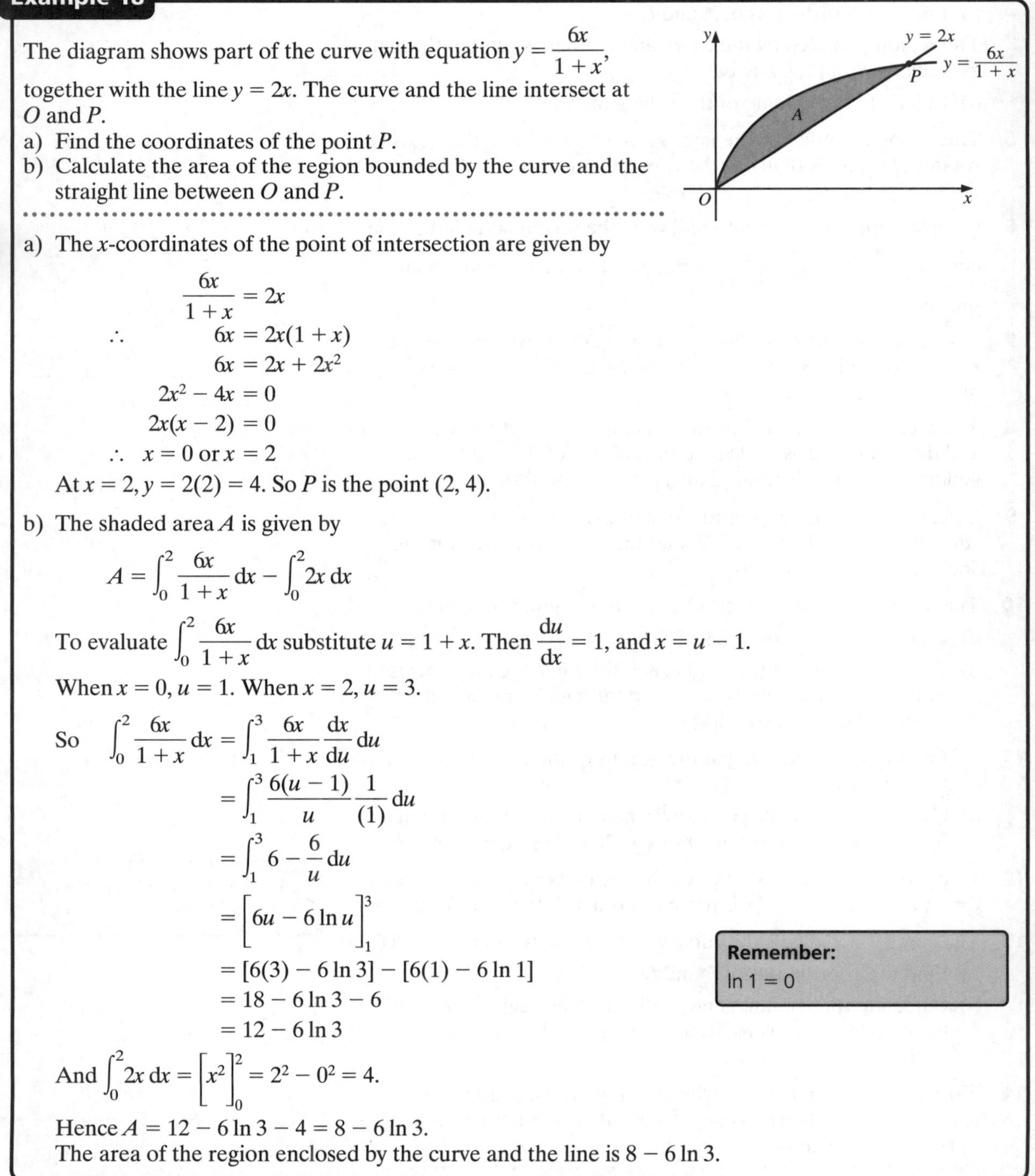

C3

a) The x-coordinates of the point of intersection are given by

$$\frac{6x}{1+x} = 2x$$

$$\therefore \quad 6x = 2x(1+x)$$

$$6x = 2x + 2x^2$$

$$2x^2 - 4x = 0$$

$$2x(x-2) = 0$$

$$\therefore \quad x = 0 \text{ or } x = 2$$

At $x = 2$, $y = 2(2) = 4$. So P is the point $(2, 4)$.

b) The shaded area A is given by

$$A = \int_0^2 \frac{6x}{1+x}\,dx - \int_0^2 2x\,dx$$

To evaluate $\int_0^2 \frac{6x}{1+x}\,dx$ substitute $u = 1 + x$. Then $\frac{du}{dx} = 1$, and $x = u - 1$.

When $x = 0$, $u = 1$. When $x = 2$, $u = 3$.

$$\text{So} \quad \int_0^2 \frac{6x}{1+x}\,dx = \int_1^3 \frac{6x}{1+x}\frac{dx}{du}\,du$$

$$= \int_1^3 \frac{6(u-1)}{u}\frac{1}{(1)}\,du$$

$$= \int_1^3 6 - \frac{6}{u}\,du$$

$$= \Big[6u - 6\ln u\Big]_1^3$$

$$= [6(3) - 6\ln 3] - [6(1) - 6\ln 1]$$

$$= 18 - 6\ln 3 - 6$$

$$= 12 - 6\ln 3$$

Remember:
$\ln 1 = 0$

And $\int_0^2 2x\,dx = \Big[x^2\Big]_0^2 = 2^2 - 0^2 = 4.$

Hence $A = 12 - 6\ln 3 - 4 = 8 - 6\ln 3$.

The area of the region enclosed by the curve and the line is $8 - 6\ln 3$.

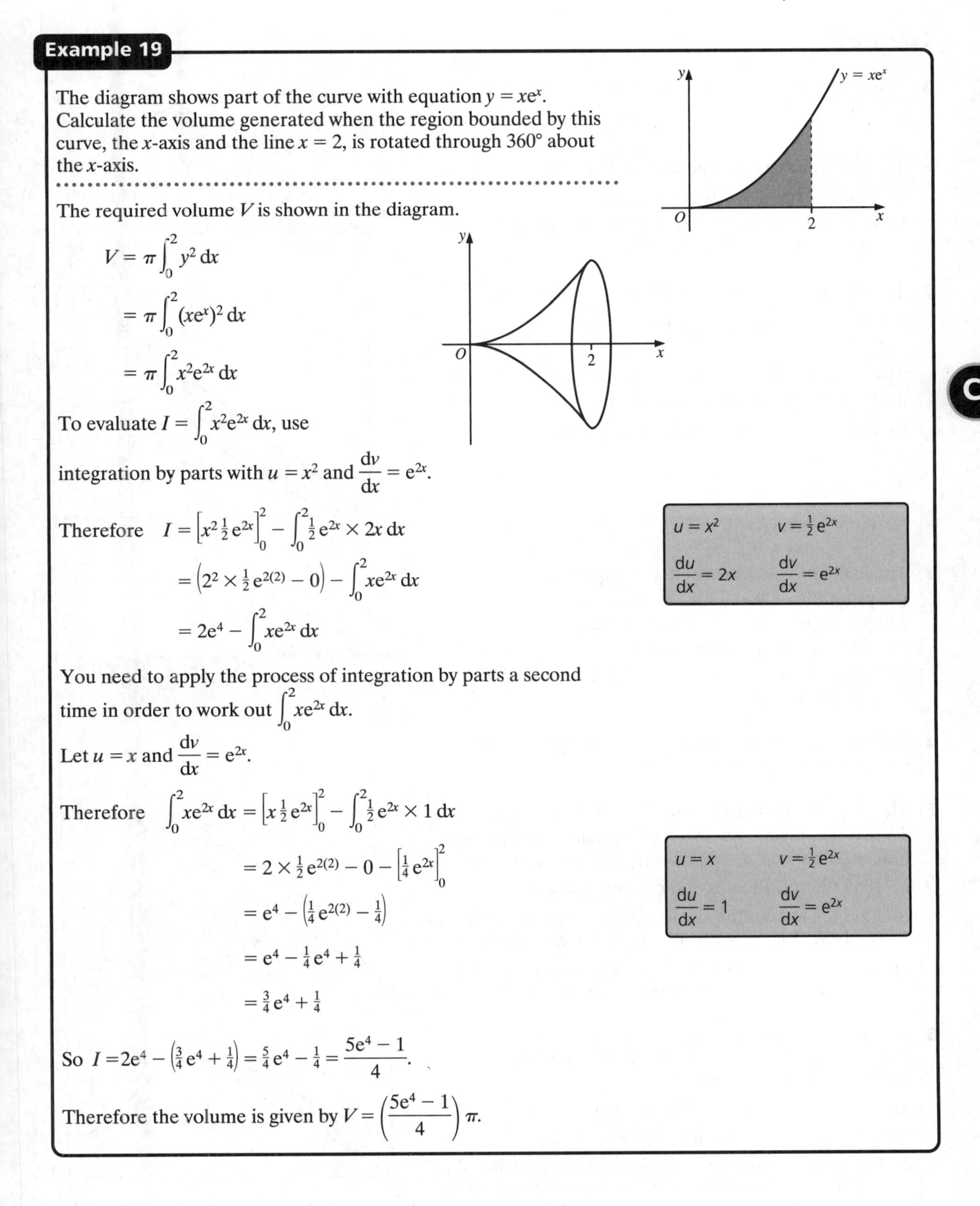

Example 19

The diagram shows part of the curve with equation $y = xe^x$.
Calculate the volume generated when the region bounded by this curve, the x-axis and the line $x = 2$, is rotated through 360° about the x-axis.

The required volume V is shown in the diagram.

$$V = \pi \int_0^2 y^2 \, dx$$

$$= \pi \int_0^2 (xe^x)^2 \, dx$$

$$= \pi \int_0^2 x^2 e^{2x} \, dx$$

To evaluate $I = \int_0^2 x^2 e^{2x} \, dx$, use integration by parts with $u = x^2$ and $\frac{dv}{dx} = e^{2x}$.

Therefore $I = \left[x^2 \tfrac{1}{2} e^{2x}\right]_0^2 - \int_0^2 \tfrac{1}{2} e^{2x} \times 2x \, dx$

$$= \left(2^2 \times \tfrac{1}{2} e^{2(2)} - 0\right) - \int_0^2 xe^{2x} \, dx$$

$$= 2e^4 - \int_0^2 xe^{2x} \, dx$$

$u = x^2$	$v = \frac{1}{2}e^{2x}$
$\frac{du}{dx} = 2x$	$\frac{dv}{dx} = e^{2x}$

You need to apply the process of integration by parts a second time in order to work out $\int_0^2 xe^{2x} \, dx$.

Let $u = x$ and $\frac{dv}{dx} = e^{2x}$.

Therefore $\int_0^2 xe^{2x} \, dx = \left[x \tfrac{1}{2} e^{2x}\right]_0^2 - \int_0^2 \tfrac{1}{2} e^{2x} \times 1 \, dx$

$$= 2 \times \tfrac{1}{2} e^{2(2)} - 0 - \left[\tfrac{1}{4} e^{2x}\right]_0^2$$

$$= e^4 - \left(\tfrac{1}{4} e^{2(2)} - \tfrac{1}{4}\right)$$

$$= e^4 - \tfrac{1}{4} e^4 + \tfrac{1}{4}$$

$$= \tfrac{3}{4} e^4 + \tfrac{1}{4}$$

$u = x$	$v = \frac{1}{2}e^{2x}$
$\frac{du}{dx} = 1$	$\frac{dv}{dx} = e^{2x}$

So $I = 2e^4 - \left(\tfrac{3}{4} e^4 + \tfrac{1}{4}\right) = \tfrac{5}{4} e^4 - \tfrac{1}{4} = \dfrac{5e^4 - 1}{4}$.

Therefore the volume is given by $V = \left(\dfrac{5e^4 - 1}{4}\right)\pi$.

Exercise 5F

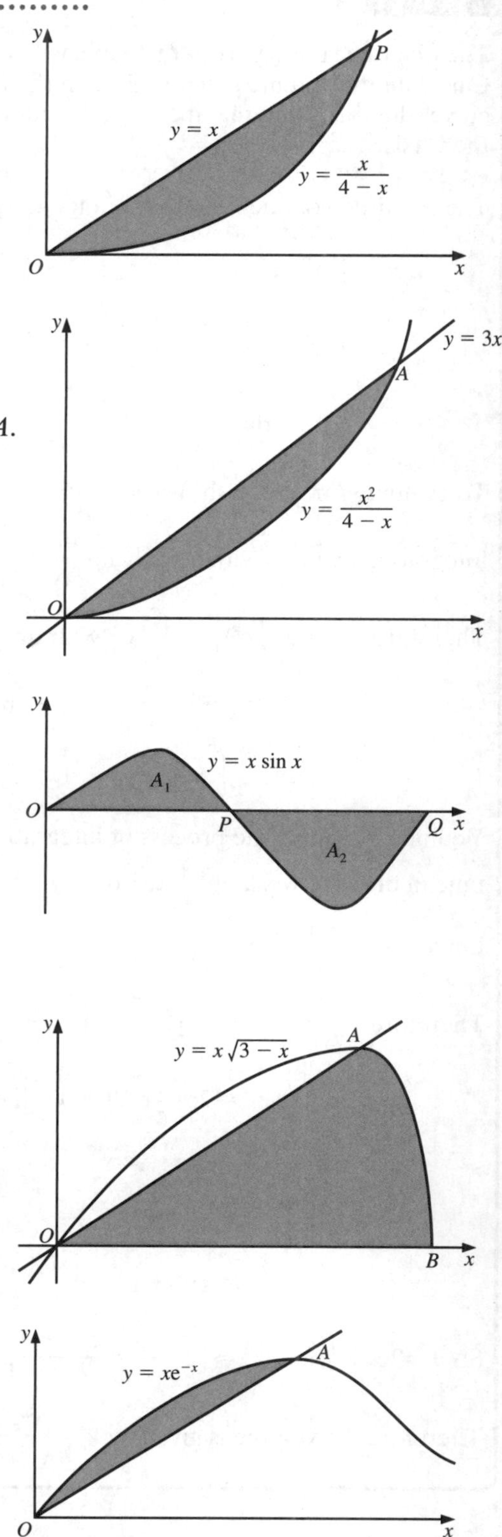

1 The diagram shows part of the curve with equation $y = \dfrac{x}{4-x}$, together with the line $y = x$.

The curve and the line intersect at O and P.

a) Find the coordinates of the point P.

b) Calculate the area of the region bounded by the curve and the line between O and P.

2 The diagram shows part of the curve with equation $y = \dfrac{x^2}{4-x}$, together with the line $y = 3x$.

The curve and the line meet at the origin and at the point A.

a) Find the coordinates of the point A.

b) Calculate the area of the shaded region.

3 The diagram shows the curve with equation $y = x \sin x$ for values of x between 0 and 2π. The curve cuts the x-axis at O, P and Q.

a) Find the coordinates of P and Q.

b) Calculate the area of each of the shaded regions, marked A_1 and A_2.

4 Find the area of the finite region bounded by the curve $y = x \ln x$, the lines $x = 1$ and $x = \mathrm{e}$.

5 The diagram shows part of the curve with equation $y = x\sqrt{3-x}$, together with a line segment, OA. The curve has a maximum at A, and crosses the x-axis at B.

a) Find the coordinates of the points A and B.

b) Find the area of the shaded region bounded by the line segment OA, the arc of the curve AB, and the x-axis.

c) Find also the volume of the solid generated when this region is rotated through 360° about the x-axis.

6 The diagram shows part of the curve with equation $y = x\mathrm{e}^{-x}$, together with a line segment OA, where A is a local maximum of the curve.

a) Find the coordinates of the point A.

b) Calculate the area of the shaded region.

c) Find also the volume of the solid generated when this region is rotated through 2π radians about the x-axis.

7 a) Sketch the curve with equation $y = (x - 1)^4(2 - x)$.

b) Calculate the area of each of the two finite regions bounded by the curve and the x-axis.

8 The curve $y = \dfrac{x}{x + 1}$ meets the curve $y = \dfrac{2}{5 - x}$ at the points A and B.

a) Find the coordinates of the points A and B.

b) Sketch the two curves on the same set of axes.

c) Calculate the area of the finite region bounded by the two curves.

9 a) Sketch the curve with equation $y = (1 - x)e^x$.

b) Calculate the area of the region in the positive quadrant bounded by the curve and the x- and y-axes.

c) Find also the volume of the solid generated when this region is rotated through 360° about the x-axis.

Summary

You should know how to ...	Check out
1 Integrate by substitution.	**1** a) Use the substitution $u = x + 5$ to find $\int x(x + 5)^7\, dx$ b) Use the substitution $u = 4 - x^2$ to find $\int_0^1 \dfrac{3x}{\sqrt{4 - x^2}}\, dx$
2 Integrate by parts.	**2** Find a) $\int xe^{3x}\, dx$ b) $\int_0^{\frac{\pi}{2}} x^2 \sin 2x\, dx$
3 Evaluate standard integrals.	**3** a) Find $\int \dfrac{1}{x^2 + 36}\, dx$ b) Evaluate $\int_1^2 \dfrac{4}{\sqrt{9 - x^2}}\, dx$
4 Evaluate volumes of rotation.	**4** a) The curve $y = x^2 + 2$ is rotated through 360° about the x-axis, between the lines $x = 1$ and $x = 3$. Calculate the volume of revolution generated. b) The curve $y = \sqrt{4 - x^2}$ is rotated through 360° about the y-axis, between the lines $y = 0$ and $y = 2$. Calculate the volume of revolution generated.

Revision exercise 5

1 a) Use integration by parts to evaluate $\int_0^{\frac{\pi}{2}} x \cos x \, dx$.

b) i) Use the substitution $t = x^2 + 4$ to show that $\int \frac{2x \, dx}{\sqrt{x^2 + 4}} = \int \frac{1}{\sqrt{t}} \, dt$.

ii) Show that $\int_0^2 \frac{2x \, dx}{\sqrt{x^2 + 4}} = 4(\sqrt{2} - 1)$. *(AQA, 2004)*

2 a) By using the chain rule, or otherwise, find $\frac{dy}{dx}$ when $y = \ln(x^2 + 9)$.

b) Show that $\int_0^3 \frac{x}{x^2 + 9} \, dx = \frac{1}{2} \ln 2$.

c) Show that $\int_0^3 \frac{x + 1}{x^2 + 9} \, dx = \frac{1}{2} \ln 2 + \frac{\pi}{12}$. *(AQA, 2004)*

C3

3 The diagram shows the graph of $y = x \ln x$ for $1 \leqslant x \leqslant 3$.

a) Differentiate $2x^2 \ln x - x^2$ with respect to x.

b) Hence, or otherwise, evaluate $\int_1^3 x \ln x \, dx$

(AQA, 2004)

4 Use integration by parts to find $\int_0^{\frac{1}{2}} xe^{2x} \, dx$. *(AQA, 2003)*

5 The graph shows the region R enclosed by the curve $y = x - \frac{1}{x}$, the x-axis and the line $x = 2$.

Find the exact volume of the solid formed when the region R is rotated through 2π radians about the x-axis.

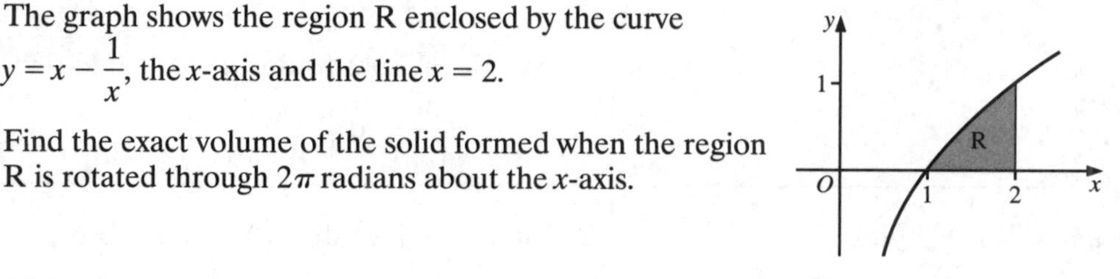

(AQA, 2002)

6 a) Express $\frac{1}{2 - x} + \frac{1}{2 + x}$ in the form $\frac{A}{4 - x^2}$ where A is a constant.

Part of the graph of $y = \frac{1}{\sqrt{4 - x^2}}$ is shown.

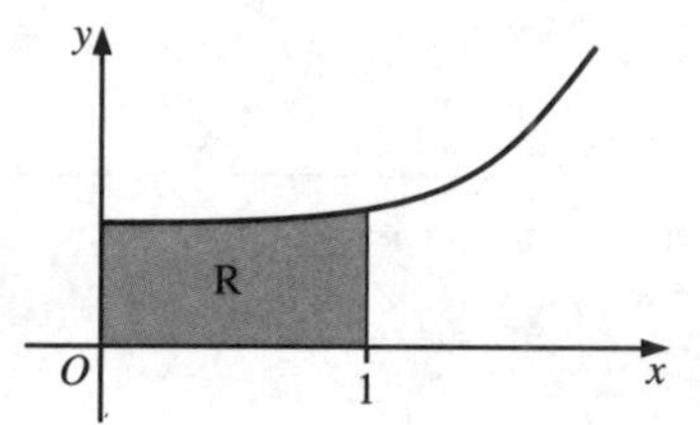

b) Using the result from part a), show that the exact volume of the solid formed when the shaded region R is rotated through 2π radians about the x-axis is $\frac{\pi \ln 3}{4}$.

c) i) By using the substitution $x = 2 \sin \theta$ show that

$$\int \frac{dx}{\sqrt{4 - x^2}} = \sin^{-1}\left(\frac{x}{2}\right) + C \text{ where } C \text{ is a constant.}$$

ii) Hence find the area of the shaded region R.

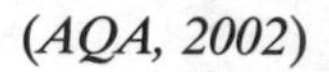
(AQA, 2002)

7 The curve with equation $y = x\sqrt{x^2 + 3}$ is sketched in the diagram. The region R, shaded in the diagram, is bounded by the curve, the x-axis and the line $x = 1$.

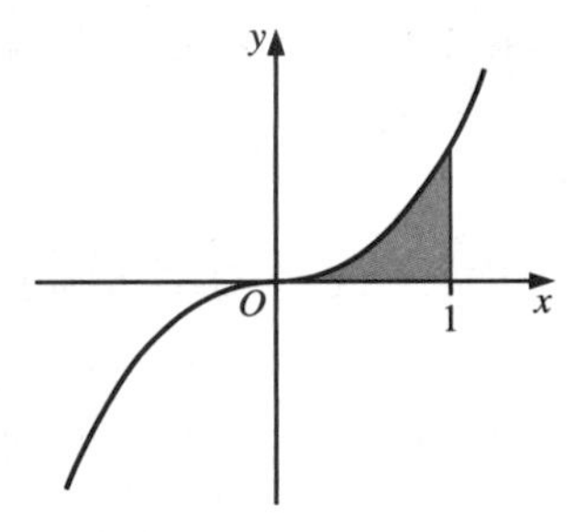

a) The region R is rotated through 2π radians about the x-axis. Find the volume of the solid generated.

b) Determine an equation of the tangent to the curve at the point where $x = 1$.

c) i) Differentiate $(x^2 + 3)^{\frac{3}{2}}$ with respect to x.

ii) Use the result from part c) i) to find the area of the region R.

(AQA, 2002)

8 The curve with equation $y = (x + 2)e^{-x}$ is sketched.

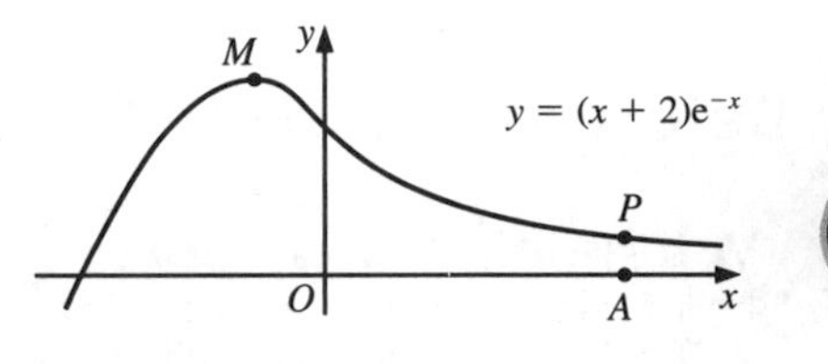

The maximum point M and the point P lie on the curve and A lies on the x-axis.
The points A and P each have coordinates equal to 3.

a) Find the value of $\dfrac{dy}{dx}$ at the point P leaving your answer in terms of e.

b) Determine the equation of the tangent to the curve at P and find the x-coordinate of the point B where this tangent crosses the x-axis.

c) Calculate the coordinates of the maximum point M.

d) Use integration by parts to find $\displaystyle\int_0^3 (x + 2)e^{-x}\,dx$.

e) The finite region bounded by the coordinate axes, the curve and the line with equation $x = 3$ is R.
Show that the line OP divides R into two regions with areas in the ratio $2e^3 - 9:5$

(AQA/AEB, 1997)

9 a) Use integration by parts to find $\displaystyle\int x \sin 2x\,dx$.

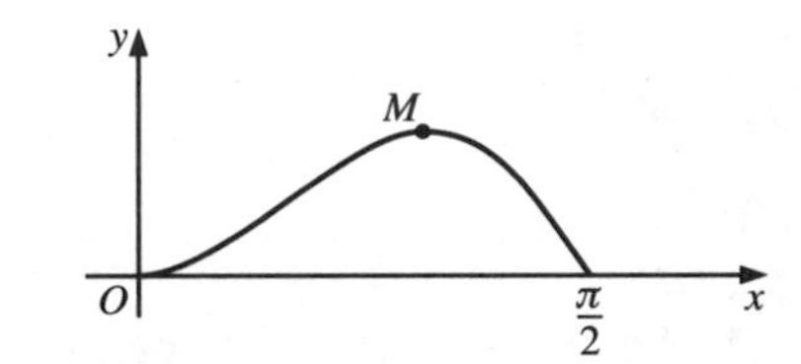

b) The graph with equation $y = x \sin 2x$ is sketched for $0 \leqslant x \leqslant \dfrac{\pi}{2}$.

i) Determine the area of the region bounded by the curve and the x-axis from $x = 0$ to $x = \dfrac{\pi}{2}$ leaving your answer in terms of π.

ii) Show that the x-coordinate of the maximum point M is the least positive root of the equation $2x + \tan 2x = 0$.

(AQA/AEB, 1998)

10 Use the substitution $u = x + 1$ to express the integral $I = \displaystyle\int_0^1 \frac{x^2}{(x + 1)^2}\,dx$ in the form $\displaystyle\int_a^b \left(1 + \frac{c}{u} + \frac{1}{u^2}\right) du$ stating the value of each of the constants a, b and c. Hence find the exact value of I.

(AQA/AEB, 2000)

11 a) Use integration by parts to find $\displaystyle\int x^3 \ln x\,dx$.

b) By means of the substitution $u = 3 + e^x$ or otherwise, evaluate $\displaystyle\int_0^{\ln 6} e^{2x}(3 + e^x)^{\frac{1}{2}}\,dx$.

(AQA/AEB, 2000)

6 Numerical methods

This chapter will show you how to

- ✦ Locate the root of an equation $f(x) = 0$
- ✦ Solve equations approximately using an iterative formula and illustrate this graphically
- ✦ Estimate definite integrals numerically using the mid-ordinate rule and Simpson's rule

Before you start

C3

You should know how to ...	Check in
1 Evaluate functions of the form $f(x)$.	**1** a) If $f(x) = x^2 - 3x - 7$, find i) $f(-1)$ ii) $f(-2)$ b) If $f(x) = 15 - x^3$, find i) $f(2.4)$ ii) $f(2.5)$ c) If $f(x) = xe^{-3x} + 2$, find i) $f(0)$ ii) $f(-1)$
2 Rearrange equations.	**2** a) Make x the subject of these formulae. i) $y = \frac{1}{2}\left(3 - \frac{1}{x}\right)$ ii) $y = \frac{1 - 3x^2}{4}$ b) Rearrange these equations to the form $f(x) = 0$. i) $7(x + 1) - 2 = 3x$ ii) $(2x + 3)^2 - 2x = 5(3 - 2x)$
3 Estimate a definite integral using the trapezium rule.	**3** Use the trapezium rule with four strips to estimate the value of the definite integral $\int_0^{\frac{\pi}{4}} \tan^2 x \, dx$.

6.1 Numerical solution of equations

In practice, most equations do not have nice algebraic solutions, and you need other methods. Many of these methods may seem clumsy when done by hand, as they frequently involve lots of small calculations.

For this reason, a computer can be a great asset with these problems. However, they can easily be done using a calculator.

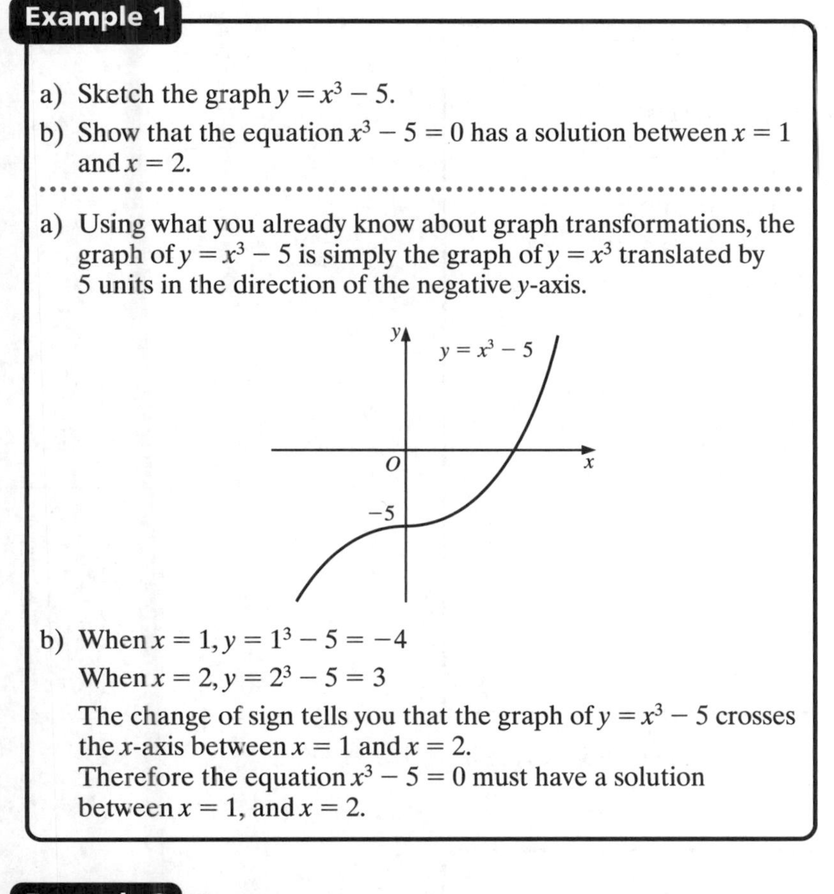

Example 1

a) Sketch the graph $y = x^3 - 5$.

b) Show that the equation $x^3 - 5 = 0$ has a solution between $x = 1$ and $x = 2$.

a) Using what you already know about graph transformations, the graph of $y = x^3 - 5$ is simply the graph of $y = x^3$ translated by 5 units in the direction of the negative y-axis.

b) When $x = 1$, $y = 1^3 - 5 = -4$

When $x = 2$, $y = 2^3 - 5 = 3$

The change of sign tells you that the graph of $y = x^3 - 5$ crosses the x-axis between $x = 1$ and $x = 2$.
Therefore the equation $x^3 - 5 = 0$ must have a solution between $x = 1$, and $x = 2$.

C3

Example 2

Show that the equation $7 + 2x - x^4 = 0$ has a root between $x = -2$ and $x = -1$.

Root is another word for solution.

Let $f(x) = 7 + 2x - x^4$

When $x = -2$, $f(-2) = 7 + 2(-2) - (-2)^4 = -13$

When $x = -1$, $f(-1) = 7 + 2(-1) - (-1)^4 = 4$

The graph of $y = f(x)$ crosses the x-axis between $x = -2$ and $x = -1$, so the equation $7 + 2x - x^4 = 0$ has a root between $x = -2$ and $x = -1$.

The graph of $y = 7 + 2x - x^4$ is sketched.

y
y = 7 + 2
7
−2 −1 O x

Example 3

Show that the equation $x - 5\sin(3x - 1) = 0$ has a root between $x = 0.3$ and $x = 0.4$, where x is measured in radians.

Let $f(x) = x - 5\sin(3x - 1)$

When $x = 0.3$, $f(0.3) = 0.3 - 5\sin(-0.1) = 0.799\,167$

When $x = 0.4$, $f(0.4) = 0.4 - 5\sin(0.2) = -0.593\,347$

The graph of $y = f(x)$ crosses the x-axis between $x = 0.3$ and $x = 0.4$, so the equation $x - 5\sin(3x - 1) = 0$ has a root between $x = 0.3$ and
$x = 0.4$.

The graph of $y = x - 5\sin(3x - 1)$ is sketched.

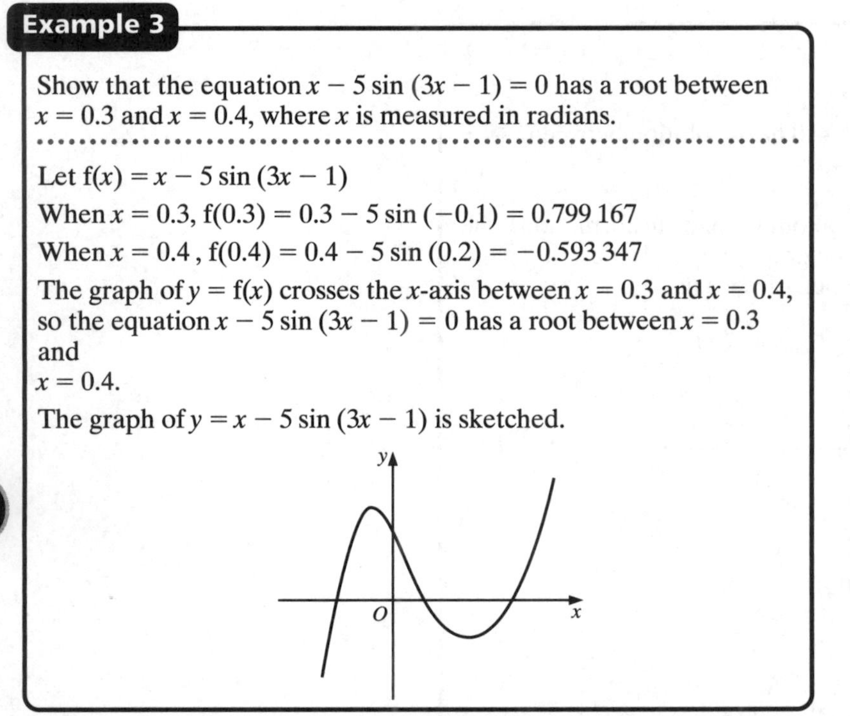

Sometimes it is necessary to rearrange an equation in order to get it into the form '$f(x) = 0$'.

Example 4

Show that the equation $x^2 = e^x - 7$ has a root between $x = 2.6$ and $x = 2.7$.

Rearranging $\quad x^2 = e^x - 7$

you get $\quad x^2 - e^x + 7 = 0$

Let $\quad f(x) = x^2 - e^x + 7$

When $x = 2.6$, $f(2.6) = 2.6^2 - e^{2.6} + 7 = 0.296\,262$

When $x = 2.7$, $f(2.7) = 2.7^2 - e^{2.7} + 7 = -0.589\,732$

The graph of $y = f(x)$ crosses the x-axis between $x = 2.6$ and $x = 2.7$, so the equation $x^2 = e^x - 7$ has a root between $x = 2.6$ and $x = 2.7$.

Example 5

a) Sketch the graphs of $y = x^2 + 1$ and $y = \dfrac{1}{x}$ on the same set of axes.

b) Use your graphs to explain why the equation $x^3 + x - 1 = 0$ has only one root.

c) Show that this root lies in the interval $0.68 < x < 0.69$.

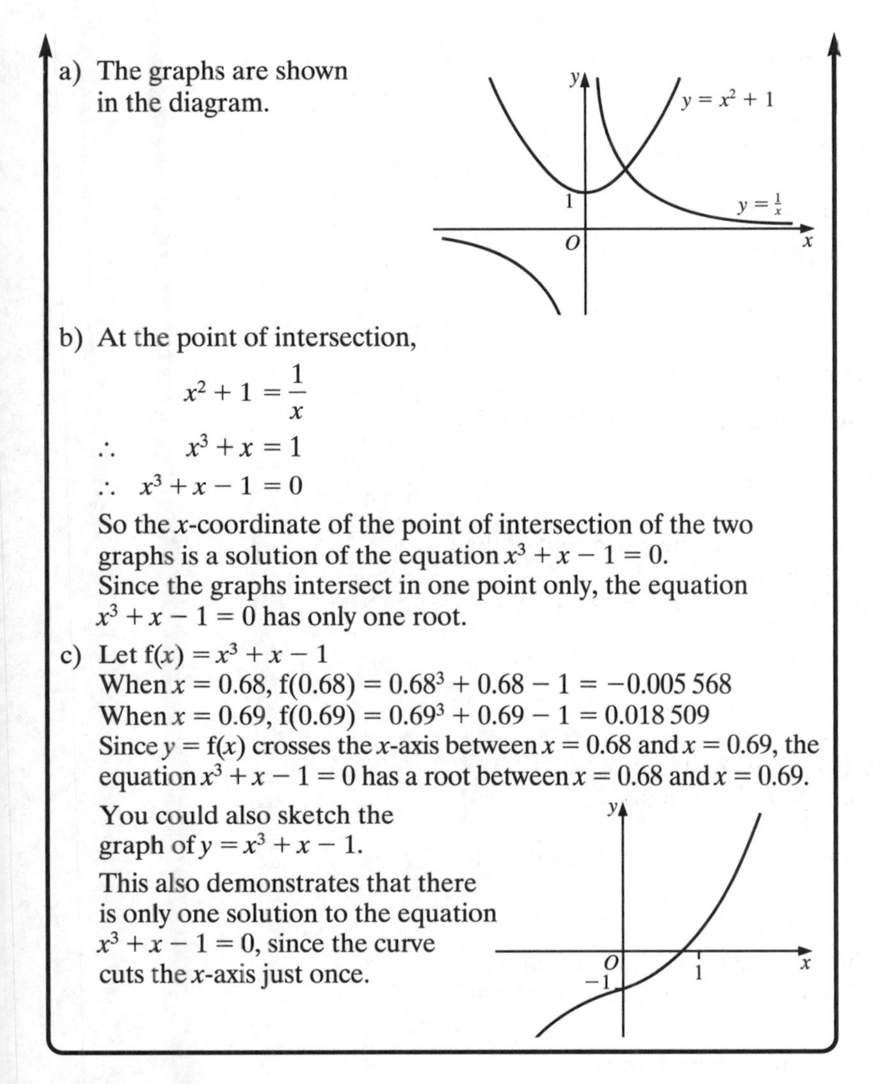

a) The graphs are shown in the diagram.

b) At the point of intersection,

$$x^2 + 1 = \frac{1}{x}$$

$$\therefore \quad x^3 + x = 1$$

$$\therefore \quad x^3 + x - 1 = 0$$

So the x-coordinate of the point of intersection of the two graphs is a solution of the equation $x^3 + x - 1 = 0$.
Since the graphs intersect in one point only, the equation $x^3 + x - 1 = 0$ has only one root.

c) Let $f(x) = x^3 + x - 1$
When $x = 0.68$, $f(0.68) = 0.68^3 + 0.68 - 1 = -0.005\,568$
When $x = 0.69$, $f(0.69) = 0.69^3 + 0.69 - 1 = 0.018\,509$
Since $y = f(x)$ crosses the x-axis between $x = 0.68$ and $x = 0.69$, the equation $x^3 + x - 1 = 0$ has a root between $x = 0.68$ and $x = 0.69$.

You could also sketch the graph of $y = x^3 + x - 1$.
This also demonstrates that there is only one solution to the equation $x^3 + x - 1 = 0$, since the curve cuts the x-axis just once.

C3

Exercise 6A

1 For each of these equations show that there is a root between the values stated.
  a) $x^3 + 2x - 7 = 0$, $x = 1, x = 2$
  b) $2x^3 + x^2 - 7x + 1 = 0$, $x = -3, x = -2$
  c) $x^4 + 2x^3 - x - 1 = 0$, $x = 0, x = 1$
  d) $4x^4 + x^2 - 4 = 0$, $x = -1, x = 0$
  e) $x^5 + x^2 = 1$, $x = 0, x = 1$
  f) $2x^5 = 7 - x$, $x = 1, x = 2$

2 Show that the equation $x = 2\sin x$, where x is measured in radians, has a root between $x = 1$ and $x = 2$.

3 Show that the equation $e^x = x^2$ has a root in the interval $-1 < x < 0$.

4 a) On the same set of axes sketch the graphs of $y = x^3$ and $y = 10 + x$.
  b) Use your graphs to explain why the equation $x^3 - x - 10 = 0$ has only one root.
  c) Show that this root lies in the interval $2.3 < x < 2.4$.

5 a) On the same set of axes sketch the graphs of $y = x^2$ and $y = \dfrac{20}{x-1}$ for $x < 0$.

b) Use your graphs to explain why the equation $x^3 - x^2 - 20 = 0$ has only one root.

c) Show that this root lies in the interval $3.09 < x < 3.10$.

6 a) On the same set of axes sketch the graphs of $y = \ln x$ and $y = 5 - \frac{1}{2}x$ for $x > 0$.

b) Use your graphs to explain why the equation $x + 2\ln x - 10 = 0$ has only one root.

c) Show that this root lies in the interval $6.3 < x < 6.4$.

7 a) On the same set of axes sketch the graphs of $y = \dfrac{1}{x}$ and $y = x + 5$.

b) Show that the positive root of the equation $\dfrac{1}{x} = x + 5$ lies between 0.1 and 0.2.

c) Use trial and improvement to estimate the negative root of the equation to the same degree of accuracy.

d) Use the quadratic formula to find these roots correct to two decimal places.

C3

6.2 Iterative methods

An **iterative** method is a process that is repeated to produce a sequence of approximations to the required solution.

If you want to solve the equation $g(x) = 0$ by an iterative method, you need a relationship of the form

$$x_{n+1} = f(x_n)$$

where x_{n+1} is a better approximation to the solution of $g(x) = 0$ than is x_n. To find such a relationship, you need to rearrange $g(x) = 0$ into the form

$$x = f(x) \quad \text{or} \quad x - f(x) = 0$$

Suppose that the graphs of $y = x$ and $y = f(x)$ are as shown.

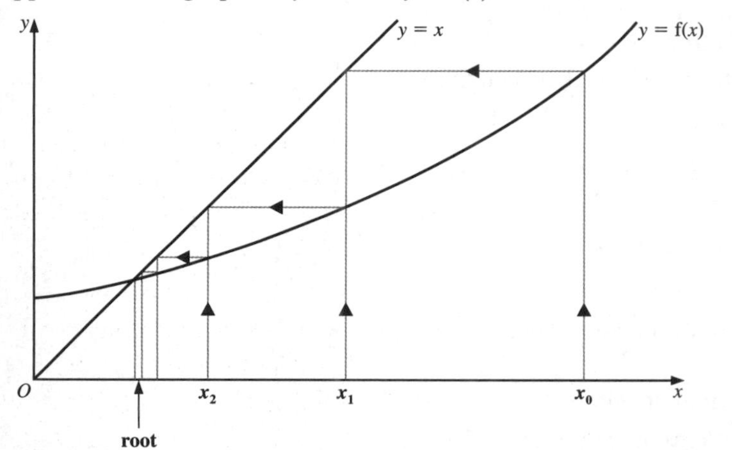

This is called a **staircase diagram**.

If your initial guess for the root of the equation is x_0, then $x_1 = f(x_0)$ gives a better approximation. And $x_2 = f(x_1)$ gives a better approximation than x_1, and so on.

By repeating the process you can obtain an increasingly accurate approximation to the root of the equation.

Suppose you want to solve the equation $x^2 + 4x - 1 = 0$ using iterative methods. Two of the forms into which the equation can be rearranged are

- $x(x + 4) - 1 = 0$ which gives $x = \dfrac{1}{x+4}$, so $f(x) = \dfrac{1}{x+4}$
- $x^2 = 1 - 4x$ which gives $x = \dfrac{1}{x} - 4$, and here $f(x) = \dfrac{1}{x} - 4$

giving the iterative formulae

$$x_{n+1} = \frac{1}{x_n + 4} \quad \text{and} \quad x_{n+1} = \frac{1}{x_n} - 4$$

If you use $x_{n+1} = \dfrac{1}{x_n + 4}$ together with an initial guess of $x_0 = 1$, you get

$$x_1 = \frac{1}{1+4} = 0.2$$

$$x_2 = \frac{1}{0.2+4} = 0.238\,095\,2381$$

$$x_3 = 0.235\,955\,0562$$

$$x_4 = 0.236\,074\,2706$$

This gives one of the solutions as 0.2361 to four decimal places. The graphs of $y = \dfrac{1}{x+4}$ and $y = x$ illustrate how x_n is converging to a root of $x^2 + 4x - 1 = 0$.

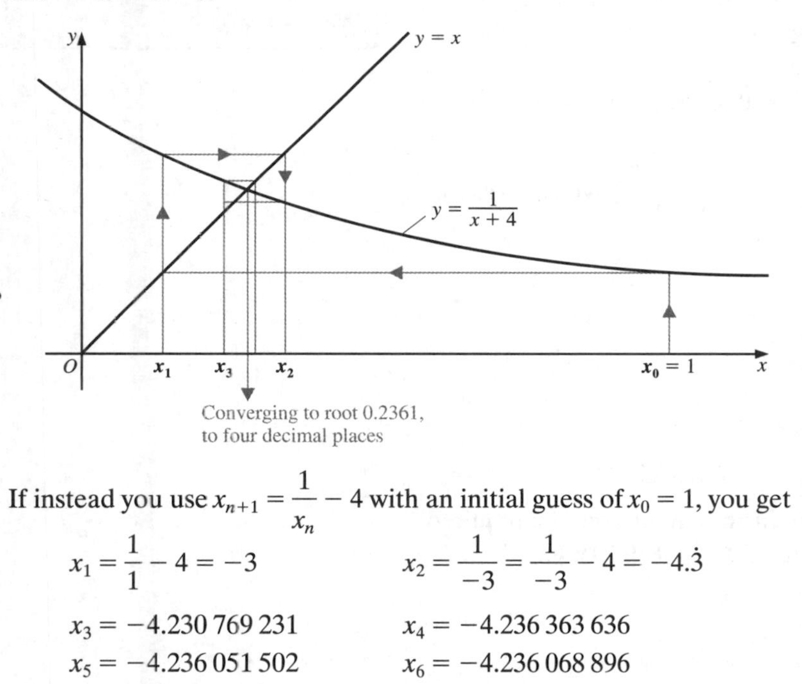

This is called a **cobweb diagram**.

If instead you use $x_{n+1} = \dfrac{1}{x_n} - 4$ with an initial guess of $x_0 = 1$, you get

$$x_1 = \frac{1}{1} - 4 = -3 \qquad x_2 = \frac{1}{-3} = \frac{1}{-3} - 4 = -4.\dot{3}$$

$$x_3 = -4.230\,769\,231 \qquad x_4 = -4.236\,363\,636$$

$$x_5 = -4.236\,051\,502 \qquad x_6 = -4.236\,068\,896$$

This gives the other solution as −4.2361, to 4 d.p.

With a graphics calculator, iteration is made very simple. In this example, and using a standard Casio calculator, you would first key

1 EXE

(the x_0 value). Then enter the iterative formula in the form

1 ÷ ANS − 4 EXE

This gives the value of x_1 as −3. Repeatedly pressing EXE gives the values x_2, x_3, …

Again, the graphs show how x_n is converging to this root.

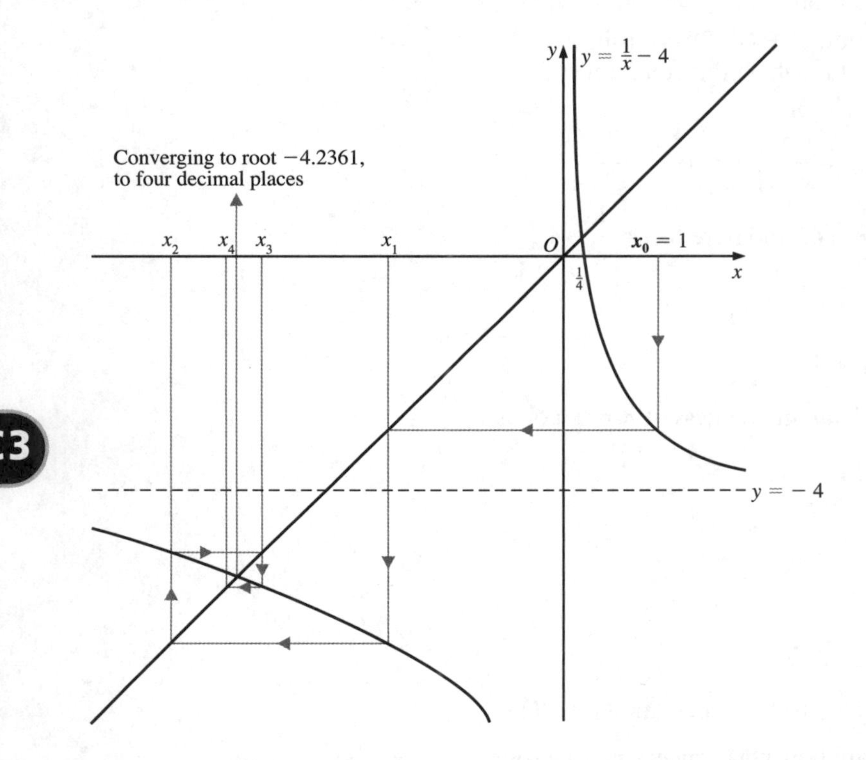

This example shows two possible arrangements of the original equation, both of which gave iterative formulae which converged to a root of the original equation. However, this is not always the case.

The equation is quadratic so you expect two solutions.

For example, the equation $x^2 + 4x + 2 = 0$ has roots between 0 and −1 and between −3 and −4.

One rearrangement of this equation is $x = -\left(\frac{x^2 + 2}{4}\right)$, which gives the iterative formula

$$x_{n+1} = -\left(\frac{x_n{}^2 + 2}{4}\right)$$

Letting $x_0 = -4$ gives

$x_1 = -4.5$ $\quad x_2 = -5.5625$

$x_3 = -8.235\,351\,563$ $\quad x_4 = -17.455\,253\,84$

It is clear that these values are getting further away from the required root. You can say that the sequence $x_0, x_1, x_2, \ldots$ is **diverging**.

The diagram illustrates geometrically what is happening in this case.

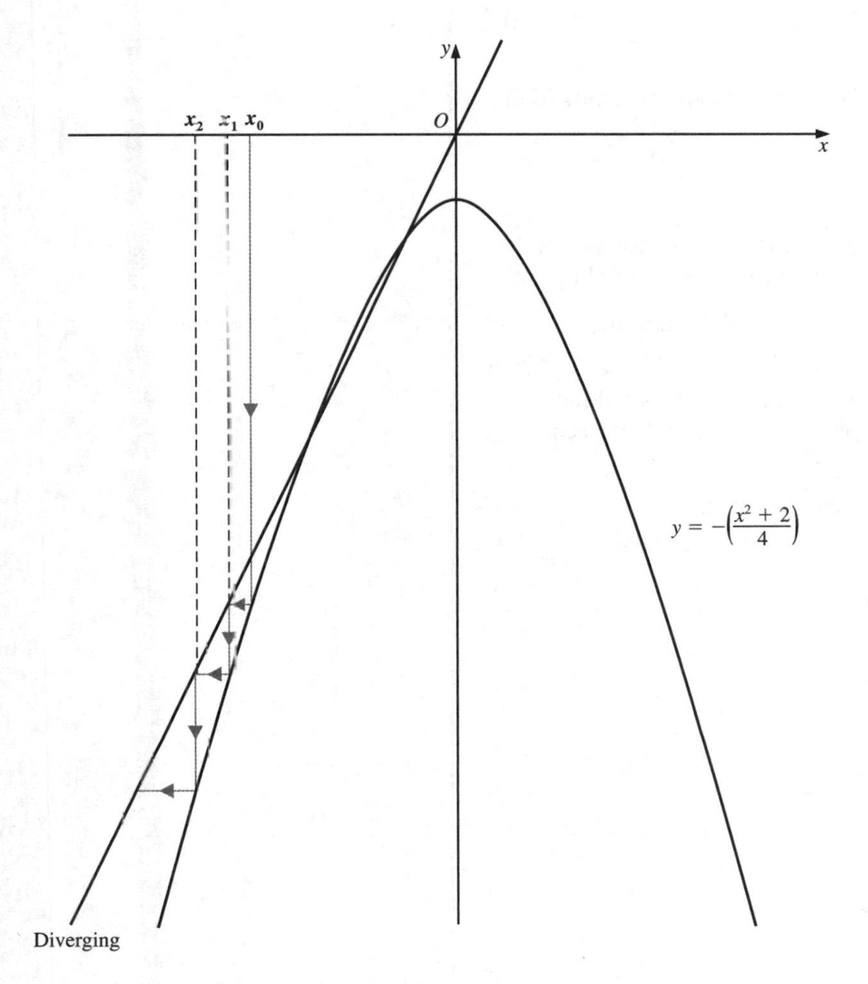

Diverging

Notice that if you had chosen x_0 to be -2, then it would have produced a **converging** sequence,

Try this for yourself.

namely -1.5

-1.0625

$-0.782\,226\,5625$

$\vdots$

$-0.585\,786\,4376.$

It is worth noting that you can't tell whether an iteration will converge or diverge without advanced analysis, which is beyond the Core modules.

Example 3

a) Show that the equation $x^3 + 7x - 2 = 0$ has a root between 0 and 1.

b) Show that the equation $x^3 + 7x - 2 = 0$ can be rearranged in the form

$$x = \frac{2}{x^2 + 7}$$

c) Use an iteration based on this rearrangement with an initial value $x_0 = 1$ to find this root correct to three decimal places.

d) Illustrate the convergence using a suitable diagram.

a) Let $f(x) = x^3 + 7x - 2$, then $f(0) = -2$ and $f(1) = 6$. Since $f(0) < 0$ and $f(1) > 0$, the graph of $y = f(x)$ must intersect the x-axis between $x = 0$ and $x = 1$.

b) Given $x^3 + 7x - 2 = 0$, rearranging leads to

$$x(x^2 + 7) - 2 = 0$$

$$\therefore \quad x = \frac{2}{x^2 + 7}$$

c) This gives the iterative formula

$$x_{n+1} = \frac{2}{x_n{}^2 + 7}$$

Letting $x_0 = 1$ gives

$x_1 = 0.25$

$x_2 = 0.283\,185\,8407$

$x_3 = 0.282\,478\,1267$

$x_4 = 0.282\,494\,0995$

The root is 0.282, to three decimal places.

d)

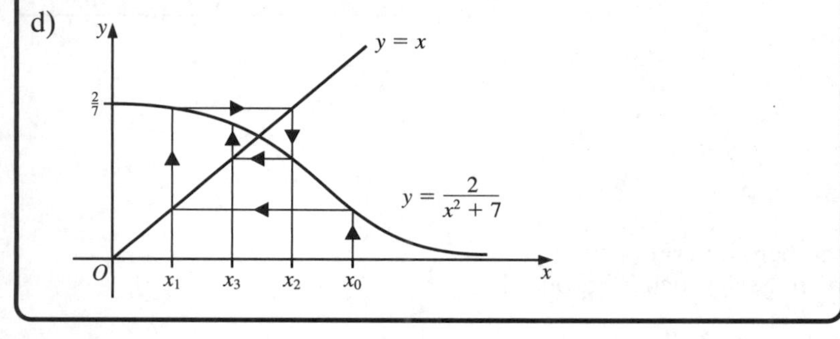

Exercise 6B

1 In each of these cases:

i) Show how the iterative formula is derived from the given equation.

ii) Determine whether the iterative formula converges to a root of the equation. If the formula converges, state the root of the equation correct to four decimal places.

a) $x^2 - 6x + 1 = 0$; $\quad x_{n+1} = \dfrac{1}{6 - x_n}, x_0 = 1$

b) $x^2 + 10x - 3 = 0$; $\quad x_{n+1} = \dfrac{3 - x_n^2}{10}, x_0 = 2$

c) $3x^2 - 6x + 1 = 0$; $\quad x_{n+1} = \dfrac{1}{6 - 3x_n}, x_0 = 1$

d) $2x^3 - x^2 + 1 = 0$; $\quad x_{n+1} = \dfrac{2x_n^3 + 1}{x_n}, x_0 = 0.5$

e) $x^4 + x - 3 = 0$; $\quad x_{n+1} = \dfrac{3}{x_n^3 + 1}, x_0 = 1$

2 a) Starting with $x_0 = 2.5$ use the iterative formula below to find x_1, x_2, x_3, x_4.

$$x_{n+1} = \frac{5}{x_n} + 1$$

b) Illustrate the convergence using a suitable diagram.

3 a) Starting with $x_0 = 1$, use the iterative formula

$$x_{n+1} = \frac{1}{x_n + 4}$$

to find x_1, x_2, x_3 and x_4.

b) Illustrate the convergence using a suitable diagram.

4 a) Starting with $x_0 = 0.5$ use the iterative formula below to find x_1, x_2, x_3, x_4.

$x_{n+1} = \cos x_n$, where x is measured in radians.

b) Illustrate the convergence using a suitable diagram.

5 a) Starting with $x_0 = 1$ use the iterative formula below to find x_1, x_2, x_3.

$$x_{n+1} = e^{x_n}$$

b) Use a suitable diagram to explain why this is diverging.

6 a) Starting with $x_0 = 2$, use the iterative formula

$$x_{n+1} = \sqrt{\left(\frac{x_n + 7}{3}\right)}$$

to find $x_1, x_2, \ldots, x_5$.

b) Find the equation which is solved by this iterative formula.

7 a) Find the equation which is solved by the iterative formula

$$x_{n+1} = 2 + \frac{1}{x_n^2}$$

b) Starting with $x_0 = 2.5$, find a solution of this equation to four decimal places.

C3

8 The sequence given by the iterative formula $x_{n+1} = \ln(2 + x_n)$ with $x_0 = 1$ converges to α. Find α correct to two decimal places.

9 The golden ratio, ψ can be found using the formula

$$x_{n+1} = \sqrt[3]{x_n(1 - x_n)}$$

Choose a suitable value for x_0 and, showing all intermediate calculations, obtain the value of ψ correct to two decimal places.

10 The sequence given by the iterative formula $x_{n+1} = 2(1 + e^{-x_n})$ $x_0 = 0$ converges to α. Find α correct to two decimal places.

11 The sequence given by the iterative formula $x_{n+1} = \frac{1}{4}\sin^{-1}(1 - x_n)$, where x is measured in radians and $x_0 = 0.5$, converges to α. Find α correct to two decimal places.

12 An iterative formula for finding the square root of any positive number N is

$$x_{n+1} = \frac{1}{2}\left(x_n + \frac{N}{x_n}\right)$$

a) Explain why the formula works.

b) Use this iterative formula to find the square root of each of these numbers, correct to three decimal places.

i) 7 ii) 15 iii) 38

6.3 Numerical integration

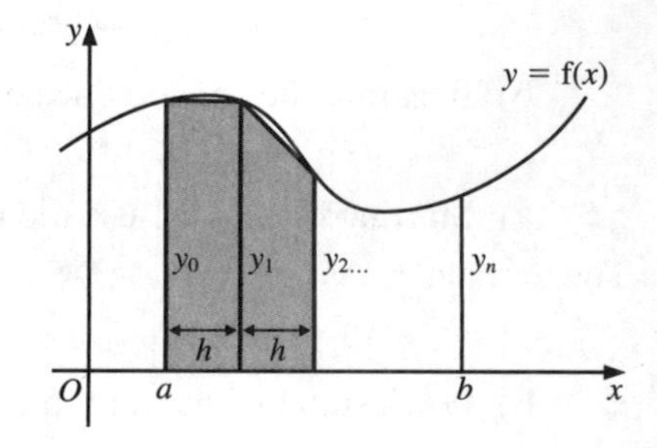

In C2 you learnt how to find an estimate for the area under a curve by using the **trapezium rule**. The method involved splitting the required area into n trapeziums, and summing up these areas. This leads to the formula

$$\int_a^b y\,dx \approx \frac{h}{2}[(y_0 + y_n) + 2(y_1 + y_2 + \ldots + y_{n-1})] \quad \text{where } h = \frac{b-a}{n}$$

This formula has its limitations, as you are splitting a curve into lots of straight lines. For curves which are not approximately linear there are two rules that give a more accurate result than this:

Mid-ordinate rule

$$\int_a^b y\,dx \approx h\,[y_{\frac{1}{2}} + y_{\frac{3}{2}} + \ldots + y_{n-\frac{3}{2}} + y_{n-\frac{1}{2}}] \quad \text{where } h = \frac{b-a}{n}$$

Simpson's rule

$$\int_a^b y\,dx \approx \frac{h}{3}[(y_0 + y_n) + 4(y_1 + y_3 + \ldots + y_{n-1}) + 2(y_2 + y_4 + \ldots + y_{n-2})]$$

where $h = \dfrac{b-a}{n}$ and n is even.

To understand how each of these three rules works, consider the function $\int_0^4 (1 + x^3)\,dx$, and use three ordinates.

Trapezium rule

Consider the case of just two trapeziums.

A is the point (0, 1), *B* is the point (2, 9) and *E* is the point (4, 65).

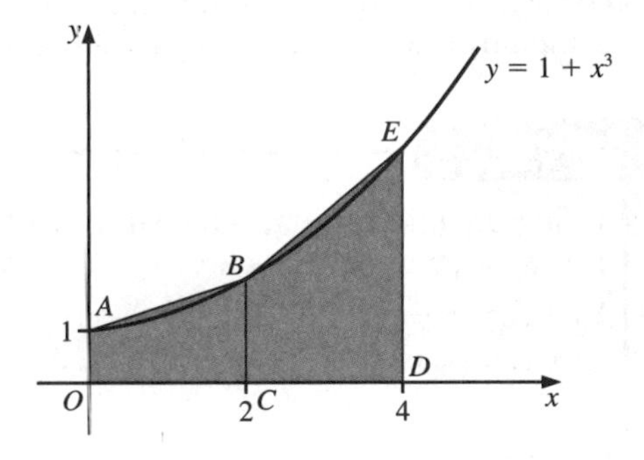

Area of *OABC* is $\frac{2}{2}(1 + 9) = 10$

Area of *BCDE* is $\frac{2}{2}(9 + 65) = 74$

The total area is 84.

Alternatively, using the formula

$$\int_a^b y\,dx \approx \frac{h}{2}[(y_0 + y_n) + 2(y_1 + y_2 + \ldots + y_{n-1})]$$

in this simple case, $h = \dfrac{4-0}{2} = 2$, $y_0 = 1$, $y_n = 65$, $y_2 = 9$.

$$\therefore \quad \int_0^4 (1 + x^3)\,dx = \frac{2}{2}[(1 + 65) + 2(9)] = 84$$

From the sketch you can see that this is an overestimate of the actual area, because of the curvature of the curve.

The actual area is given by $\displaystyle\int_0^4 (1 + x^3)\,dx = \left[x + \frac{x^4}{4}\right]_0^4$

$$= 4 + 64 = 68$$

Mid-ordinate rule

The mid-ordinate rule tries to allow for the curvature of the curve by averaging out the values. So, instead of splitting the area into trapeziums the mid-ordinate rule splits the area into rectangles. Each rectangle is centred at the middle of a strip.

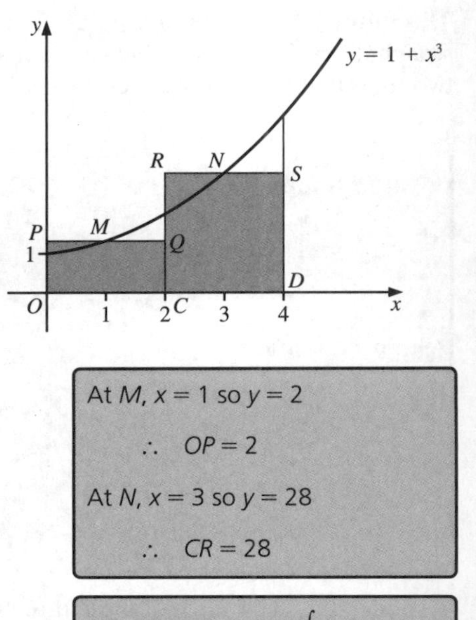

Remember: The lines $x = 0$, $x = 2$ and $x = 4$ are **ordinates**. In this diagram there are three ordinates, but only two strips.

M is the point (1, 2) and N is the point (3, 28).

Area of $OPQC$ is $2 \times 2 = 4$

Area of $CRSD$ is $2 \times 28 = 56$

At M, $x = 1$ so $y = 2$

$\therefore \quad OP = 2$

At N, $x = 3$ so $y = 28$

$\therefore \quad CR = 28$

C3

The total area is 60.

Using the formula

$$\int_a^b y \, dx \approx h\left[y_{\frac{1}{2}} + y_{\frac{3}{2}} + \ldots + y_{n-\frac{3}{2}} + y_{n-\frac{1}{2}}\right]$$

in this simple case, $h = \dfrac{4-0}{2} = 2$, as before, $y_{\frac{1}{2}} = 2, y_{\frac{3}{2}} = 28$

$$\therefore \quad \int_0^4 (1 + x^3) \, dx = 2[2 + 28] = 60$$

It is clear from the sketch that this is an underestimate, and this is verified by the calculation which shows the true value to be 68.

The actual value of $\int (1 + x^3) \, dx$ is 68. The trapezium rule gives an estimate of 84. The mid-ordinate rule gives an estimate of 60. An estimate using Simpson's rule is given on page 132.

Example 4

Using the mid-ordinate rule with two strips, find an approximation for the area under the curve $y = \sqrt{1 + x}$ between $x = 2$ and $x = 10$.

$$A = \int_2^{10} y \, dx = h\left[y_{\frac{1}{2}} + y_{\frac{3}{2}}\right]$$

$$h = \frac{10 - 2}{2} = 4 \text{ and}$$

$$x_{\frac{1}{2}} = x_0 + \tfrac{1}{2}h = 2 + \tfrac{1}{2}(4) = 4$$

$$x_{\frac{3}{2}} = x_0 + \tfrac{3}{2}h = 2 + \tfrac{3}{2}(4) = 8$$

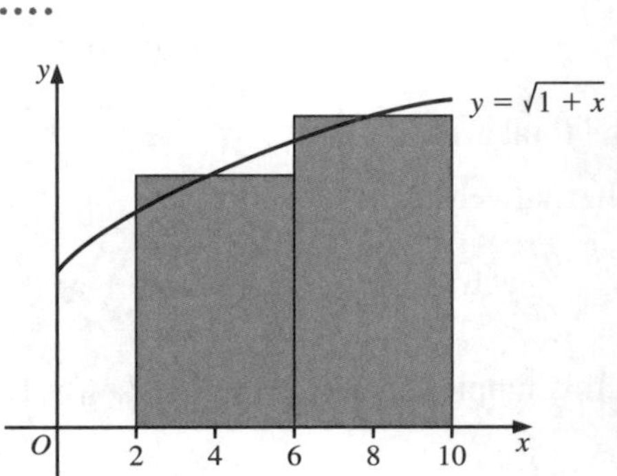

So:

x_i	$x_{\frac{1}{2}} = 4$	$x_{\frac{3}{2}} = 8$
y_i	$\sqrt{5}$	3

By the mid-ordinate rule

$A = 4(\sqrt{5} + 3)$

$A \approx 20.9$ (3 significant figures)

An estimate of the area is 20.9.

Example 5

a) Using the mid-ordinate rule with four strips, find an approximate value for $I = \int_0^{\frac{\pi}{3}} \cos x \, dx$, where x is measured in radians.

b) Calculate the exact value of I.

c) Calculate the percentage error involved when taking the answer to a) as an approximation to I.

a) $h = \dfrac{\frac{\pi}{3} - 0}{4} = \dfrac{\pi}{12}$ and

$$x_{\frac{1}{2}} = x_0 + \tfrac{1}{2}h = 0 + \tfrac{1}{2}\left(\frac{\pi}{12}\right) = \frac{\pi}{24}$$

$$x_{\frac{3}{2}} = x_0 + \tfrac{3}{2}h = 0 + \tfrac{3}{2}\left(\frac{\pi}{12}\right) = \frac{\pi}{8}$$

$$x_{\frac{5}{2}} = x_0 + \tfrac{5}{2}h = 0 + \tfrac{5}{2}\left(\frac{\pi}{12}\right) = \frac{5\pi}{24}$$

$$x_{\frac{7}{2}} = x_0 + \tfrac{7}{2}h = 0 + \tfrac{7}{2}\left(\frac{\pi}{12}\right) = \frac{7\pi}{24}$$

y
y = cos x
$\frac{\pi}{24}$ $\frac{\pi}{8}$ $\frac{5\pi}{25}$ $\frac{7\pi}{24}$
O $\frac{\pi}{12}$ $\frac{\pi}{6}$ $\frac{\pi}{4}$ $\frac{\pi}{3}$ $\frac{\pi}{2}$ x

So:

x_i	$x_{\frac{1}{2}} = \frac{\pi}{24}$	$x_{\frac{3}{2}} = \frac{\pi}{8}$	$x_{\frac{5}{2}} = \frac{5\pi}{24}$	$x_{\frac{7}{2}} = \frac{7\pi}{24}$
y_i	0.991 44	0.923 88	0.793 35	0.608 76

$y_i = \cos x_i$

By the mid-ordinate rule $I = \dfrac{\pi}{12}(0.991\,44 + 0.923\,88 + 0.793\,35 + 0.608\,76)$

$\therefore \quad I \approx 0.8685$ (4 decimal places)

An estimate of I is 0.8685.

b) $I = \int_0^{\frac{\pi}{3}} \cos x \, dx$

$= \Big[\sin x\Big]_0^{\frac{\pi}{3}}$

$= \sin\left(\frac{\pi}{3}\right) - \sin(0)$

$= \dfrac{\sqrt{3}}{2}$ (or 0.866025403)

c) Percentage error is $\dfrac{0.8685 - \frac{\sqrt{3}}{2}}{\frac{\sqrt{3}}{2}} \times 100 = 0.286$

The percentage error is 0.286% (3 significant figures)

C3

Simpson's rule

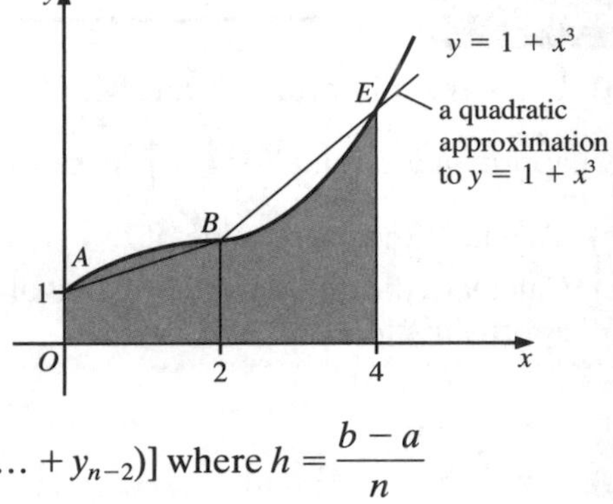

Both the trapezium rule and the mid-ordinate rule use straight lines to approximate a curve.

Simpson's rule works by fitting a quadratic function to the curve, and this quadratic passes through the mid-point and each of the two end points. By using a curve the rule is able to bring the curvature of the curve into the calculation.

Simpson's rule states that:

$$\int_a^b y\,dx \approx \frac{h}{3}[(y_0 + y_n) + 4(y_1 + y_2 + \ldots + y_{n-1}) + 2(y_2 + y_4 + \ldots + y_{n-2})] \text{ where } h = \frac{b-a}{n}$$

or

$$\int_a^b y\,dx \approx \frac{h}{3}[(\text{first ordinate} + \text{last ordinate}) + 4(\text{sum of odd ordinates}) + 2\,(\text{sum of even ordinates})]$$

C3

For three ordinates Simpson's rule is simply

$$\int_a^b y\,dx \approx \frac{h}{3}(y_0 + 4y_1 + y_2) \text{ where } h = \frac{b-a}{n}$$

because $y_n = y_2$ and there are no other even ordinates.

Returning to the example $\int_0^4 (1 + x^3)$ using two strips,

A is the point (0, 1), B is the point (2, 9) and E is the point (4, 65).

So $y_0 = 1, y_1 = 9$ and $y_2 = 65$; and $h = \frac{4-0}{2} = 2$.

$$\text{Area} = \tfrac{2}{3}[1 + 4(9) + 65] = \tfrac{2}{3}[102] = 68$$

The total area is 68, which is 'spot on'. In fact Simpson's rule will always give the exact value for any cubic graph. It is less accurate for other functions or polynomials of higher order, but it is still more accurate than the trapezium rule or the mid-ordinate rule for polynomials.

Example 6

Using Simpson's rule with two strips, find an approximation for the area under the curve $y = \frac{1}{x}$ between $x = 1$ and $x = 2$.

Two strips include three ordinates. Therefore,

$$h = \frac{2-1}{3-1} = \frac{1}{2}$$

and:

x_i	1	1.5	2
y_i	1	$0.\dot{6}$	0.5

It is often useful to draw a sketch:

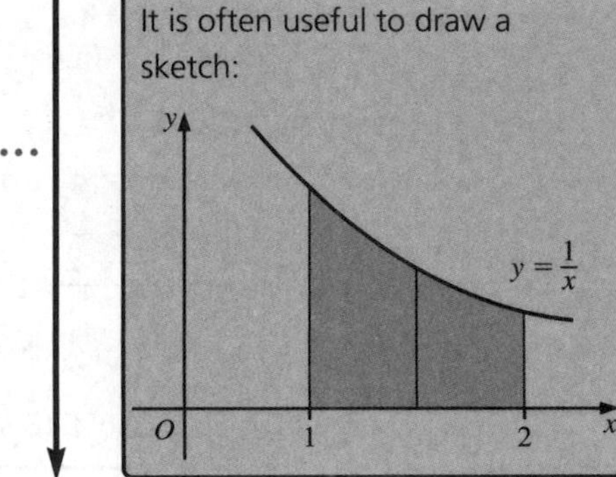

By Simpson's rule,

$$A = \approx \frac{h}{3}(y_0 + 4y_1 + y_2)$$

$$A \approx \tfrac{1}{3}\left(\tfrac{1}{2}\right)[1 + 4(0.\dot{6}) + 0.5]$$

$$A \approx 0.694 \text{ (3 significant figures)}$$

An estimate of the area is 0.694.

You could find the percentage error in the approximation, as in Example 5 on page 131. You should find it is about 0.1%.

Remember to read questions carefully. The next example gives you the number of *ordinates*, not strips.

Example 7

Using Simpson's rule with five ordinates, find an approximation for the area under the curve $f(x) = e^{-2x}$ between $x = 1$ and $x = 3$.

Now $h = \dfrac{3-1}{5-1} = \dfrac{2}{4} = \dfrac{1}{2}$

x_i	1	1.5	2	2.5	3
y_i	e^{-2}	e^{-3}	e^{-4}	e^{-5}	e^{-6}

By Simpson's rule,

$$A \approx \tfrac{1}{3}\left(\tfrac{1}{2}\right)[e^{-2} + 4e^{-3} + 4e^{-5} + 2e^{-4} + e^{-6}]$$

$$A \approx 0.067 \text{ (2 significant figures)}$$

An estimate of the area is 0.067.

A sketch is particularly useful in avoiding the confusion between ordinates and strips.

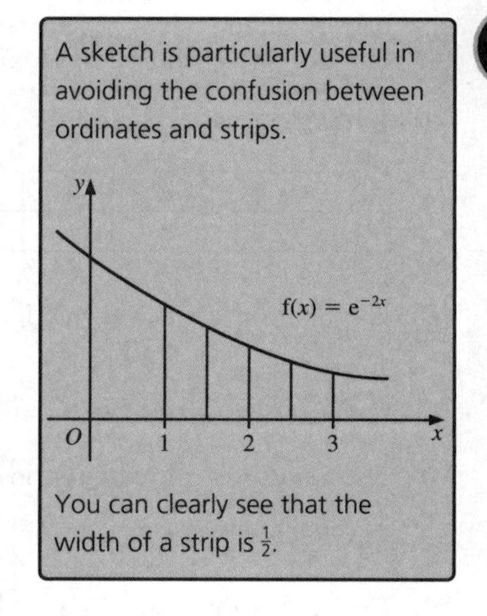

You can clearly see that the width of a strip is $\frac{1}{2}$.

Whichever method you use, you can improve the accuracy of your estimate for the area by increasing the number of strips you consider.

Exercise 6C

Unless stated otherwise, give the answers in this exercise correct to three significant figures.

1 Find an approximate value for $\int_0^1 \sqrt{1 + x^2}\,dx$ using the mid-ordinate rule with 2 strips.

2 Use the mid-ordinate rule with 4 strips to find an approximate value for $\displaystyle\int_4^{12} \frac{1}{1 + \sqrt{x}}\,dx$.

3 Use the mid-ordinate rule with 4 strips to find an approximate value for $\int_1^9 \sqrt{\ln x}\,dx$

4 Find an approximate value for $\int_0^{\frac{\pi}{3}} \frac{1}{1+\sin x}\,dx$ using the mid-ordinate rule with 5 strips, where x is measured in radians.

5 Using the mid-ordinate rule with 3 strips, find an approximation for the area under the curve $y = e^{\sin x}$ between $x = 0$ and $x = 6$, where x is measured in radians.

6 Using the mid-ordinate rule with 3 strips, find an approximation for the area under the curve $y = \sqrt{1+\tan x}$ between $x = 0$ and $x = \frac{\pi}{3}$, where x is measured in radians.

7 Given that $y = f(x)$ and that for the given values of x the corresponding values of y are shown in the table below, use Simpson's rule with 4 strips to find an approximate value for $\int_0^8 f(x)\,dx$.

x	0	2	4	6	8
$f(x)$	1	3	9	15	30

Remember:
Check whether the question talks about ordinates or strips.

8 Use Simpson's rule with 4 strips to find an approximate value for $\int_0^2 \frac{1}{\sqrt{1+x^3}}\,dx$.

9 Use Simpson's rule with 6 strips to find an approximate value for $\int_0^6 \sqrt{36-x^2}\,dx$.

10 Find an approximate value for $\int_1^2 e^{x^2}\,dx$ using Simpson's rule with 5 ordinates.

11 Find an approximate value for $\int_{\frac{\pi}{6}}^{\frac{\pi}{2}} \sqrt{\cos x}\,dx$ using Simpson's rule with 3 ordinates, where x is measured in radians.

12 Use Simpson's rule with 4 strips to find an approximate value for $\int_1^{21} \frac{1}{1+\ln x}\,dx$.

13 a) Sketch the graph of $y = \frac{1}{x}$ for $x > 0$.

b) Use Simpson's rule with 7 ordinates to estimate the value of $\int_1^7 \frac{1}{x}\,dx$.

c) Calculate, to three decimal places, the percentage error involved when taking the answer to part b) as an approximation to $\ln 7$.

14 a) Given that $I = \int_0^{\pi} \sin x \, dx$, where x is measured in radians,

i) estimate, to three decimal places, the value of I using the mid-ordinate rule with 4 strips,

ii) estimate, to three decimal places, the value of I using Simpson's rule with 2 strips.

b) Calculate the exact value of I.

c) Calculate, to one decimal place, the percentage error involved with each of the estimates in part a).

15 By evaluating $\int_0^1 \frac{1}{1+x^2} \, dx$, and using Simpson's rule with 11 ordinates, show that $\pi \approx 3.141\,593$.

16 a) Given that the parabola $y = ax^2 + bx + c$ passes through the points $(-h, y_0)$, $(0, y_1)$ and (h, y_2), show that

i) $c = y_1$

ii) $a = \dfrac{y_0 - 2y_1 + y_2}{2h^2}$

b) Deduce that $\int_{-h}^{h} (ax^2 + bx + c) \, dx = \frac{h}{3}(y_0 + 4y_1 + y_2)$.

C3

Summary

You should know how to ...	Check out
1 Locate the roots of an equation between two values of x.	**1** a) Show that $3x^2 + 2x - 7 = 0$ has a root α $1 < \alpha < 2$. b) Show that $x^2 - 1 = 3\sqrt{x}$ has a root β $2.1 < \beta < 2.5$.
2 Use iteration to estimate the root of an equation.	**2** a) Show that $3x^2 + 2x - 7 = 0$ can be expressed as $x = \sqrt{\frac{7-2x}{3}}$. b) Taking $x_0 = 1$, find the root α, $1 < \alpha < 2$, to two decimal places. c) Illustrate the convergence of the iteration using a cobweb diagram.
3 Use numerical methods to estimate a definite integral.	**3** a) Use the mid-ordinate rule with four strips to estimate $\int_2^4 \sqrt{x^2 - 1} \, dx$. b) Use Simpson's rule with four strips to estimate $\int_1^3 x^2 \tan^{-1} x \, dx$.

Revision exercise 6

1 a) Show without using a calculator that the equation $x^3 - 15$ has a root in the interval $2 \leqslant x \leqslant 3$.

b) i) Show that the equation $x = \frac{2x}{3} + \frac{5}{x^2}$ can be rearranged to give the equation $x^3 - 15 = 0$.

ii) Use the iterative formula $x_{n+1} = \frac{2x_n}{3} + \frac{5}{x_n^{\,2}}$ starting with $x_1 = 3$, to find the values of x_2, x_3 and x_4 giving your answers to six decimal places.

iii) The graphs of $y = \frac{2x}{3} + \frac{5}{x^2}$ and $y = x$ are sketched in the diagram.

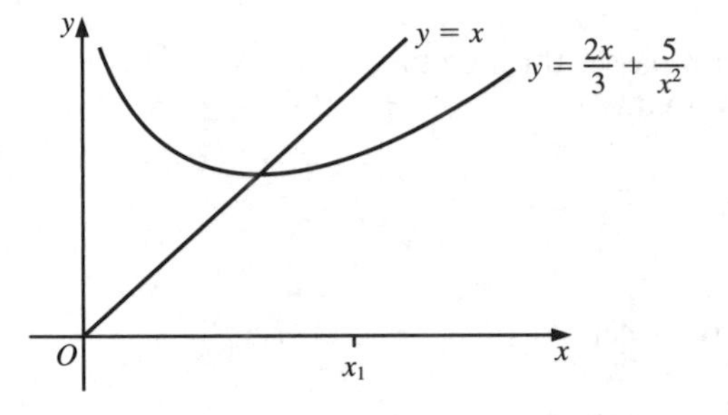

On a copy of the graphs, draw a staircase diagram to illustrate the convergence of the sequence $x_1, x_2, x_3, \ldots$

iv) Write down the **exact** value to which this sequence converges. *(AQA, 2004)*

2 A sequence is defined by $x_{n+1} = \sqrt{(x_n + 12)}$, $x_1 = 2$.

a) Find the values of x_2, x_3 and x_4 giving your answers to 3 decimal places.

b) Given that the limit of the sequence is L,

i) show that L must satisfy the equation $L^2 - L - 12 = 0$

ii) find the value of L.

c) The graphs of $y = \sqrt{(x + 12)}$ and $y = x$ are sketched in the diagram.

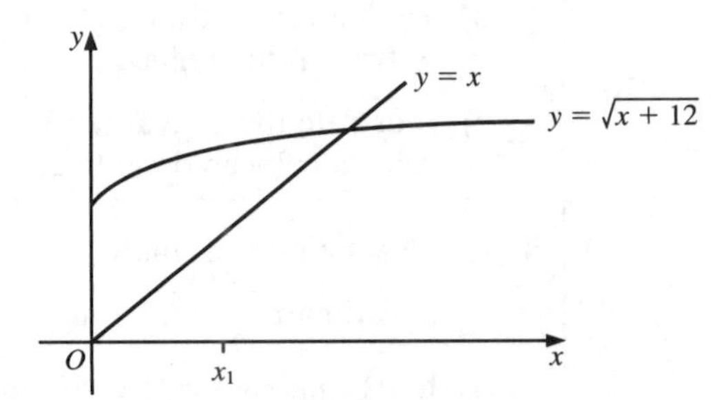

On a copy of the sketch, draw a cobweb or staircase diagram to show how convergence takes place. *(AQA, 2004)*

3 Use Simpson's rule with five ordinates (four equal strips) to find an approximation to the integral $\int_0^2 \ln(x^2+1)\,dx$, giving your answer to 3 decimal places. *(AQA, 2003)*

4 a) The diagram shows the graph of $y = \sin^{-1} x$.

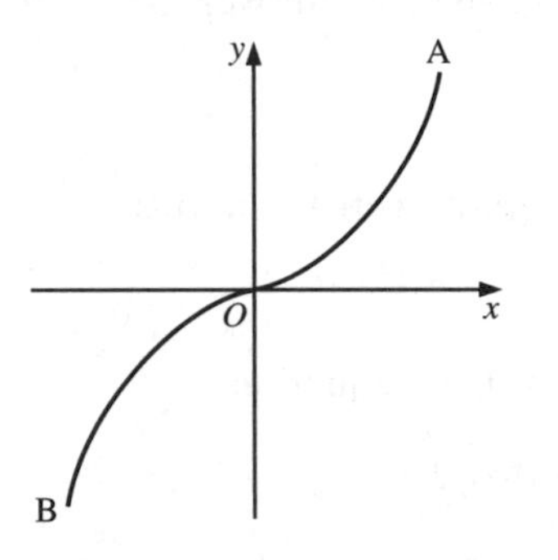

Write down the coordinates of the end points A and B.

b) Use the mid-ordinate rule, with five strips of equal width, to estimate the value of $\int_0^1 \sin^{-1} x\,dx$. Give your answer to three decimal places. *(AQA, 2003)*

5 The diagram shows a sketch of the curve $y = x\cos x$, $0 \leqslant x \leqslant \frac{\pi}{2}$.

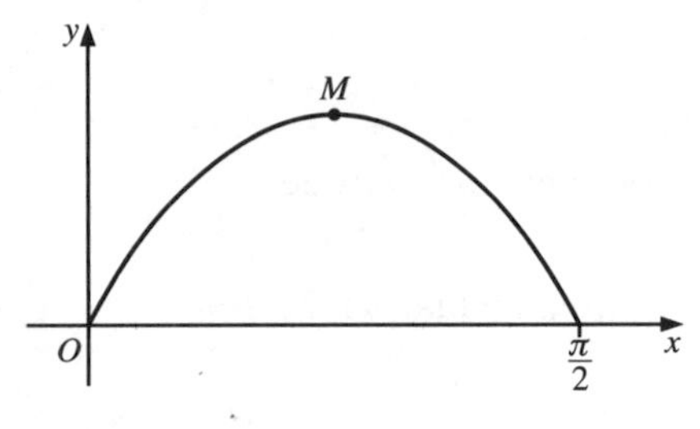

The maximum point is M.

a) i) Find $\frac{dy}{dx}$.

ii) Show that the x-coordinate of M satisfies the equation $x = \tan^{-1}\frac{1}{x}$.

iii) It is estimated that the relevant root of the above equation is approximately 0.9. Use the iterative formula $x_{n+1} = \tan^{-1}\frac{1}{x_n}$ starting with $x_1 = 0.9$ to obtain the root correct to two decimal places.

b) Find the area of the region bounded by the curve and the x-axis. *(AQA, 2003)*

6 The volume $V\,\text{cm}^3$ of liquid in a container when the depth is x cm is given by $V = (2x^2 + x^3)^{\frac{1}{2}}$. An attempt is made to find the value of x when $V = 10$.

a) Show that a possible iterative formula that can be used to find x is

$$x_{n+1} = \frac{10}{\sqrt{2 + x_n}}.$$

b) Use the value $x_1 = 4$ in the iterative formula above to find the value of x_3 giving your answer to four significant figures. *(AQA, 2002)*

7 a) Use Simpson's rule with five ordinates (four strips) to find an approximation to the integral $I = \int_0^1 x\cos x \, dx$ giving your answer to five decimal places.

b) Use integration by parts to find the value of I, giving your answer to five decimal places. *(AQA, 2002)*

8 a) i) Draw on the same diagram sketches of the graphs with equations $y = 5e^{2x}$ and $y = \frac{4}{x}$ for $x > 0$.

ii) Explain why this diagram shows that, for $x > 0$, the equation $5e^{2x} - \frac{4}{x} = 0$ has just one root, α, and show that $0.3 < \alpha < 0.4$.

b) Show, using calculus, that $y = 5e^{2x} - \frac{4}{x}$ is an increasing function of x for $x > 0$. *(AQA, 2001)*

9 a) By taking logarithms, solve the equation $0.6 = 2^{-x}$ giving your answer to three significant figures.

b) A sequence is defined by $x_{n+1} = 2^{-x_n}$, $x_1 = 0.6$.

i) Find the values of x_2, x_3, x_4 and x_5, giving your answers to three significant figures.

ii) Given that the root converges to α write down an equation in x for which α is a root. *(AQA, 2001)*

C3 Practice Paper

90 minutes *75 marks* *You may use a calculator*

1 a) Find $\frac{dy}{dx}$ when i) $y = \sin 3x$ *(2 marks)*

ii) $y = e^{2x} \sin 3x$ *(3 marks)*

b) Given that $y = \frac{e^{2x}}{\sin 3x}$ show that $\frac{dy}{dx} = e^{2x} \operatorname{cosec} 3x\,(2 - 3 \cot 3x)$ *(5 marks)*

2 The diagram shows part of the graph of the function

$f(x) = \frac{5}{\sqrt{9 + x^2}}$.

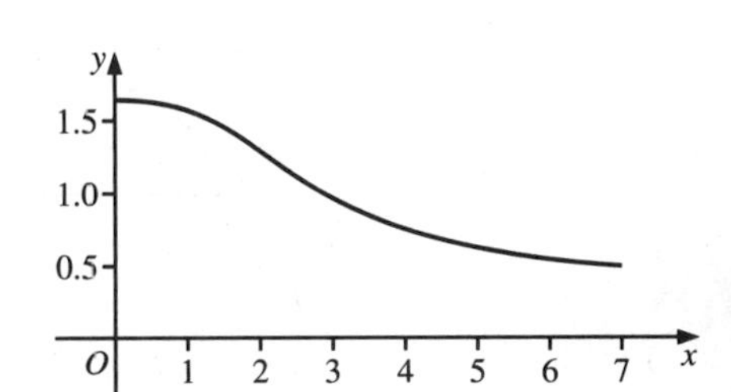

The region R is defined by the curve, the x- and y-axes and the line $x = 6$.

a) i) Copy and complete the table.

x	1	3	5
f(x)		1.1785	

(2 marks)

ii) Use the mid-ordinate rule with three ordinates to estimate the area of R. *(3 marks)*

b) Calculate the volume of revolution formed when the region R is rotated through 360° about the x-axis. Give your answer to two significant figures. *(5 marks)*

3 a) Find $\int \cos 2x \, dx$ *(2 marks)*

b) Find $\int_0^{\frac{\pi}{6}} x \cos 2x \, dx$ *(6 marks)*

4 a) Describe a sequence of two geometrical transformations that maps the graph of $y = e^x$ onto the graph of $y = e^{-3x}$. *(3 marks)*

b) i) Show that the curve $y = e^{-3x} + 2x$ has a stationary point at $y = \frac{2}{3}(1 + \ln \frac{3}{2})$. *(7 marks)*

ii) Determine the nature of this stationary point. *(2 marks)*

5 The functions f and g are defined with their respective domains by

$$f(x) = \frac{1}{x - 2} \text{ for all real } x, x \neq 2$$

$$g(x) = 4 - 5x \text{ for all real } x.$$

a) i) Find the function fg(x). *(2 marks)*

ii) State the domain and range of fg(x). *(2 marks)*

b) Show that the inverse function f^{-1} is given by $f^{-1}(x) = \frac{1}{x} + 2$. *(2 marks)*

c) Solve the equation $f^{-1}(x) = g^{-1}(x)$. *(5 marks)*

6 a) Using the substitution $u = 3x^2 + 2$, or otherwise, find $\int \frac{2x}{\sqrt{3x^2 + 2}}\,dx$. *(4 marks)*

b) Show that $\int_1^3 \frac{2x}{3x^2 + 2}\,dx = p \ln q$ where p and q are rational numbers. *(6 marks)*

7 The diagram shows part of the curve $y = 2 \ln 5x$. The curve crosses the x-axis at $x = a$.

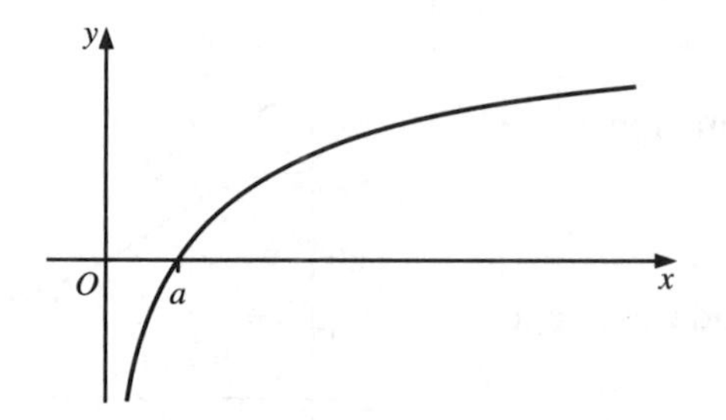

a) Find the value of a. *(2 marks)*

b) i) By drawing a suitable line on a copy of the diagram, show that the equation $2 \ln 5x - 3x = 0$ has two roots. *(2 marks)*

ii) Show that one of these roots lies between $x = 0.3$ and $x = 0.4$. *(2 marks)*

c) Solve the equation $2 \ln 5x - 3 = 0$ giving your answer to three decimal places. *(3 marks)*

d) i) Sketch the graph of $y = |2 \ln 5x|$. *(2 marks)*

ii) Solve the equation $|2 \ln 5x| - 3 = 0$ giving your answers to three decimal places. *(3 marks)*

7 Algebra and functions

This chapter will show you how to

- ✦ Use the factor and remainder theorems with $(ax + b)$
- ✦ Simplify rational functions using a variety of techniques

Before you start

You should know how to ...	Check in
1 Use the Factor Theorem.	**1** The cubic polynomial $p(x)$ is defined as $p(x) = 2x^3 - 5x^2 - x + 6$ a) Show that $(x - 2)$ is a factor of $p(x)$. b) Find $p(-1)$. What can you conclude?
2 Use the Remainder Theorem.	**2** a) Find the remainder when $x^2 - 3x - 5$ is divided by $x - 3$. b) When $x^3 - 3x^2 - 9x + d$ is divided by $x + 2$, there is a remainder of 5. Find the value of d.
3 Factorise polynomials up to degree 3.	**3** Factorise these polynomials. a) $6 - 9x$ b) $10x - 5x^2$ c) $2x^2 + 5x - 3$ d) $x^3 + 2x^2 - 5x - 6$

C4

7.1 The Remainder Theorem

When a polynomial, $f(x)$, is divided by $(x - a)$ the remainder is $f(a)$.

For example, the remainder when $f(x) = x^3 - 3x^2 + 2x + 5$ is divided by $(x - 4)$ is given by

$$f(4) = 4^3 - 3(4)^2 + 2(4) + 5 = 29.$$

This result can be extended to include division by any linear factor of the form $ax + b$.

So the **Remainder Theorem** becomes:

> When the polynomial $f(x)$ is divided by $ax + b$, the remainder is $f\left(-\frac{b}{a}\right)$.

To prove this, write

$$f(x) = (ax + b)(\text{quotient}) + (\text{remainder})$$

When $x = -\frac{b}{a}$, $f\left(-\frac{b}{a}\right) = \left[a\left(-\frac{b}{a}\right) + b\right](\text{quotient}) + (\text{remainder})$

$$= (0)(\text{quotient}) + (\text{remainder})$$
$$= \text{remainder}$$

as required.

Example 1

Find the remainder when the polynomial $4x^3 - 6x^2 + 5x - 2$ is divided by $(2x - 3)$.

Let $f(x) = 4x^3 - 6x^2 + 5x - 2$

By the Remainder Theorem, the remainder when $f(x)$ is divided by $(2x - 3)$ is $f\left(\frac{3}{2}\right)$.

Now

$$f\left(\tfrac{3}{2}\right) = 4\left(\tfrac{3}{2}\right)^3 - 6\left(\tfrac{3}{2}\right)^2 + 5\left(\tfrac{3}{2}\right) - 2$$
$$= \tfrac{27}{2} - \tfrac{27}{2} + \tfrac{15}{2} - 2$$
$$= 5\tfrac{1}{2}$$

Therefore the remainder is $5\frac{1}{2}$.

C4

You may be given the remainder and asked to find one of the constants in the function.

Example 2

The cubic $3x^3 + ax^2 + 3x + 5$ leaves a remainder of 3 when divided by $(3x + 2)$. Find the value of the constant a.

Let $f(x) = 3x^3 + ax^2 + 3x + 5$.

By the Remainder Theorem, the remainder when $f(x)$ is divided by $(3x + 2)$ is $f\left(-\frac{2}{3}\right)$.

Therefore

$$3\left(-\tfrac{2}{3}\right)^3 + a\left(-\tfrac{2}{3}\right)^2 + 3\left(-\tfrac{2}{3}\right) + 5 = 3$$
$$\therefore \quad -\frac{8}{9} + \frac{4a}{9} - 2 + 5 = 3$$
$$\frac{4a}{9} = \frac{8}{9}$$
$$\therefore \quad a = 2$$

Exercise 7A

1 Use the Remainder Theorem to find the remainder when

a) $2x^3 + x^2 + 5x - 4$ is divided by $2x - 1$

b) $4x^3 - 2x^2 + 3x - 1$ is divided by $2x + 3$

c) $8x^3 + 4x + 3$ is divided by $2x - 1$

d) $18x^3 - 3x^2 + 2x + 1$ is divided by $3x - 2$

e) $8x^3 + 4x^2 - 3$ is divided by $4x + 3$

f) $4x^4 - 3x^2 + x + 2$ is divided by $2x + 3$

2 The cubic $6x^3 + 5x^2 + ax + 3$ leaves a remainder of 1 when divided by $(3x - 2)$. Find the value of the constant a.

3 The cubic $ax^3 + 7x^2 - x - 2$ leaves a remainder of 4 when divided by $(2x + 3)$. Find the value of the constant a.

4 The expression $cx^3 - 9x^2 + 16x - 1$ leaves a remainder of 2 when divided by $(5x - 1)$. Find the value of the constant c.

5 The remainder when $4x^3 + bx^2 - x - 6$ is divided by $(4x + 3)$ is -3. Find the value of the constant b.

6 The remainder when $3x^3 - x^2 + 5x + c$ is divided by $(3x + 5)$ is 9. Find the value of the constant c.

7 The expression $2x^3 + 3x^2 + ax + b$ leaves a remainder of 18 when divided by $(x + 3)$ and a remainder of 2 when divided by $(x - 1)$. Find the values of the constants a and b.

8 The expression $ax^3 + bx^2 + 2x + 5$ leaves a remainder of 4 when divided by $(3x + 1)$ and a remainder of -2 when divided by $(x + 1)$. Find the values of the constants a and b.

9 The expression $4x^3 + bx^2 + cx + 5$ leaves a remainder of 6 when divided by $(4x - 1)$ and a remainder of 15 when divided by $(x + 2)$. Find the values of the constants b and c.

C4

7.2 The Factor Theorem

In C1 you learnt how to factorise a cubic expression like $x^3 + 4x^2 - 7x - 10$ using the Factor Theorem. This stated that if $f(a) = 0$ then $(x - a)$ is a factor of the polynomial $f(x)$.

You start by checking some values of x which are factors of 10:

$f(1) = 1 + 4 - 7 - 10 = -12 \quad \therefore \quad (x - 1)$ is not a factor

$f(-1) = -1 + 4 + 7 - 10 = 0 \quad \therefore \quad (x + 1)$ is a factor

So $x^3 + 4x^2 - 7x - 10 = (x + 1)(ax^2 + bx + c)$

Multiply out the brackets to find the coefficient of each x-term.

Looking at the coefficient of x^3 gives $a = 1$.

Looking at the coefficient of x^2 gives $b + a = 4$, hence $b = 3$.

Looking at the coefficient of x gives $c + b = -7$, hence $c = -10$.

And if you check the coefficient of the term independent of x, you see that $c = -10$ is correct.

So

$$\begin{aligned} x^3 + 4x^2 - 7x - 10 &= (x + 1)(x^2 + 3x - 10) \\ &= (x + 1)(x + 5)(x - 2) \end{aligned}$$

This form of the Factor Theorem works for cubics where the coefficient of x^3 is 1. This result can be generalised so that it also applies to cubics where the coefficient of x^3 is not 1.

If $ax - b$ is a factor of the polynomial f(x), then there is a zero remainder when f(x) is divided by $ax - b$.

Hence, by the Remainder Theorem $f\left(\frac{b}{a}\right) = 0$.

> The general form of the Factor Theorem is:
>
> If $ax - b$ is a factor of the polynomial f(x), then $f\left(\frac{b}{a}\right) = 0$.

C4

Example 3

Factorise the cubic $f(x) = 2x^3 - 5x^2 - 19x + 42$. Hence solve the equation $f(x) = 0$, and sketch the graph of $y = f(x)$.

You know that if f(x) has a linear factor $ax + b$, the constant b will be a factor of 42. In other words, one of $\pm1, \pm2, \pm3, \pm6, \pm7, \pm14, \pm21$ or ±42.

Start by evaluating $f(+1)$ and $f(-1)$. Since $f(-1) = 54$ and $f(1) = 20$, $x \pm 1$ are not factors.

Next, evaluate $f(+2)$ and $f(-2)$. Since $f(-2) = 44$, $x + 2$ is not a factor. But $f(2) = 0$, therefore $x - 2$ is a factor. So,

$$f(x) = (x - 2)(ax^2 + bx + c)$$

If you compare coefficients, you get $a = 2$, $b = -1$ and $c = -21$. Therefore,

> The coefficient of x^3 is a.
> The coefficient of x^2 is $(b - 2a)$.
> The coefficient of x is $(c - 2b)$.
> The constant is $-2c$.

$$\begin{aligned} 2x^3 - 5x^2 - 19x + 42 &= (x - 2)(2x^2 - x - 21) \\ &= (x - 2)(2x - 7)(x + 3) \end{aligned}$$

Since $f(x) = (x + 3)(x - 2)(2x - 7)$, solving $f(x) = 0$ gives $x = -3$, $x = 2$ or $x = \frac{7}{2}$. Therefore, the curve cuts the x-axis at $x = -3$, $x = 2$ and $x = \frac{7}{2}$. Also, $f(0) = 42$. Therefore, the curve cuts the y-axis at 42. The sketch is shown in the diagram.

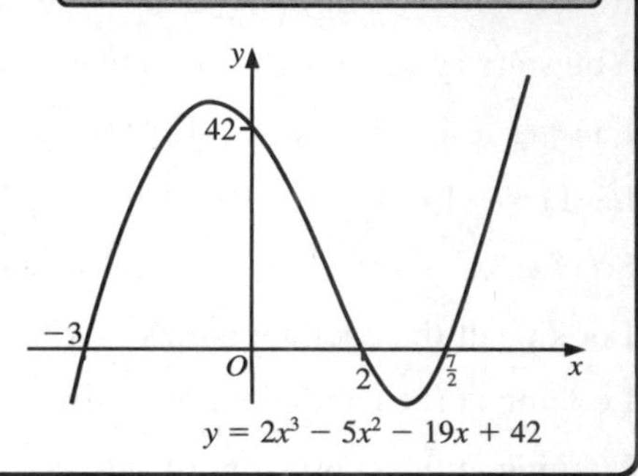

Example 4

Given $(2x^2 + 3x + a)(x - 3) = 2x^3 + bx^2 + cx - 12$, find the values of the constants a, b and c. Hence show that, with these values of a, b and c, the equation $x^3 + bx^2 + cx - 12 = 0$ has only one real solution, and find it.

So $(2x^2 + 3x + a)(x - 3) = 2x^3 + 3x^2 + ax - 6x^2 - 9x - 3a$

$2x^3 - 3x^2 + (a - 9)x - 3a = 2x^3 + bx^2 + cx - 12$

Coefficients of x^2: $b = -3$

Constant: $-3a = -12$, so $a = 4$

Coefficients of x: $a - 9 = c$, so $c = -5$

So $2x^3 + bx^2 + cx - 12 = 2x^3 - 3x^2 - 5x - 12 = (2x^2 + 3x + 4)(x - 3)$.

One solution of the equation $(2x^2 + 3x + 4)(x - 3) = 0$ is $x = 3$.

To solve the quadratic $2x^2 + 3x + 4 = 0$ you first need to calculate the discriminant:

$$D = b^2 - 4ac = 3^2 - 4(2)(4) = -23$$

Since the discriminant is negative the quadratic equation $2x^2 + 3x + 4 = 0$ has no real solutions.

The only solution to the equation $(2x^2 + 3x + 4)(x - 3) = 0$ is $x = 3$.

Multiply out.

Equate coefficients.

Remember:
You learnt in C1 that the type of roots an equation has depends on the value of the discriminant.

C4

Exercise 7B

1 Factorise each of these expressions.

a) $x^3 + 2x^2 - 5x - 6$ b) $x^3 - 5x^2 + 2x + 8$

c) $x^3 - 6x^2 + 11x - 6$ d) $x^3 - 6x^2 + 3x + 10$

e) $x^3 + 3x^2 - 9x + 5$ f) $x^3 - 6x^2 + 12x - 8$

2 Factorise each of these expressions.

a) $2x^3 - x^2 - 2x + 1$ b) $2x^3 - x^2 - 5x - 2$

c) $3x^3 + 2x^2 - 7x + 2$ d) $2x^3 + 7x^2 + 2x - 3$

e) $3x^3 - x^2 + 3x - 1$ f) $5x^3 + 14x^2 + 7x - 2$

3 Find the real solutions to each of these equations.

a) $2x^3 - 3x^2 - 5x + 6 = 0$ b) $3x^3 - 20x^2 + 29x + 12 = 0$

c) $2x^3 + x^2 = 16x + 15$ d) $5x^3 + 23x = 34x^2 - 6$

e) $x^3 + 8x^2 = 2 - 11x$ f) $x^3 + 30x = 10x^2 + 27$

4 Find the real solutions to each of these equations.

a) $2x^3 + x^2 - 13x + 6 = 0$ b) $3x^3 + 23x = 16x^2 + 6$

c) $2x^3 - 7x = 7x^2 - 12$ d) $2x^3 + 12x^2 + 3x + 18 = 0$

e) $10x^3 + 1 = x(7x + 4)$ f) $6(x^3 + 1) = 17x^2 + 5x$

5 By first factorising the expression $f(x) = x^3 - 5x^2 + 2x - 10$, show that the equation $f(x) = 0$ has only one real solution, and state its value.

6 Given

$$(x + a)(x + 3)(x - 2) = x^3 + bx^2 + cx - 30$$

find the values of the constants a, b and c. Hence, with these values of b and c, solve the equation $x^3 + bx^2 + cx - 30 = 0$.

7 Given

$$(x + c)(x + d)(x - 1) = x^3 + 2x^2 - 13x + e$$

find the possible values of the constants c, d and e. With this value of e, solve the equation $x^3 + 2x^2 - 13x + e = 0$.

8 Find the values of the constants a, b and c for which

$$(x - 4)(x - 2)(x + a) = x^3 - 7x^2 + bx + c$$

Taking these values for b and c, solve the equation $x^3 - 7x^2 + bx + c = 0$.

9 Given

$$(x^2 + a)(x - 4) = x^3 + bx^2 + cx - 20$$

find the values of the constants a, b and c. Hence show that, with these values of b and c, the equation $x^3 + bx^2 + cx - 20 = 0$ has only one real solution.

C4

10 Find the values of the constants b and c for which

$$(x^2 + b)(x + c) = x^3 - 3x^2 + bx - 15$$

With this value of b, solve the equation $x^3 - 3x^2 + bx - 15 = 0$.

11 Express $(2x - 3)(x^2 - 5x - 1) + 7$ as the product of linear factors. Hence solve the equation $(2x - 3)(x^2 - 5x - 1) = -7$.

12 Given $\mathrm{p}(x) \equiv (x^2 - 12)(x + 1) + 4x$, express $\mathrm{p}(x)$ as a product of linear factors. Hence solve the equation $\mathrm{p}(x) = 0$.

13 Find the values of the constants a, b and c for which

$$(x + 5)(x - 4)(x - a) = x^3 + bx^2 - 23x + c$$

Taking these values for b and c, solve the equation

$$x^3 + bx^2 - 23x + c = 0$$

14 Given

$$(2x + c)(x + c)(x - 4) = 2x^3 + dx^2 + ex - 36$$

where c, d and e are constants and $c > 0$, find the values of c, d and e. With these values of d and e, solve the equation

$$2x^3 + dx^2 + ex = 36$$

15 Given that $a > b$ and

$$(3x + 1)(ax - 1)(bx - 1) = 24x^3 - 10x^2 + cx + 1$$

find the values of the constants a, b and c. With this value of c, solve the equation

$$24x^3 - 10x^2 + cx + 1 = 0$$

7.3 Algebraic fractions

You can add and subtract algebraic fractions in the same way that you add and subtract numerical fractions. To find the sum or difference of two numerical fractions, each fraction must be expressed in terms of the same denominator, called the lowest common denominator. For example, consider

$$\frac{1}{3}+\frac{1}{6}$$

The lowest common denominator is 6, since 3 is a factor of 6. Therefore,

$$\frac{1}{3}+\frac{1}{6}=\frac{2}{6}+\frac{1}{6}=\frac{3}{6}=\frac{1}{2}$$

When the only common factor of the denominators is 1, then find their product and use the product as the common denominator. For example,

$$\frac{1}{3}+\frac{1}{7}$$

The lowest common denominator is $3 \times 7 = 21$, since 3 and 7 have no common factor other than 1. Therefore,

$$\frac{1}{3}+\frac{1}{7}=\frac{7}{21}+\frac{3}{21}=\frac{10}{21}$$

A numerical fraction whose numerator is greater than or equal to the denominator is called an **improper fraction**. For example, $\frac{7}{5}$ is an improper fraction which can be written as $1\frac{2}{5}$.

C4

An algebraic fraction is called improper if the degree of the numerator is greater than or equal to the degree of the denominator.

Remember: The degree of an algebraic expression f(x) is the highest power of x in it.

For example,

$$\frac{x^2+2x-8}{3x+7}$$

is an improper fraction because the degree of the numerator is two and the degree of the denominator is one.

You will learn some techniques for simplifying improper algebraic fractions on page 160.

Example 5

Express $\dfrac{4}{x+6}-\dfrac{2}{x+7}$ as a single fraction.

The lowest common denominator is $(x+6)(x+7)$. Therefore,

$$\begin{aligned}\frac{4}{x+6}-\frac{2}{x+7}&=\frac{4(x+7)-2(x+6)}{(x+6)(x+7)}\\&=\frac{4x+28-2x-12}{(x+6)(x+7)}\\&=\frac{2x+16}{(x+6)(x+7)}\end{aligned}$$

$$\therefore\quad \frac{4}{x+6}-\frac{2}{x+7}=\frac{2(x+8)}{(x+6)(x+7)}$$

If the denominators of the fractions to be added are of different degree, look carefully to see if they have a common factor.

Example 6

Express $\dfrac{2x}{x^2 + 3x + 2} + \dfrac{3}{x + 1}$ as a single fraction.

In this case, $(x^2 + 3x + 2)(x + 1)$ is a common denominator but there is a much simpler common denominator, namely $(x^2 + 3x + 2)$, since

$$x^2 + 3x + 2 = (x + 1)(x + 2)$$

In other words, $(x + 1)$ is a factor of $x^2 + 3x + 2$. Therefore,

$$\frac{2x}{x^2 + 3x + 2} + \frac{3}{x + 1} = \frac{2x}{(x + 1)(x + 2)} + \frac{3(x + 2)}{(x + 1)(x + 2)}$$

$$= \frac{2x + 3(x + 2)}{(x + 1)(x + 2)}$$

$$\therefore \quad \frac{2x}{x^2 + 3x + 2} + \frac{3}{x + 1} = \frac{5x + 6}{(x + 1)(x + 2)}$$

C4

You may need to add or subtract algebraic fractions in order to solve an equation.

Example 7

Express $3\frac{3}{5} + \dfrac{4}{x + 7}$ as a single fraction. Hence solve the equation

$$3\tfrac{3}{5} + \frac{4}{x + 7} = \frac{8}{5 - x}$$

Writing $3\frac{3}{5}$ as an improper fraction gives $\frac{18}{5}$. Therefore,

$$3\tfrac{3}{5} + \frac{4}{x + 7} = \frac{18}{5} + \frac{4}{x + 7}$$

$$= \frac{18(x + 7) + 4(5)}{5(x + 7)}$$

$$= \frac{18x + 146}{5(x + 7)}$$

Using this result, the equation

$$3\tfrac{3}{5} + \frac{4}{x + 7} = \frac{8}{5 - x}$$

becomes

$$\frac{18x + 146}{5(x + 7)} = \frac{8}{5 - x}$$

Cross-multiplying and simplifying give

$$(18x + 146)(5 - x) = 40(x + 7)$$

$$\therefore \quad 18x^2 + 96x - 450 = 0$$

Dividing by 6 and factorising give

$$3x^2 + 16x - 75 = 0$$

$$\therefore \quad (3x + 25)(x - 3) = 0$$

Solving gives $x = -\frac{25}{3}$ or $x = 3$.

In the next example you can find the constants by simplifying the algebraic fractions.

Example 8

Given $\dfrac{a}{(x-2)} - \dfrac{6}{(2x+3)} = \dfrac{b}{(x-2)(2x+3)}$, where a and b are both constants, find the values of a and b.

Now $\dfrac{a}{(x-2)} - \dfrac{6}{(2x+3)} = \dfrac{b}{(x-2)(2x+3)}$

Multiply each side by $(x-2)(2x+3)$ to get

$$\frac{a(x-2)(2x+3)}{(x-2)} - \frac{6(x-2)(2x+3)}{(2x+3)} = \frac{b(x-2)(2x+3)}{(x-2)(2x+3)}$$

Cancelling gives

$$a(2x + 3) - 6(x - 2) = b$$

Equating coefficients of x and the constant:

$0(x)$: $\quad 2a - 6 = 0$

$\therefore \quad a = 3$

$0(1)$: $\quad 3a + 12 = b$

$\therefore \quad b = 21$

So $a = 3$ and $b = 21$.

$0(x)$ is shorthand for 'coefficients of x'.
$0(1)$ is shorthand for 'the constant'.

C4

Exercise 7C

1 Express each of these as a single fraction.

a) $\dfrac{3}{x-1} + \dfrac{2}{x+3}$

b) $\dfrac{2}{x+4} - \dfrac{1}{x-3}$

c) $\dfrac{4}{x-5} - \dfrac{2}{x+4}$

d) $\dfrac{x}{x-3} - \dfrac{2}{x+2}$

e) $\dfrac{2x}{x+2} - \dfrac{1}{x-2}$

f) $\dfrac{5}{2x-3} - \dfrac{3}{2x+5}$

g) $\dfrac{2x+1}{x-2} + \dfrac{3x}{x+4}$

h) $\dfrac{5x}{x^2+3} + \dfrac{6}{2x-5}$

2 Express each of these as a single fraction.

a) $\dfrac{x}{x^2+3x-4}+\dfrac{2}{x-1}$ b) $\dfrac{5}{x^2+6x+8}-\dfrac{4}{x+2}$

c) $\dfrac{3}{x^2-3x+2}+\dfrac{2}{x-1}$ d) $\dfrac{4x}{x^2-4x+4}+\dfrac{3}{x-2}$

e) $\dfrac{3}{x^2+5x+6}-\dfrac{2}{x+3}-\dfrac{1}{x+2}$

f) $\dfrac{5x}{2x^2+3x-5}-\dfrac{1}{x-1}-\dfrac{2}{2x+5}$

3 Show that

$$\frac{3}{x-2}+\frac{4}{x-1}=\frac{7x-11}{(x-2)(x-1)}$$

Hence solve the equation

$$\frac{3}{x-2}+\frac{4}{x-1}=\frac{7}{x}$$

4 Express $\dfrac{2}{x-3}+\dfrac{1}{2x+1}$ as a single fraction. Hence solve the equation

$$\frac{2}{x-3}+\frac{1}{2x+1}=\frac{5}{2x-3}$$

5 Express $\dfrac{3}{x-2}+\dfrac{x}{x+1}$ as a single fraction. Hence solve the equation

$$\frac{3}{x-2}+\frac{x}{x+1}=2\tfrac{1}{7}$$

6 Given

$$\frac{a}{6x-1}-\frac{1}{3x+1}=\frac{b}{(6x-1)(3x+1)}$$

where a and b are both constants, find the values of a and b.

7 Find the values of the constants A and B for which

$$\frac{A}{x-5}+\frac{B}{x+5}=\frac{2x}{x^2-25}$$

8 Given

$$\frac{6}{x+3}+\frac{P}{x-1}=\frac{Qx}{(x+3)(x-1)}$$

where P and Q are both constants, find the values of P and Q.

9 Find the values of the constants a and b for which

$$\frac{a}{x+b}+\frac{2}{x-4}=\frac{3x}{(x+b)(x-4)}$$

7.4 Rational functions

Sometimes a fraction will be given in the form $\frac{f(x)}{p(x)}$ where $f(x)$ and $p(x)$ are both polynomials. An example of this is

$$\frac{x^2 - 4x}{x^2 - 5x + 4}$$

Such an expression is called a **rational function**. In this case, factorising the numerator and the denominator, and then cancelling the common factor, allows you to simplify the rational function.

$$\frac{x^2 - 4x}{x^2 - 5x + 4} = \frac{x(x-4)}{(x-1)(x-4)} = \frac{x}{x-1}$$

A rational number is any number in fraction form $\frac{a}{b}$, where a and b are integers.

Example 9

Simplify each of these rational functions.

a) $\frac{2x - 4}{x^2 + x - 6}$ b) $\frac{2x^2 - 3x + 1}{x^2 - 1}$ c) $\frac{4x^2 + 12x + 9}{6x^2 + 5x - 6}$

a) $\frac{2x - 4}{x^2 + x - 6} = \frac{2(x-2)}{(x-2)(x+3)}$
$= \frac{2}{x+3}$

b) $\frac{2x^2 - 3x + 1}{x^2 - 1} = \frac{(2x-1)(x-1)}{(x+1)(x-1)}$
$= \frac{2x-1}{x+1}$

c) $\frac{4x^2 + 12x + 9}{6x^2 + 5x - 6} = \frac{(2x+3)(2x+3)}{(2x+3)(3x-2)}$
$= \frac{2x+3}{3x-2}$

You may need to simplify a rational function in order to solve an equation.

Example 10

a) Simplify $\frac{6x^2 - 3x}{6x^2 + x - 2}$.

b) Hence solve the equation $\frac{6x^2 - 3x}{6x^2 + x - 2} = 5$.

a) $\frac{6x^2 - 3x}{6x^2 + x - 2} = \frac{3x(2x-1)}{(3x+2)(2x-1)}$
$= \frac{3x}{3x+2}$

C4

b) $\dfrac{3x}{3x+2} = 5$

Multiply each side by $(3x + 2)$ to get

$$3x = 5(3x+2)$$
$$\therefore \quad 3x = 15x + 10$$
$$\therefore \quad -10 = 12x$$
$$\therefore \quad x = -\frac{5}{6}$$

The solution is $x = -\frac{5}{6}$.

Exercise 7D

1 Simplify each of these rational functions.

a) $\dfrac{x-3}{x^2-x-6}$ b) $\dfrac{2x+2}{x^2+6x+5}$

c) $\dfrac{x^2-2x}{x^2-7x+10}$ d) $\dfrac{3x^2+3x}{x^2-2x-3}$

e) $\dfrac{x^2+7x+10}{x^2+5x+6}$ f) $\dfrac{x^2-6x-7}{x^2-x-2}$

g) $\dfrac{x^2-4}{x^2-8x+12}$ h) $\dfrac{x^2-8x+15}{x^2-x-6}$

C4

2 Simplify each of these rational functions.

a) $\dfrac{2x+1}{2x^2+5x+2}$ b) $\dfrac{6x-9}{4x^2-9}$

c) $\dfrac{3x^2+2x}{3x^2+17x+10}$ d) $\dfrac{6x^2-10x}{3x^2-8x+5}$

e) $\dfrac{3x^2+10x+3}{2x^2+5x-3}$ f) $\dfrac{4x^2-1}{4x^2+4x+1}$

g) $\dfrac{5x^2+7x+2}{4x^2+x-3}$ h) $\dfrac{4x^2-25}{2x^2+3x-5}$

3 a) Simplify $\dfrac{10x^2-5x}{2x^2+7x-4}$.

b) Hence solve the equation $\dfrac{10x^2-5x}{2x^2+7x-4} = 3$.

4 a) Simplify $\dfrac{5x^2+9x-2}{3x^2+11x+10}$.

b) Hence solve the equation $\dfrac{5x^2+9x-2}{3x^2+11x+10} = 2$.

5 a) Simplify $\dfrac{9x^2-4}{3x^2-11x+6}$.

b) Hence solve the equation $\dfrac{9x^2-4}{3x^2-11x+6} = 7$.

7.5 Algebraic division

In C1 you learnt how to apply the method of algebraic long division to find the quotient and remainder when one polynomial is divided by another polynomial. For example, when $x^2 - 3x + 6$ is divided by $(x + 2)$ the quotient is $(x - 5)$ and the remainder is 16.

Check this for yourself!

Now you can learn another method for performing the same operation.

Since $x^2 - 3x + 6$ is a quadratic function and $(x + 2)$ is a linear function, the result of the division will be a linear function.

That is,

$$\frac{x^2 - 3x + 6}{x + 2} = Ax + B + \frac{C}{x + 2},$$

where $(Ax + B)$ is the linear quotient, and C is the remainder.

Multiply throughout by $(x + 2)$:

$$x^2 - 3x + 6 = (Ax + B)(x + 2) + C$$

Expand the bracket and equate coefficients of x^2, x and the constant:

$$\begin{aligned} 0(x^2)&: \quad 1 = A \\ 0(x)&: \quad -3 = 2A + B \\ \therefore& \quad B = -5 \\ 0(1)&: \quad 6 = 2B + C \\ \therefore& \quad C = 16 \end{aligned}$$

So $\dfrac{x^2 - 3x + 6}{x + 2} = x - 5 + \dfrac{16}{x + 2}$

When a polynomial of degree n is divided by a polynomial of degree m, the quotient will be a polynomial of degree $n - m$.

Example 11

Find the values of the constants A, B and C for which $\dfrac{3x^2 - 2x + 7}{x - 4} = Ax + B + \dfrac{C}{x - 4}$.

$$\frac{3x^2 - 2x + 7}{x - 4} = Ax + B + \frac{C}{x - 4}$$

Multiply each side by $(x - 4)$:

$$3x^2 - 2x + 7 = (Ax + B)(x - 4) + C$$

Expand the bracket and equate coefficients of x^2, x and the constant:

$$\begin{aligned} 0(x^2)&: \quad 3 = A \\ 0(x)&: \quad -2 = -4A + B \\ &\therefore \; B = 10 \\ 0(1)&: \quad 7 = -4B + C \\ &\therefore \; C = 47 \end{aligned}$$

So $\dfrac{3x^2 - 2x + 7}{x - 4} = 3x + 10 + \dfrac{47}{x - 4}$

Remember:
$0(x^2)$ just means coefficients of x^2.

You could use the identity symbol $\equiv$ instead of $=$ in the final statement, as it is true for all $x (x \neq 4)$.

When both functions are linear the quotient will be a constant, as in Example 12.

Example 12

Find the values of the constants A and B for which

$\frac{3x+4}{x-1} = A + \frac{B}{x-1}$

$$\frac{3x+4}{x-1} = A + \frac{B}{x-1}$$

Multiply each side by $(x-1)$:

$$3x + 4 = A(x-1) + B$$

Expand the brackets and equate coefficients of x and the constants:

$0(x)$: $\quad 3 = A$

$0(1)$: $\quad 4 = -A + B$

$\therefore \quad B = 7$

So $\frac{3x+4}{x-1} = 3 + \frac{7}{x-1}$

C4

You can apply this method in the same way to polynomials of higher degree.

In general,

- ✦ when a polynomial of degree n is divided by a polynomial also of degree n, the quotient is a constant.
- ✦ when a polynomial of degree n is divided by a polynomial of degree m, where $m < n$, the quotient is a polynomial of degree $n - m$.

Example 13

Find the values of the constants A, B and C for which

$$\frac{2x^2}{(x+5)(x-3)} = A + \frac{Bx+C}{(x+5)(x-3)}$$

Here both polynomials are of degree 2. So the quotient is a constant, A.

$$\frac{2x^2}{(x+5)(x-3)} = A + \frac{Bx+C}{(x+5)(x-3)}$$

Multiply each side by $(x+5)(x-3)$:

$$2x^2 = A(x+5)(x-3) + Bx + C$$

Expand the bracket and equate coefficients of x^2, x and the constants:

$0(x^2)$: $\quad 2 = A$

$0(x)$: $\quad 0 = 2A + B$

$\therefore \quad B = -4$

$0(1)$: $\quad 0 = -15A + C$

$\therefore \quad C = 30$

So $\dfrac{2x^2}{(x+5)(x-3)} = 2 + \dfrac{30-4x}{(x+5)(x-3)}$

Example 14

Find the values of the constants A, B, C and D for which

$$\frac{2x^3 - 3x^2 - 2x + 2}{(x-2)} = Ax^2 + Bx + C + \frac{D}{(x-2)}$$

$$\frac{2x^3 - 3x^2 - 2x + 2}{(x-2)} = Ax^2 + Bx + C + \frac{D}{(x-2)}$$

Multiply each side by $(x-2)$:

$$2x^3 - 3x^2 - 2x + 2 = (Ax^2 + Bx + C)(x-2) + D$$

Expand the bracket and equate coefficients of x^3, x^2, x and the constants:

$0(x^3)$: $\quad 2 = A$

$0(x^2)$: $\quad -3 = -2A + B$

$\therefore \quad B = 1$

$0(x)$: $\quad -2 = -2B + C$

$\therefore \quad C = 0$

$0(1)$: $\quad 2 = -2C + D$

$\therefore \quad D = 2$

So $\dfrac{2x^3 - 3x^2 - 2x + 2}{(x-2)} = 2x^2 + x + \dfrac{2}{(x-2)}$

Here a polynomial of degree 3 is divided by a linear factor of degree 1, so the quotient is of degree $3 - 1 = 2$ i.e. $Ax^2 + Bx + C$.

Exercise 7E

1 Find the values of the constants A and B for each of these:

a) $\dfrac{2x+3}{x-2} = A + \dfrac{B}{x-2}$ $\qquad$ b) $\dfrac{3x-4}{x+1} = A + \dfrac{B}{x+1}$

c) $\dfrac{5x-3}{x+2} = A + \dfrac{B}{x+2}$ $\qquad$ d) $\dfrac{4x+3}{2x-1} = A + \dfrac{B}{2x-1}$

2 Find the values of the constants A, B and C for each of these identities.

a) $$\frac{2x^2 - 3x + 5}{x - 3} = Ax + B + \frac{C}{x - 3}$$

b) $$\frac{x^2 + 5x - 7}{x + 4} = Ax + B + \frac{C}{x + 4}$$

c) $$\frac{4x^2 + 12x - 3}{x + 6} = Ax + B + \frac{C}{x + 6}$$

3 Find the values of the constants A, B and C for each of these identities.

a) $$\frac{x^2}{(x + 4)(x - 2)} = A + \frac{Bx + C}{(x + 4)(x - 2)}$$

b) $$\frac{2x^2}{(x + 6)(x - 3)} = A + \frac{Bx + C}{(x + 6)(x - 3)}$$

c) $$\frac{4x^2}{(x - 1)(x - 5)} = A + \frac{Bx + C}{(x - 1)(x - 5)}$$

4 Find the values of the constants A, B, C and D for each of these identities.

C4

a) $$\frac{2x^3 - x^2 - 3x + 1}{(x - 2)} = Ax^2 + Bx + C + \frac{D}{(x - 2)}$$

b) $$\frac{3x^3 + x^2 + 4x + 9}{(x + 1)} = Ax^2 + Bx + C + \frac{D}{(x + 1)}$$

c) $$\frac{x^3 - 5x^2 - 7x + 4}{(x - 3)} = Ax^2 + Bx + C + \frac{D}{(x - 3)}$$

5 Find the values of the constants in each of these identities.

a) $$\frac{4x^2}{(x + 1)(x - 3)} = A + \frac{Bx + C}{(x + 1)(x - 3)}$$

b) $$\frac{x^3 - 2x^2 + 4x + 3}{(x - 2)} = Ax^2 + Bx + C + \frac{D}{(x - 2)}$$

c) $$\frac{5x^2 + 6x - 8}{x + 3} = Ax + B + \frac{C}{x + 3}$$

d) $$\frac{8x + 3}{2x + 5} = A + \frac{B}{2x + 5}$$

e) $$\frac{3x^3 - 2x^2 + 5x - 2}{(x - 1)} = Ax^2 + Bx + C + \frac{D}{(x - 1)}$$

f) $$\frac{4 - 7x}{x + 5} = A + \frac{B}{x + 5}$$

7.6 Partial fractions

In Example 5 (page 147), the algebraic fraction $\frac{4}{x+6} - \frac{2}{x+7}$ is expressed as a single fraction, namely $\frac{2(x+8)}{(x+6)(x+7)}$. Now let's look at how you can express $\frac{2(x+8)}{(x+6)(x+7)}$ as $\frac{4}{x+6} - \frac{2}{x+7}$. This reverse process is called expressing $\frac{2(x+8)}{(x+6)(x+7)}$ in **partial fractions.**

There are basically four different types of algebraic fraction that can be expressed in partial fractions.

1 Denominator with linear factors

In this type of fraction, each linear factor $(ax + b)$ in the denominator has a corresponding partial fraction of the form

$$\frac{A}{(ax+b)}$$

where a, b and A are constants.

C4

Example 15

Express $\frac{7x+8}{(x+4)(x-6)}$ in partial fractions.

$$\frac{7x+8}{(x+4)(x-6)} = \frac{A}{x+4} + \frac{B}{x-6}$$

Multiply throughout by $(x + 4)(x - 6)$:

$$7x + 8 = A(x - 6) + B(x + 4) \qquad [1]$$

There are two techniques for finding the constants A and B. First, expanding the right-hand side of identity [1] gives

$$7x + 8 = Ax - 6A + Bx + 4B$$

Now you can compare the coefficients of the x terms and the constants to find the constants A and B.

$0(x)$: $\quad 7 = A + B \qquad [2]$

$0(1)$: $\quad 8 = -6A + 4B$

$\therefore \quad 4 = -3A + 2B \qquad [3]$

You could use ≡ instead of = in statement [1], as it is an identity. You must use = in [2] and [3], however, as these are (simultaneous) equations.

Solving [2] and [3] simultaneously gives $A = 2$ and $B = 5$.

Second, if you let $x = 6$, the constant A will be eliminated from the equation, and you can find the constant B:

$$7(6) + 8 = A(6 - 6) + B(6 + 4)$$

$\therefore \quad 50 = 10B$

$\therefore \quad B = 5$

Then, if you let $x = -4$, you can find the constant A:

$$7(-4) + 8 = A(-4 - 6) + B(-4 + 4)$$
$$\therefore \quad -20 = -10A$$
$$\therefore \quad A = 2, \text{ as before.}$$

So:

$$\frac{7x + 8}{(x + 4)(x - 6)} = \frac{2}{x + 4} + \frac{5}{x - 6}$$

In fact it is easier to use a combination of the two techniques shown in Example 15, as you will see later in Examples 17 to 20.

Example 16

Express $\dfrac{9x^2 + 34x + 14}{(x + 2)(x^2 - x - 12)}$ in partial fractions.

Although the denominator appears to have only one linear factor, this is not the case since

$$x^2 - x - 12 = (x + 3)(x - 4)$$

Therefore,

$$\frac{9x^2 + 34x + 14}{(x + 2)(x^2 - x - 12)} = \frac{9x^2 + 34x + 14}{(x + 2)(x + 3)(x - 4)}$$

Assume that

$$\frac{9x^2 + 34x + 14}{(x + 2)(x + 3)(x - 4)} = \frac{A}{x + 2} + \frac{B}{x + 3} + \frac{C}{x - 4}$$

Multiply throughout by $(x + 2)(x + 3)(x - 4)$:

$$9x^2 + 34x + 14 = A(x + 3)(x - 4) + B(x + 2)(x - 4) + C(x + 2)(x + 3)$$

Let $x = -2$: $9(-2)^2 + 34(-2) + 14 = A(-2 + 3)(-2 - 4)$

$$\therefore \quad -18 = -6A$$
$$\therefore \quad A = 3$$

Let $x = -3$: $9(-3)^2 + 34(-3) + 14 = B(-3 + 2)(-3 - 4)$

$$\therefore \quad -7 = 7B$$
$$\therefore \quad B = -1$$

Let $x = 4$: $9(4)^2 + 34(4) + 14 = C(4 + 2)(4 + 3)$

$$\therefore \quad 294 = 42C$$
$$\therefore \quad C = 7$$

Therefore,

$$\frac{9x^2 + 34x + 14}{(x + 2)(x + 3)(x - 4)} = \frac{3}{x + 2} - \frac{1}{x + 3} + \frac{7}{x - 4}$$

2 Denominator with a repeated factor

In this type of fraction, each repeated linear factor $(ax + b)^2$ in the denominator has corresponding partial fractions of the form

$$\frac{A}{ax+b} + \frac{B}{(ax+b)^2}$$

where a, b, A and B are constants.

Example 17

Express $\dfrac{2x-5}{(x-4)^2}$ in partial fractions.

$(x - 4)$ is a repeated linear factor.

Assume that $\dfrac{2x-5}{(x-4)^2} = \dfrac{A}{x-4} + \dfrac{B}{(x-4)^2}$

Multiply each side by $(x - 4)^2$:

$$2x - 5 = A(x - 4) + B$$

Both techniques described in Example 15 are used here.

Equate coefficients in x:

$0(x)$: $2 = A$

Let $x = 4$: $2(4) - 5 = B$

$$\therefore \quad B = 3$$

Therefore $\dfrac{2x-5}{(x-4)^2} = \dfrac{2}{x-4} + \dfrac{3}{(x-4)^2}$

A fraction may have several factors in the denominator, one of which is repeated.

Example 18

Express $\dfrac{2x^2+29x-11}{(2x+1)(x-2)^2}$ in partial fractions.

Assume that

$$\frac{2x^2+29x-11}{(2x+1)(x-2)^2} = \frac{A}{2x+1} + \frac{B}{x-2} + \frac{C}{(x-2)^2}$$

Multiply throughout by $(2x + 1)(x - 2)^2$:

$$2x^2 + 29x - 11 = A(x-2)^2 + B(2x+1)(x-2) + C(2x+1)$$

Let $x = -\frac{1}{2}$: $2\left(-\frac{1}{2}\right)^2 + 29\left(-\frac{1}{2}\right) - 11 = A\left(-\frac{1}{2} - 2\right)^2$

$$\therefore \quad -25 = \frac{25}{4}A$$

$$\therefore \quad A = -4$$

Let $x = 2$: $2(2)^2 + 29(2) - 11 = C(4 + 1)$

$$\therefore\quad 55 = 5C$$

$$\therefore\quad C = 11$$

Comparing the coefficients of the x^2 terms gives

$$2 = A + 2B \qquad [1]$$

Substituting $A = -4$ into [1] gives $B = 3$. Therefore,

$$\frac{2x^2 + 29x - 11}{(2x + 1)(x - 2)^2} = -\frac{4}{2x + 1} + \frac{3}{x - 2} + \frac{11}{(x - 2)^2}$$

3 Improper fractions

You know that:

- When a polynomial of degree n is divided by a polynomial also of degree n, the quotient is a constant.
- When a polynomial of degree n is divided by a polynomial of degree m, where $m < n$ the quotient is a polynomial of degree $n - m$.

You can use these two facts when expressing improper algebraic fractions in partial fractions.

C4

Example 19

Express $\dfrac{5x^2 - 71}{(x + 5)(x - 4)}$ in partial fractions.

The degree of $5x^2 - 71$ is 2 and the degree of $(x + 5)(x - 4)$ is also 2, so the quotient is a constant. Write the quotient as A and assume that:

$$\frac{5x^2 - 71}{(x + 5)(x - 4)} = A + \frac{B}{x + 5} + \frac{C}{x - 4}$$

Multiplying throughout by $(x + 5)(x - 4)$ gives

$$5x^2 - 71 = A(x + 5)(x - 4) + B(x - 4) + C(x + 5)$$

Comparing coefficients of the x^2 terms gives $A = 5$.

Let $x = -5$: $5(-5)^2 - 71 = B(-5 - 4)$

$$\therefore\quad 54 = -9B$$

$$\therefore\quad B = -6$$

Let $x = 4$: $5(4)^2 - 71 = C(4 + 5)$

$$\therefore\quad 9 = 9C$$

$$\therefore\quad C = 1$$

So:

$$\frac{5x^2 - 71}{(x + 5)(x - 4)} = 5 - \frac{6}{x + 5} + \frac{1}{x - 4}$$

The next example shows you how to extend the method to harder questions, but is more difficult than you can expect in the exam.

Example 20

Express $\dfrac{2x^3 - 2x^2 - 26x - 18}{(x+2)(x+1)(x-3)}$ in partial fractions.

> The numerator and denominator are both of degree 3, so the quotient is a constant.

Assume that $\dfrac{2x^3 - 2x^2 - 26x - 18}{(x+2)(x+1)(x-3)} = A + \dfrac{B}{x+2} + \dfrac{C}{x+1} + \dfrac{D}{x-3}$

Multiplying each side by $(x+2)(x+1)(x-3)$ gives

$$\begin{aligned} 2x^3 - 2x^2 - 26x - 18 = &A(x+2)(x+1)(x-3) \\ &+ B(x+1)(x-3) \\ &+ C(x+2)(x-3) \\ &+ D(x+2)(x+1) \end{aligned}$$

Equating coefficients in x^3 gives

$0(x^3)$: $2 = A$

Let $x = -2$: $2(-2)^3 - 2(-2)^2 - 26(-2) - 18 = B(-1)(-5)$

$$-16 - 8 + 52 - 18 = 5B$$
$$10 = 5B$$
$$\therefore \quad B = 2$$

Let $x = -1$: $2(-1)^3 - 2(-1)^2 - 26(-1) - 18 = C(1)(-4)$

$$-2 - 2 + 26 - 18 = -4C$$
$$4 = -4C$$
$$\therefore \quad C = -1$$

Let $x = 3$: $2(3)^3 - 2(3)^2 - 26(3) - 18 = D(5)(4)$

$$\therefore \quad 54 - 18 - 78 - 18 = 20D$$
$$\therefore \quad -60 = 20D$$
$$\therefore \quad D = -3$$

Therefore $\dfrac{2x^3 - 2x^2 - 26x - 18}{(x+2)(x+1)(x-3)} \equiv 2 + \dfrac{2}{x+2} - \dfrac{1}{x+1} - \dfrac{3}{x-3}$

C4

Exercise 7F

In Questions **1** to **12**, write each expression in partial fractions.

1 ***Denominator with two distinct linear factors***

a) $\dfrac{3x-1}{(x+3)(x-2)}$ b) $\dfrac{5x+6}{(x+4)(x-3)}$

c) $\dfrac{2x+1}{(x+2)(x+1)}$ d) $\dfrac{9-8x}{(2x-1)(3-x)}$

e) $\dfrac{7x+16}{x^2+2x-8}$ f) $\dfrac{5x-1}{2x^2+x-10}$

2 ***Denominator with three distinct linear factors***

a) $\dfrac{2}{(x+3)(x+2)(x+1)}$ b) $\dfrac{x^2-9x+2}{(x+1)(x-1)(x-2)}$

c) $\dfrac{x+1}{(x+3)(x+2)(x-1)}$ d) $\dfrac{2x^2-7x+1}{(2x+1)(2x-1)(x-2)}$

e) $\dfrac{7x^2+39x+56}{(x+4)(x+3)(2x+5)}$ f) $\dfrac{2x^2+11x+3}{x(3x+1)(x+3)}$

3 ***Denominator with a repeated factor***

a) $\dfrac{2x+3}{(x+2)^2}$ b) $\dfrac{4x-9}{(x-3)^2}$

c) $\dfrac{3x-14}{x^2-8x+16}$ d) $\dfrac{5x+7}{(x+1)^2(x+2)}$

e) $\dfrac{6x^2-11x+13}{(x-2)^2(x+3)}$ f) $\dfrac{7+5x-6x^2}{(2x+1)^2(x+2)}$

4 ***Improper fractions***

a) $\dfrac{x^2+7x-14}{(x+5)(x-3)}$ b) $\dfrac{2x^2+x-5}{(x+2)(x+1)}$

c) $\dfrac{8x^2-2x-9}{(2x+1)(2x-3)}$ d) $\dfrac{12x^2-15x+6}{(3x-1)(2x-3)}$

e) $\dfrac{x^3+4x^2-x-17}{(x+3)(x-2)}$ f) $\dfrac{3x^3-10x^2+2x-1}{(3x-1)(x-3)}$

5 $\dfrac{4x+7}{(x+1)(x+3)}$ **6** $\dfrac{x^2+29x+2}{(3x+5)(2x+1)(x-3)}$

7 $\dfrac{1-2x}{(x+1)^2}$ **8** $\dfrac{4x^2-7x-5}{(x-3)^2(x+2)}$

9 $\dfrac{3x^2+7x+11}{(x^2+3x+3)(x-1)}$ **10** $\dfrac{x^2+3x-34}{x^2+2x-15}$

11 $\dfrac{x+10}{(2x-1)(x-4)}$ **12** $\dfrac{x^2+9x+17}{(x+4)^2(x+3)}$

13 a) Express $\dfrac{1}{(x-5)(x-4)}$ in partial fractions.

b) Hence prove that

$$\frac{1}{(x-5)^2(x-4)^2}=\frac{1}{(x-5)^2}+\frac{1}{(x-4)^2}-\frac{2}{x-5}+\frac{2}{x-4}$$

14 a) Express $\dfrac{7}{(x+5)(x-2)}$ in partial fractions.

b) Hence prove that

$$\frac{1}{(x+5)^2(x-2)^2}=\frac{1}{343}\left[\frac{7}{(x+5)^2}+\frac{7}{(x-2)^2}+\frac{2}{x+5}-\frac{2}{x-2}\right]$$

C4

Summary

You should know how to ...	Check out
1 Use the Factor Theorem with more demanding problems.	**1** The polynomial $p(x)$ is given as $$p(x) = 8x^3 + 18x^2 + x - 6$$ a) Find $p(\frac{1}{2})$. What can you conclude? b) Show that $(4x + 3)$ is a factor of $p(x)$. c) Write $p(x)$ as a product of three linear factors.
2 Use the Remainder Theorem with more demanding problems.	**2** The polynomial $p(x) = 9x^3 - 3x^2 + ax + 5$ is known to have a remainder of 3, when divided by $3x + 1$. Find the value of a.
3 Simplify a rational function.	**3** Simplify $\dfrac{3x^2 + 9x}{4x^3 + 12x^2 - x - 3}$
4 Divide polynomials, including the use of partial fractions.	**4** a) Express $\dfrac{5x + 3}{x + 1}$ in the form $A + \dfrac{B}{x + 1}$. b) Express $\dfrac{4x^2 - 7x - 22}{(2x + 1)(x - 3)}$ in the form $A + \dfrac{B}{2x + 1} + \dfrac{C}{x - 3}$.

C4

Revision exercise 7

1 A polynomial is defined by $p(x) = 4x^3 - 5x^2 + 2$.
Find the remainder when $p(x)$ is divided by $(2x + 1)$. *(AQA, 2002)*

2 a) Given that $(2x - 1)$ is a factor of $p(x) = 6x^3 - kx^2 - 6x + 8$ use the Factor Theorem to show that $k = 23$.
b) Express $p(x)$ as the product of three linear factors. *(AQA, 2001)*

3 A curve C has the equation $y = \dfrac{2x + 1}{x + 2}$ $x \neq -2$.
Express the equation of C in the form $y = A + \dfrac{B}{x + 2}$. *(AQA, 2004)*

4 a) Given that $(x + 2)$ is a factor of $p(x) = 6x^3 + kx^2 - 9x + 2$ show that $k = 7$.
b) Find the value of $p(\frac{1}{2})$ and hence show that $(2x - 1)$ is a factor of $p(x)$.
c) Express $p(x)$ as a product of three linear factors.
d) Hence find the value of θ, in radians, in the interval $0 < \theta < 2\pi$ for which $6\sin^3\theta + 7\sin^2\theta - 9\sin\theta + 2 = 0$. *(AQA, 2004)*

Remember:
$\sin^3\theta$ just means $(\sin\theta)^3$.

5 a) Given that $(x-2)$ is a factor of $p(x) = 4x^3 - kx + 10$ show that $k = 21$.

b) Use the Factor Theorem to show that $(2x-1)$ is also a factor of $p(x)$.

c) Write $p(x)$ as a product of three linear factors.

d) Hence find the value of θ, in radians, in the interval $-\pi < \theta < \pi$ for which $4\tan^3\theta - 21\tan\theta + 10 = 0$. *(AQA, 2003)*

6 a) Given that $(2x+1)$ is a factor of $f(x) = 8x^3 - 6x^2 + ax + b$ where a and b are integers, show that $a - 2b = -5$.

b) Given also that $(x-1)$ is a factor of $f(x)$, show that $a = -3$ and find the value of b.

c) Write $f(x)$ as a product of three linear factors.

d) Find, in radians, the value of θ in the interval $0 \leqslant \theta \leqslant 2\pi$ for which $8\cos^3\theta - 3\cos\theta = 5 - 6\sin^2\theta$. *(AQA, 2002)*

7 Express $\dfrac{x^2+1}{x^2(x-1)}$ in partial fractions.

8 a) Express each of the following in partial fractions.

i) $\dfrac{x-8}{(x-2)(x-4)}$ ii) $\dfrac{12}{(x-2)(x-4)}$

b) Hence prove that

$$\frac{(x-8)^2}{(x-2)^2(x-4)^2} = \frac{9}{(x-2)^2} + \frac{4}{(x-4)^2} + \frac{6}{x-2} - \frac{6}{x-4}$$

C4

8 Binomial series

This chapter will show you how to

- ✦ Find binomial expansions for any number n
- ✦ Use binomial expansions to find numerical approximations
- ✦ Apply binomial expansions to rational functions

Before you start

You should know how to ...	Check in
1 Calculate binomial coefficients.	**1** a) Find the value of $\binom{7}{3}$. b) Find the coefficient of x^3 in the expansion of $(2 - 5x)^7$.
2 Find the binomial expansion of $(a + bx)^n$ for a positive integer n.	**2** a) Obtain the full expansion of $(1 + x)^5$. b) Obtain the full expansion of $(2 - 5x)^4$.
3 Express a rational function in terms of partial fractions.	**3** a) Express $\dfrac{4x - 7}{(x + 2)(2x - 1)}$ in the form $\dfrac{A}{x + 2} + \dfrac{B}{2x - 1}$, where A and B are integers. b) Express $\dfrac{2 - 7x}{(1 + x)(1 - 2x)^2}$ in the form $\dfrac{A}{1 + x} + \dfrac{B}{1 - 2x} + \dfrac{C}{(1 - 2x)^2}$

8.1 Binomial expansion when n is not a positive integer

In C2 you learnt that

$$(1 + x)^n = 1 + nx + \frac{n(n - 1)x^2}{2!} + \frac{n(n - 1)(n - 2)x^3}{3!} + \ldots + x^n$$

This is called the binomial expansion of $(1 + x)^n$.

with the expansion terminating at x^n when n is a positive integer.

For example,

$$(1 + x)^5 = 1 + 5x + \frac{5 \times 4x^2}{2!} + \frac{5 \times 4 \times 3x^3}{3!} + \frac{5 \times 4 \times 3 \times 2x^4}{4!} + x^5$$

$$= 1 + 5x + 10x^2 + 10x^3 + 5x^4 + x^5$$

However, if n is not a positive integer then the expansion does not terminate and is only valid for $-1 < x < 1$, that is, $|x| < 1$. In this case, the expansion gives only an approximation to $(1 + x)^n$.

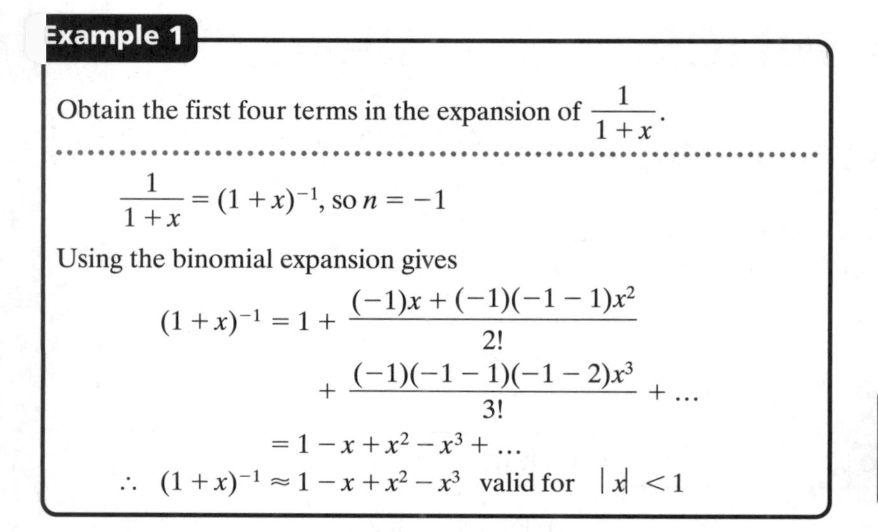

Example 1

Obtain the first four terms in the expansion of $\dfrac{1}{1+x}$.

$$\frac{1}{1+x} = (1+x)^{-1}, \text{ so } n = -1$$

Using the binomial expansion gives

$$(1+x)^{-1} = 1 + \frac{(-1)x + (-1)(-1-1)x^2}{2!} + \frac{(-1)(-1-1)(-1-2)x^3}{3!} + \dots$$

$$= 1 - x + x^2 - x^3 + \dots$$

$$\therefore \quad (1+x)^{-1} \approx 1 - x + x^2 - x^3 \quad \text{valid for } |x| < 1$$

Remember:
$n! = n(n-1)(n-2) \dots 3 \times 2 \times 1$
(called n factorial)

Usually the terms after the x^3 term are so small that they can be ignored.

Example 2

Given that x is so small that x^4 and higher powers of x can be neglected, show that $\sqrt{1-4x} \approx 1 - 2x - 2x^2 - 4x^3$, giving the range of values of x for which the expansion is valid.

C4

$$\sqrt{1-4x} = (1-4x)^{\frac{1}{2}}$$

Using the binomial expansion:

$n = \frac{1}{2}$
'x' $= -4x$

$$(1-4x)^{\frac{1}{2}} = 1 + \tfrac{1}{2}(-4x) + \frac{\frac{1}{2}\left(-\frac{1}{2}\right)}{2!}(-4x)^2 + \frac{\frac{1}{2}\left(-\frac{1}{2}\right)\left(-\frac{3}{2}\right)}{3!}(-4x)^3 + \dots$$

$$= 1 - 2x - 2x^2 - 4x^3 + \dots$$

The expansion is valid for $|-4x| < 1$, i.e. $|x| < \frac{1}{4}$.

8.2 The expansion of $(a + x)^n$

Sometimes a bracket does not begin with a term '1'. In this case you need to factorise the expression before attempting to expand it.

Example 3

Obtain the first four terms in the expansion of $\dfrac{1}{3-x}$, giving the range of values of x for which the expansion is valid.

$$\frac{1}{3-x} = (3-x)^{-1}$$

$$= \left[3\left(1 - \frac{x}{3}\right)\right]^{-1}$$

$$= 3^{-1}\left[1 - \frac{x}{3}\right]^{-1}$$
$$= \frac{1}{3}\left[1 - \frac{x}{3}\right]^{-1}$$

The expression is now in the form $k(1 + x)^n$, where 'x' $= -\frac{x}{3}$ and $n = -1$.

Using the binomial expansion gives

$$\frac{1}{3 - x} = \frac{1}{3}\left[1 + (-1)\left(-\frac{x}{3}\right) + \frac{(-1)(-2)}{2!}\left(-\frac{x}{3}\right)^2 + \frac{(-1)(-2)(-3)}{3!}\left(-\frac{x}{3}\right)^3 + \ldots\right]$$
$$= \tfrac{1}{3}\left[1 + \tfrac{1}{3}x + \tfrac{1}{9}x^2 + \tfrac{1}{27}x^3 + \ldots\right]$$
$$= \tfrac{1}{3} + \tfrac{1}{9}x + \tfrac{1}{27}x^2 + \tfrac{1}{81}x^3 + \ldots$$

So $\frac{1}{3 - x} = \frac{1}{3} + \frac{1}{9}x + \frac{1}{27}x^2 + \frac{1}{81}x^3 + \ldots$

The expansion is valid for $\left|\frac{x}{3}\right| < 1$, that is, $|x| < 3$.

Example 4

Find a quadratic approximation for $(2 + 3x)^{-2}$, giving your answer in the form $a + bx + cx^2$. State the range of values of x for which the expansion is valid.

$$(2 + 3x)^{-2} = \left[2\left(1 + \frac{3x}{2}\right)\right]^{-2}$$
$$= 2^{-2}\left[1 + \frac{3x}{2}\right]^{-2}$$
$$= \frac{1}{4}\left[1 + \frac{3x}{2}\right]^{-2}$$

First factorise the expression.

Using the binomial expansion gives

$$(2 + 3x)^{-2} = \frac{1}{4}\left[1 - 2\left(\frac{3x}{2}\right) + \frac{(-2)(-3)}{2!}\left(\frac{3x}{2}\right)^2 + \ldots\right]$$
$$= \frac{1}{4}\left[1 - 3x + \frac{27}{4}x^2 + \ldots\right]$$
$$= \tfrac{1}{4} - \tfrac{3}{4}x + \tfrac{27}{16}x^2 + \ldots$$

The question asks for the answer in the form $a + bx + cx^2$, so you only need to expand as far as the x^2 term.

So $(2 + 3x)^{-2} \approx \frac{1}{4} - \frac{3}{4}x + \frac{27}{16}x^2 + \ldots$

The expansion is valid for $\left|\frac{3x}{2}\right| < 1$, i.e. $|x| < \frac{2}{3}$.

It is also possible to use expansions to find approximate values for calculations. This is illustrated in the next example.

C4

Example 5

Given that x is so small that x^3 and higher powers of x can be neglected, show that

$$\frac{1}{\sqrt{1+x}} = 1 - \frac{x}{2} + \frac{3x^2}{8}$$

By letting $x = \frac{1}{4}$, find a rational approximation for $\sqrt{5}$.

$$\frac{1}{\sqrt{1+x}} = (1+x)^{-\frac{1}{2}}$$

Using the binomial expansion gives

$$(1+x)^{-\frac{1}{2}} = 1 + \left(-\frac{1}{2}\right)x + \frac{\left(-\frac{1}{2}\right)\left(-\frac{1}{2}-1\right)x^2}{2!} + \ldots$$

$$\approx 1 - \frac{x}{2} + \frac{3x^2}{8}$$

as required. This expansion is valid for $|x| < 1$.

Let $x = \frac{1}{4}$, then

$$\frac{1}{\sqrt{1+\frac{1}{4}}} \approx 1 - \frac{1}{2}\left(\frac{1}{4}\right) + \frac{3}{8}\left(\frac{1}{4}\right)^2$$

$$\therefore \quad \sqrt{\frac{4}{5}} \approx \frac{115}{128}$$

$$\therefore \quad \frac{2}{\sqrt{5}} \approx \frac{115}{128}$$

$$\therefore \quad \sqrt{5} \approx \frac{2(128)}{115} = \frac{256}{115}$$

A rational approximation for $\sqrt{5}$ is $\frac{256}{115}$. If you use your calculator to convert this fraction to a decimal, you get 2.226…. If you use the calculator to calculate $\sqrt{5}$ you get $\approx$2.236… so the approximation is only accurate to one decimal place.

The final example shows how you can combine two binomial expansions to solve a more advanced problem.

Example 6

Obtain the expansion of $\dfrac{(1+x)^3}{(2-x)}$ up to and including the term in x^3. Hence evaluate $(1.2)^3$ giving your answer to two decimal places.

$$\frac{(1+x)^3}{(2-x)} = (1+x)^3(2-x)^{-1}$$

Expanding $(1+x)^3$ gives

$$(1+x)^3 = 1 + \binom{3}{1}x + \binom{3}{2}x^2 + x^3$$

$$\therefore \quad (1+x)^3 = 1 + 3x + 3x^2 + x^3, \text{ valid for all } x$$

You need the binomial expansion of $(2 - x)^{-1}$ so factorise out the 2 first.

$$(2-x)^{-1} = 2^{-1}\left[1 + \left(-\frac{x}{2}\right)\right]^{-1}$$

$$= \frac{1}{2}\left[1 + (-1)\left(-\frac{x}{2}\right) + \frac{(-1)(-1-1)}{2!}\left(-\frac{x}{2}\right)^2 + \frac{(-1)(-1-1)(-1-2)}{3!}\left(-\frac{x}{2}\right)^3 + \ldots\right]$$

$$= \frac{1}{2}\left(1 + \frac{x}{2} + \frac{x^2}{4} + \frac{x^3}{8} + \ldots\right)$$

$\therefore\quad (2-x)^{-1} = \dfrac{1}{2} + \dfrac{x}{4} + \dfrac{x^2}{8} + \dfrac{x^3}{16} + \ldots$ valid for $\left|\dfrac{x}{2}\right| < 1$, that is, $|x| < 2$

Therefore,

$$(1+x)^3(2-x)^{-1} = (1 + 3x + 3x^2 + x^3)\left(\frac{1}{2} + \frac{x}{4} + \frac{x^2}{8} + \frac{x^3}{16} + \ldots\right)$$

$$= \frac{1}{2} + \frac{7x}{4} + \frac{19x^2}{8} + \frac{27x^3}{16} + \ldots$$

$$\therefore\quad \frac{(1+x)^3}{(2-x)} \approx \frac{1}{2} + \frac{7x}{4} + \frac{19x^2}{8} + \frac{27x^3}{16} \quad \text{valid for } |x| < 2$$

Let $x = 0.2$ (which lies in the valid range), then

$$\frac{(1+0.2)^3}{(2-0.2)} \approx \frac{1}{2} + \frac{7(0.2)}{4} + \frac{19(0.2)^2}{8} + \frac{27(0.2)^3}{16}$$

$$\therefore\quad \frac{(1.2)^3}{1.8} \approx 0.9585$$

$$\therefore\quad (1.2)^3 \approx 1.8 \times 0.9585 = 1.73 \quad \text{(2 decimal places)}$$

You could have simply used a calculator to work out 1.2^3. This problem just illustrates the validity of the binomial expansion. You can check that $1.2^3 = 1.728$.

C4

Exercise 8A

Remember: $\binom{n}{r}$ is shorthand for $\dfrac{n!}{r!(n-r)!}$ and is often found on calculators as ${}^nC_r$ or ${}_nC_r$.

1 Obtain the first four terms in the binomial expansion of each of these, and state the range of values of x for which each is valid.

a) $(1+x)^{-2}$ b) $(1+x)^{\frac{1}{2}}$ c) $(1+2x)^{-3}$

d) $(1-3x)^{-2}$ e) $(1-3x)^{-\frac{1}{3}}$ f) $(1+x^2)^{-3}$

g) $\dfrac{1}{1-3x}$ h) $\sqrt{1-6x}$

2 Obtain the expansion of each of these up to and including the term in x^3, giving the range of values of x for which each is valid.

a) $(2+x)^{-1}$ b) $(4+x)^{\frac{1}{2}}$ c) $(9-4x)^{-\frac{1}{2}}$

d) $(8+3x)^{\frac{2}{3}}$ e) $\dfrac{1}{(2-x)^3}$ f) $\sqrt{4-x}$

g) $\dfrac{1}{6+x}$ h) $\dfrac{1}{(3-2x)^2}$

3 Given x is so small that x^3 and higher powers of x may be neglected, obtain a quadratic approximation to each expression.

a) $\dfrac{1+x}{1-x}$ b) $\dfrac{1+3x}{2+x}$ c) $\dfrac{x}{4-x}$

d) $\dfrac{2+5x}{(1-3x)^2}$ e) $(4-x^2)\sqrt{4-x}$ f) $\dfrac{7-3x}{\sqrt{1-x}}$

g) $(4-3x)^2\sqrt{1-6x}$ h) $\dfrac{(2-x^2)^2}{(4+x)^3}$

4 a) Expand each of these in ascending powers of x up to and including the term in x^2.

i) $(1+2x)^4$ ii) $\sqrt{1-x}$

b) Hence obtain a quadratic approximation to $(1+2x)^4\sqrt{1-x}$ which is valid for small values of x.

5 a) Given x is small obtain a quadratic approximation to each of these.

i) $\sqrt{1+2x}$ ii) $\dfrac{1}{(1-x)^4}$

b) Hence obtain the first three terms in the series expansion of

$$\frac{\sqrt{1+2x}}{(1-x)^4}$$

stating the range of values of x for which the series is valid.

C4

6 a) Expand each of these expressions in ascending powers of x up to and including the term in x^3.

i) $\dfrac{1}{1-x}$ ii) $\dfrac{1}{1-2x}$

b) Hence obtain a cubic approximation to

$$\frac{1}{(1-x)(1-2x)}$$

which is valid for small values of x.

7 a) Given x is small, find a cubic approximation to each of these.

i) $\sqrt[3]{1-3x}$ ii) $\dfrac{1}{1-4x}$

b) Hence obtain the first four terms in the series expansion of

$$\frac{\sqrt[3]{1-3x}}{1-4x}$$

stating the range of values of x for which the series is valid.

8 Expand $(1+x)^{\frac{1}{2}}$ in ascending powers of x up to and including the term in x^3. Hence evaluate $\sqrt{1.01}$ giving your answer to eight decimal places.

9 Expand $(1-2x)^{\frac{1}{4}}$ in ascending powers of x up to and including the term in x^2. Hence evaluate $\sqrt[4]{0.998}$ giving your answer to six decimal places.

10 Expand $(4 - 3x)^{-\frac{1}{2}}$ in ascending powers of x up to and including the term in x^2. Hence evaluate $\dfrac{1}{\sqrt{3.97}}$ giving your answer to four decimal places.

11 a) Show that $\left(1 + \dfrac{x}{25}\right)^{\frac{1}{2}} = 1 + \dfrac{x}{50} - \dfrac{x^2}{5000} + \dfrac{x^3}{250\,000} - \dots$

b) By substituting $x = 1$ into your answer to part a), deduce that $\sqrt{26} \approx 5.099\,02$.

12 a) Write down the first four terms in the binomial expansion of $\sqrt{1 + \dfrac{x}{100}}$.

b) By substituting a suitable value for x into your answer to part a), deduce that $\sqrt{102} \approx 10.099\,505$.

13 a) Show that $\left(1 + \dfrac{x}{125}\right)^{\frac{1}{3}} = 1 + \dfrac{x}{375} - \dfrac{x^2}{140\,625} + \dots$

b) By substituting a suitable value of x into your answer to part a), deduce that $\sqrt[3]{126} \approx 5.013$.

14 a) Write down the first three terms in the binomial expansion of $\sqrt[4]{1 + \dfrac{x}{16}}$.

b) Deduce that $\sqrt[4]{15} \approx 1.968$.

8.3 Applications of partial fractions to series expansions

You are now in a position to combine the techniques of partial fractions, which you learnt in Chapter 7, with the current work on binomial expansions. The next example shows you how this is done.

Example 7

Express

$$\frac{2x + 5}{(x + 1)(x + 2)}$$

in partial fractions and hence find the first four terms in the series expansion of

$$\frac{2x + 5}{(x + 1)(x + 2)}$$

State the values of x for which the expansion is valid.

Assume that

$$\frac{2x + 5}{(x + 1)(x + 2)} = \frac{A}{x + 1} + \frac{B}{x + 2}$$

Multiplying throughout by $(x + 1)(x + 2)$ gives

$$2x + 5 = A(x + 2) + B(x + 1)$$

Let $x = -1$: $2(-1) + 5 = A(-1 + 2)$
$\therefore\ A = 3$

Let $x = -2$: $2(-2) + 5 = B(-2 + 1)$
$\therefore\ B = -1$

Therefore,

$$\frac{2x+5}{(x+1)(x+2)} = \frac{3}{x+1} - \frac{1}{x+2}$$

To obtain the series expansion, you need the series expansion of both $\frac{3}{x+1}$ and $\frac{1}{x+2}$.

Now $\frac{3}{x+1} = 3(1+x)^{-1}$ and using the binomial expansion for $(1+x)^{-1}$ gives

$$\begin{aligned}3(1+x)^{-1} &= 3\left[1 + (-1)x + \frac{(-1)(-1-1)}{2!}x^2 + \frac{(-1)(-1-1)(-1-2)}{3!}x^3 + \ldots\right]\\ &= 3(1 - x + x^2 - x^3 + \ldots)\\ &= 3 - 3x + 3x^2 - 3x^3 + \ldots\end{aligned}$$

This expansion is valid for $-1 < x < 1$. [1]

Also, $\frac{1}{x+2} = (2+x)^{-1}$ and using the binomial expansion for $(2+x)^{-1}$ gives

$$\begin{aligned}(2+x)^{-1} &= \left[2\left(1+\frac{x}{2}\right)\right]^{-1}\\ &= 2^{-1}\left(1+\frac{x}{2}\right)^{-1}\\ &= \frac{1}{2}\left[1 + (-1)\left(\frac{x}{2}\right) + \frac{(-1)(-1-1)}{2!}\left(\frac{x}{2}\right)^2 + \frac{(-1)(-1-1)(-1-2)}{3!}\left(\frac{x}{2}\right)^3 + \ldots\right]\\ &= \frac{1}{2}\left(1 - \frac{x}{2} + \frac{x^2}{4} - \frac{x^3}{8} + \ldots\right)\\ &= \frac{1}{2} - \frac{x}{4} + \frac{x^2}{8} - \frac{x^3}{16} + \ldots\end{aligned}$$

This expansion is valid for $-1 < \frac{x}{2} < 1$, in other words for $-2 < x < 2$. [2]

Therefore,

$$\begin{aligned}\frac{3}{x+1} - \frac{1}{x+2} &= (3 - 3x + 3x^2 - 3x^3 + \ldots) - \left(\frac{1}{2} - \frac{x}{4} + \frac{x^2}{8} - \frac{x^3}{16} + \ldots\right)\\ &= \frac{5}{2} - \frac{11}{4}x + \frac{23}{8}x^2 - \frac{47}{16}x^3 \ldots\end{aligned}$$

This expansion is valid for those values of x which satisfy [1] and [2], namely $-1 < x < 1$.

Example 8

a) Express $\dfrac{2x^2 + 24}{(2x + 1)(x - 3)^2}$ in partial fractions.

b) Hence find the first three terms in the series expansion of

$$\frac{2x^2 + 24}{(2x + 1)(x - 3)^2}$$

stating the values of x for which the expansion is valid.

---

a) $\dfrac{2x^2 + 24}{(2x + 1)(x - 3)^2} = \dfrac{A}{2x + 1} + \dfrac{B}{x - 3} + \dfrac{C}{(x - 3)^2}$

Multiplying each side by $(2x + 1)(x - 3)^2$ gives

$$2x^2 + 24 = A(x - 3)^2 + B(2x + 1)(x - 3) + C(2x + 1)$$

Let $x = 3$: $2(3)^2 + 24 = C(7)$

$$18 + 24 = 7C$$
$$42 = 7C$$
$$\therefore \quad C = 6$$

Let $x = -\frac{1}{2}$: $2\left(-\frac{1}{2}\right)^2 + 24 = A\left(-\frac{7}{2}\right)^2$

$$\therefore \quad \tfrac{1}{2} + 24 = \tfrac{49}{4}A$$
$$\therefore \quad \tfrac{49}{2} = \tfrac{49}{4}A$$
$$\therefore \quad A = 2$$

Let $x = 0$: $24 = 9A - 3B + C$

$$24 = 18 - 3B + 6$$
$$\therefore \quad B = 0$$

Therefore $\dfrac{2x^2 + 24}{(2x + 1)(x - 3)^2} = \dfrac{2}{2x + 1} + \dfrac{6}{(x - 3)^2}$

b) $\dfrac{2}{2x + 1} = 2[2x + 1]^{-1}$

$$= 2[1 + 2x]^{-1}$$

Using the binomial expansion gives

$$2[1 + 2x]^{-1} = 2\left[1 + (-1)(2x) + \frac{(-1)(-2)}{2!}(2x)^2 + \ldots\right]$$
$$= 2[1 - 2x + 4x^2 + \ldots]$$
$$= 2 - 4x + 8x^2 + \ldots$$

This expansion is valid for $|2x| < 1$, that is, $|x| < \frac{1}{2}$. [1]

C4

Also $\dfrac{6}{(x-3)^2} = \dfrac{6}{(3-x)^2}$

$$= 6[3-x]^{-2}$$

$$= 6\left[3\left(1-\frac{x}{3}\right)\right]^{-2}$$

$$= 6 \times (3)^{-2} \times \left[1-\frac{x}{3}\right]^{-2}$$

$$= \frac{2}{3}\left[1-\frac{x}{3}\right]^{-2}$$

Using the binomial expansion gives

$$\frac{2}{3}\left[1-\frac{x}{3}\right]^{-2} = \frac{2}{3}\left[1-2\left(-\frac{x}{3}\right)+\frac{(-2)(-3)}{2!}\left(-\frac{x}{3}\right)^2+\ldots\right]$$

$$= \frac{2}{3}\left[1+\frac{2}{3}x+\frac{1}{3}x^2+\ldots\right]$$

$$= \frac{2}{3}+\frac{4}{9}x+\frac{2}{9}x^2+\ldots$$

This expansion is valid for $\left|\dfrac{x}{3}\right| < 1$, that is, $|x| < 3$. [2]

C4

Therefore $\dfrac{2x^2+24}{(2x+1)(x-3)^2} = (2-4x+8x^2+\ldots)$

$$+\left(\tfrac{2}{3}-\tfrac{4}{9}x+\tfrac{2}{9}x^2+\ldots\right)$$

$$= \tfrac{8}{3}-\tfrac{40}{9}x+\tfrac{74}{9}x^2+\ldots$$

The expansion is valid for those values of x which satisfy [1] and [2], namely $|x| < \frac{1}{2}$.

Exercise 8B

1 Express each of these in partial fractions. Hence obtain the series expansion for each expression, giving all terms up to and including the term in x^3, and stating the values of x for which the expansion is valid.

a) $\dfrac{2-x}{(1+x)(1-2x)}$

b) $\dfrac{1+7x}{(1+3x)(1-x)}$

c) $\dfrac{1+x}{(1+3x)(1+5x)}$

d) $\dfrac{7+x}{(1+x)(2-x)}$

e) $\dfrac{1}{(2+x)(3+x)}$

f) $\dfrac{4-x+3x^2}{(1+x^2)(1-x)}$

2 Express each of these in partial fractions. Hence obtain the series expansion for each expression, giving all terms up to and including the term in x^2, and stating the values of x for which the expansion is valid.

a) $\dfrac{1 - x - x^2}{(1 - 2x)(1 - x)^2}$ b) $\dfrac{4x^2 + 5x + 5}{(1 - 3x)(1 + x)^2}$

c) $\dfrac{2x^2 + 2x + 1}{(1 + x)(1 + 2x)^2}$ d) $\dfrac{4 - 23x}{(2 + x)(1 - 2x)^2}$

3 a) Given $\dfrac{6x^3 - 30x^2 + 23x + 5}{(1 + 3x)(1 - x)(2 - x)} = A + \dfrac{B}{1 + 3x} + \dfrac{C}{1 - x} + \dfrac{D}{2 - x}$, find the values of the constants A, B, C and D.

b) Hence obtain the series expansion for $\dfrac{6x^3 - 30x^2 + 23x + 5}{(1 + 3x)(1 - x)(2 - x)}$, giving all terms up to and including the term in x^2, and stating the values of x for which the expansion is valid.

4 a) Given $\dfrac{24x^3 - 16x - 3}{(3 + 2x)(1 + 2x)^2} = A + \dfrac{B}{3 + 2x} + \dfrac{C}{1 + 2x} + \dfrac{D}{(1 + 2x)^2}$, find the values of the constants A, B, C and D.

b) Hence obtain the series expansion for $\dfrac{24x^3 - 16x - 3}{(3 + 2x)(1 + 2x)^2}$, giving all terms up to and including the term in x^2, and stating the values of x for which the expansion is valid.

Summary

You should know how to ...	Check out
1 Obtain a binomial expansion with fractional or negative n.	**1** a) Expand $\sqrt{1 - x}$ in a binomial series up to and including the term in x^3. b) Obtain the binomial expansion of $\dfrac{1}{(2x + 3)^3}$ up to and including the term in x^3. State the range of values of x for which this expansion is valid.
2 Use a binomial expansion to make a numerical approximation.	**2** a) Obtain the binomial expansion of $(1 - x)^{\frac{1}{3}}$ up to and including the term in x^2. b) By putting $x = \frac{1}{9}$, obtain a rational approximation to $\sqrt[3]{9}$.
3 Use partial fractions to expand a rational function.	**3** Obtain the binomial expansion of $\dfrac{5 - 4x}{(1 + x)(2 - x)}$ up to and including the term in x^2.

Revision exercise 8

1 a) Obtain the binomial expansion of $(1 + x)^{\frac{1}{3}}$ as far as the term in x^2.

b) Hence, or otherwise, find the series expansion of $(8 + 4x)^{\frac{1}{3}}$ as far as the term in x^2. *(AQA, 2004)*

2 a) Obtain the binomial expansion in ascending powers of x up to and including the term in x^3 of the following giving each term in its simplest form.

i) $(1 + x)^{-1}$ ii) $(1 + 4x)^{\frac{1}{2}}$

b) Hence show that, for small values of x, $2(1 + 4x)^{\frac{1}{2}} + \dfrac{4}{1 + x} \approx 6 + kx^3$ where k is a constant to be found. *(AQA, 2004)*

3 a) Express $\dfrac{8 + 3x}{(1 + 3x)(2 - x)}$ in the form $\dfrac{A}{1 + 3x} + \dfrac{B}{2 - x}$.

b) Obtain the first three terms in the expansion of $\dfrac{1}{1 + 3x}$ in ascending powers of x.

c) Show that the first three terms in the expansion of $\dfrac{1}{2 - x}$ in ascending powers of x are $\dfrac{1}{2} + \dfrac{x}{4} + \dfrac{x^2}{8}$.

d) Hence, or otherwise, obtain the first three terms in the expansion of $\dfrac{8 + 3x}{(1 + 3x)(2 - x)}$ in ascending powers of x.

e) State the range of values of x for which the expansion in part d) is valid. *(AQA, 2004)*

4 a) Obtain the binomial expansion of $(1 + 2x)^{-2}$ in ascending powers of x up to and including the term in x^3.

b) State the range of values of x for which the full expansion is valid. *(AQA, 2003)*

5 a) Obtain the binomial expansion of $(1 + x)^{\frac{1}{2}}$ as far as the term in x^2.

b) i) Hence, or otherwise, find the series expansion of $(4 + 2x)^{\frac{1}{2}}$ as far as the term in x^2.

ii) Find the range of values of x for which the expansion is valid. *(AQA, 2003)*

6 a) Express $\dfrac{4 - x}{(1 - x)(2 + x)}$ in the form $\dfrac{A}{1 - x} + \dfrac{B}{2 + x}$.

b) i) Show the first three terms in the expansion of $\dfrac{1}{2 + x}$ in ascending powers of x are $\dfrac{1}{2} - \dfrac{x}{4} + \dfrac{x^2}{8}$.

ii) Obtain also the first three terms in the expansion of $\dfrac{1}{1 - x}$ in ascending powers of x.

c) Hence, or otherwise, obtain the first three terms in the expansion of $\dfrac{4 - x}{(1 - x)(2 + x)}$ in ascending powers of x. *(AQA, 2002)*

7 $f(x) = \dfrac{2x}{(1+2x)^3}, x \neq -\frac{1}{2}$

Obtain the first three terms of the binomial expansion of $f(x)$ in ascending powers of x. *(AQA, 2002)*

8 The function f is given by $f(x) = \dfrac{9}{(1+2x)(4-x)}$

a) Express $f(x)$ in partial fractions.

b) i) Show that the first three terms in the expansion of $\dfrac{1}{4-x}$ in ascending powers of x are $\dfrac{1}{4} + \dfrac{x}{16} + \dfrac{x^2}{64}$.

ii) Obtain a similar expansion for $\dfrac{1}{1+2x}$.

iii) Hence, or otherwise, obtain the first three terms in the expansion $f(x)$ in ascending powers of x.

iv) Find the range of values of x for which the expansion of $f(x)$ in ascending powers of x is valid. *(AQA, 2001)*

9 a) Determine the binomial expansion of $\left(1 - \dfrac{x}{10}\right)^{-3}$ in ascending powers of x, up to and including the term in x^3.

b) Show that the coefficient of x^n in this expansion is $K(n+1)(n+2) \times \dfrac{1}{10^n}$ for a rational number K whose value is to be determined.

c) Determine the value of $\left(\dfrac{1}{0.99}\right)^3$ correct to fourteen decimal places. *(AQA, 2001)*

9 Trigonometry and compound angles

This chapter will show you how to

- ✦ Understand and use compound angle identities
- ✦ Understand and use multiple angle identities

Before you start

You should know how to ...	Check in
1 Solve trigonometric equations using degrees and radians.	**1** a) Find an acute angle in degrees for which $2\cos(x + 10°) = 1$. b) Find an acute angle in radians for which $\tan 3x = 1$.
2 Sketch and use trigonometric graphs.	**2** a) Sketch the graph of $y = \sin 2x$ for $-180° \leqslant x \leqslant 360°$. b) Illustrate in your graph all solutions to the equation $\sin 2x = 0.6$.
3 Integrate trigonometric functions.	**3** a) Find $\int \sin 2x \, dx$. b) Find the value of $\int_0^{\frac{\pi}{4}} \sin 2x \, dx$.

C4

9.1 Compound angles

It is quite easy to show that

$$\sin(A + B) \neq \sin A + \sin B$$
$$\cos(A + B) \neq \cos A + \cos B$$
$$\tan(A + B) \neq \tan A + \tan B$$

For example,

$$\sin(40° + 50°) = \sin 90° = 1$$

However,

$$\sin 40° + \sin 50° = 0.643 + 0.766 = 1.409 \quad (3 \text{ d.p.})$$

Fortunately, there are identities for $\sin(A + B)$, $\cos(A + B)$ and $\tan(A + B)$.

This is a **demonstration** of a particular case, not a **proof**.

For any angles A and B,

$$\sin(A+B) = \sin A \cos B + \cos A \sin B$$

$$\cos(A+B) = \cos A \cos B - \sin A \sin B$$

You could use the identity symbol $\equiv$ instead of $=$ for these formulae.

Proof

This can be proved for acute angles A and B, where $A + B < 90°$.

To prove the results for **any** angles, you would need to use vectors and matrices.

In the diagram,

$$\sin(A+B) = \frac{PQ}{OP}$$

$$= \frac{PT + TQ}{OP}$$

$$= \frac{PT + RS}{OP}$$

$$= \frac{PT}{OP} + \frac{RS}{OP}$$

$$= \frac{PT}{PR} \times \frac{PR}{OP} + \frac{RS}{OR} \times \frac{OR}{OP}$$

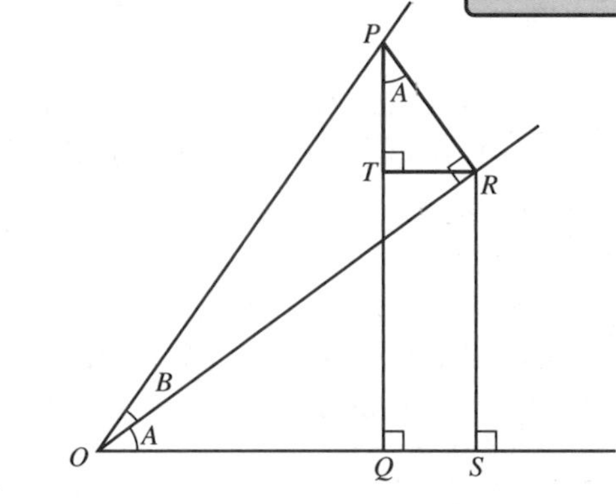

Therefore,

$$\sin(A+B) = \cos A \sin B + \sin A \cos B$$

as required.

From the same diagram,

$$\cos(A+B) = \frac{OQ}{OP}$$

$$= \frac{OS - TR}{OP}$$

$$= \frac{OS}{OP} - \frac{TR}{OP}$$

$$= \frac{OS}{OR} \times \frac{OR}{OP} - \frac{TR}{PR} \times \frac{PR}{OP}$$

These proofs will not be set in the examination.

Therefore,

$$\cos(A+B) = \cos A \cos B - \sin A \sin B$$

as required.

For any angles A and B,

$$\sin(A-B) = \sin A \cos B - \cos A \sin B$$

$$\cos(A-B) = \cos A \cos B + \sin A \sin B$$

Proof

Again, this can be proved for acute angles A and B.

You know that

$$\sin(A+B) = \sin A\cos B + \cos A\sin B \quad \text{and}$$
$$\cos(A+B) = \cos A\cos B - \sin A\sin B$$

Replace B with $-B$, noting that

- $\cos(-B) = \cos B$
- $\sin(-B) = -\sin B$

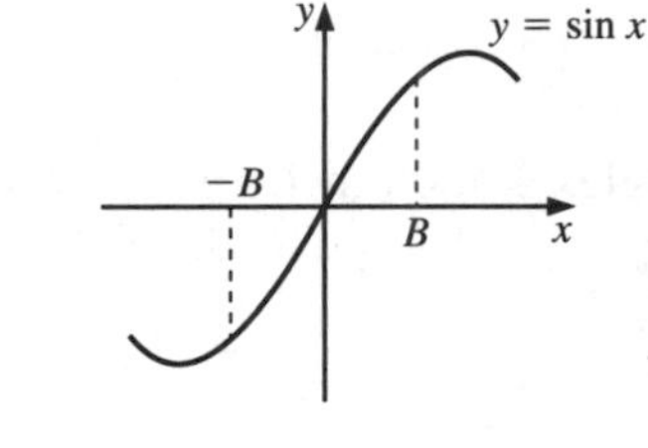

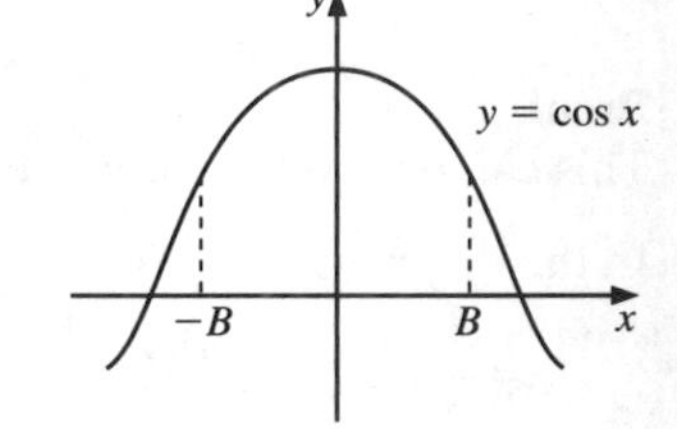

So:

$$\sin(A-B) \equiv \sin(A+(-B)) = \sin A\cos(-B) + \cos A\sin(-B)$$
$$= \sin A\cos B - \cos A\sin B$$

as required.

Also:

$$\cos(A-B) \equiv \cos(A+(-B)) = \cos A\cos(-B) - \sin A\sin(-B)$$
$$= \cos A\cos B + \sin A\sin B$$

as required.

C4

For any angles A and B,

$$\tan(A \pm B) = \frac{\tan A \pm \tan B}{1 \mp \tan A\tan B}$$

Note that $\mp$ in the denominator means that if you want $\tan(A+B)$ you need $\tan A + \tan B$ in the numerator, $1 - \tan A\tan B$ in the denominator, and vice versa.

Proof

You know that $\tan\theta = \dfrac{\sin\theta}{\cos\theta}$, so

$$\text{LHS} \equiv \tan(A \pm B) = \frac{\sin(A \pm B)}{\cos(A \pm B)}$$
$$= \frac{\sin A\cos B \pm \cos A\sin B}{\cos A\cos B \mp \sin A\sin B}$$

You want $1 \mp \tan A\tan B$ in the denominator, so divide the numerator and the denominator by $\cos A\cos B$.

$$\tan(A \pm B) = \frac{\left(\dfrac{\sin A\cos B}{\cos A\cos B} \pm \dfrac{\cos A\sin B}{\cos A\cos B}\right)}{\left(\dfrac{\cos A\cos B}{\cos A\cos B} \mp \dfrac{\sin A\sin B}{\cos A\cos B}\right)}$$

$$= \frac{\left(\dfrac{\sin A}{\cos A} \pm \dfrac{\sin B}{\cos B}\right)}{\left(1 \mp \dfrac{\sin A}{\cos A}\cdot\dfrac{\sin B}{\cos B}\right)}$$

$$= \frac{\tan A \pm \tan B}{1 \mp \tan A\tan B}$$

as required.

Example 1

Given that A and B are acute angles and that $\sin A = \dfrac{1}{\sqrt{3}}$ and that $\cos B = \dfrac{3}{5}$, evaluate

a) $\sin(A + B)$

b) $\tan(A - B)$

c) $\operatorname{cosec}(A + B)$.

When $\sin A = \dfrac{1}{\sqrt{3}}$ (A acute), you can see from the right-angled triangle that

$$x^2 = (\sqrt{3})^2 - 1^2 \quad \therefore \quad x = \sqrt{2}$$

Therefore,

$$\cos A = \frac{\sqrt{2}}{\sqrt{3}} = \sqrt{\frac{2}{3}} \quad \text{and} \quad \tan A = \frac{1}{\sqrt{2}}$$

$\sqrt{3}$ 1 A x

When $\cos B = \frac{3}{5}$ (B acute), you can see from the right-angled triangle that $y = 4$. Therefore,

$$\sin B = \frac{4}{5} \quad \text{and} \quad \tan B = \frac{4}{3}$$

5 y B 3

a) $$\sin(A + B) = \sin A \cos B + \cos A \sin B \qquad [1]$$

So: $$\sin(A + B) = \frac{1}{\sqrt{3}} \times \frac{3}{5} + \frac{\sqrt{2}}{\sqrt{3}} \times \frac{4}{5} = \frac{3 + 4\sqrt{2}}{5\sqrt{3}}$$

b) $$\tan(A - B) = \frac{\tan A - \tan B}{1 + \tan A \tan B} = \frac{\frac{1}{\sqrt{2}} - \frac{4}{3}}{1 + \left(\frac{1}{\sqrt{2}}\right)\left(\frac{4}{3}\right)} = \frac{3 - 4\sqrt{2}}{3\sqrt{2} + 4}$$

$$\therefore \quad \tan(A - B) = \frac{3 - 4\sqrt{2}}{4 + 3\sqrt{2}}$$

c) $$\operatorname{cosec}(A + B) = \frac{1}{\sin(A + B)} = \frac{1}{\left(\frac{3 + 4\sqrt{2}}{5\sqrt{3}}\right)}$$

$$\therefore \quad \operatorname{cosec}(A + B) = \frac{5\sqrt{3}}{3 + 4\sqrt{2}}$$

$\sin A = \dfrac{1}{\sqrt{3}}$ $\quad$ $\sin B = \dfrac{4}{5}$

$\cos A = \sqrt{\dfrac{2}{3}}$ $\quad$ $\cos B = \dfrac{3}{5}$

$\tan A = \dfrac{1}{\sqrt{2}}$ $\quad$ $\tan B = \dfrac{4}{3}$

C4

Questions on trigonometric ratios will not be set requiring surd form. Examples 2, 3 and 4 are included to provide interest and challenge.

Example 2

Given that $\tan(\theta - 30°) = 2$, prove that $\tan\theta = -8 - 5\sqrt{3}$.

$$\tan(\theta - 30°) = \frac{\tan\theta - \tan 30°}{1 + \tan\theta\tan 30°}$$

$$= \frac{\tan\theta - \frac{1}{\sqrt{3}}}{1 + \tan\theta\left(\frac{1}{\sqrt{3}}\right)}$$

$$= \frac{\sqrt{3}\tan\theta - 1}{\sqrt{3} + \tan\theta}$$

Since $\tan(\theta - 30°) = 2$,

$$\frac{\sqrt{3}\tan\theta - 1}{\sqrt{3} + \tan\theta} = 2$$

$$\therefore \quad \sqrt{3}\tan\theta - 1 = 2(\sqrt{3} + \tan\theta)$$

$$\tan\theta(\sqrt{3} - 2) = 2\sqrt{3} + 1$$

$$\therefore \quad \tan\theta = \frac{2\sqrt{3} + 1}{\sqrt{3} - 2}$$

Rationalise the denominator by multiplying top and bottom by $(\sqrt{3} + 2)$:

$$\tan\theta = \frac{(2\sqrt{3} + 1)(\sqrt{3} + 2)}{(\sqrt{3} - 2)(\sqrt{3} + 2)}$$

$$= \frac{6 + 4\sqrt{3} + \sqrt{3} + 2}{3 - 4}$$

$$= \frac{8 + 5\sqrt{3}}{-1}$$

$$= -(8 + 5\sqrt{3})$$

So $\tan\theta = -8 - 5\sqrt{3}$, as required.

If you want to find a rational expression for a trigonometric function of an angle, see if you can express it in terms of angles whose sin, cos or tan you *do* know the expression for.

C4

Example 3

Prove that $\sin 15° = \dfrac{\sqrt{3}-1}{2\sqrt{2}}$.

$$\begin{aligned}\sin 15° &= \sin(45° - 30°)\\ &= \sin 45° \cos 30° - \cos 45° \sin 30°\\ &= \left(\frac{1}{\sqrt{2}}\right)\left(\frac{\sqrt{3}}{2}\right) - \left(\frac{1}{\sqrt{2}}\right)\left(\frac{1}{2}\right)\\ &= \frac{\sqrt{3}}{2\sqrt{2}} - \frac{1}{2\sqrt{2}}\\ &= \frac{\sqrt{3}-1}{2\sqrt{2}}\end{aligned}$$

So $\sin 15° = \dfrac{\sqrt{3}-1}{2\sqrt{2}}$.

You can express 15° in terms of 45° and 30°, whose sin and cos are standard identities.

C4

You can use the compound-angle formulae to solve trigonometric equations.

Example 4

Solve the equation $2\sin(30° + \theta) + 2\cos(60° + \theta) = \sqrt{3}$, for $-180° \leqslant \theta \leqslant 180°$.

Using the compound formula for $\sin(A + B)$,

$$\begin{aligned}2\sin(30° + \theta) &= 2(\sin 30° \cos\theta + \cos 30° \sin\theta)\\ &= 2\left(\frac{1}{2}\cos\theta + \frac{\sqrt{3}}{2}\sin\theta\right)\\ &= \cos\theta + \sqrt{3}\sin\theta\end{aligned}$$

Using the compound formula for $\cos(A + B)$,

$$\begin{aligned}2\cos(60° + \theta) &= 2(\cos 60° \cos\theta - \sin 60° \sin\theta)\\ &= 2\left(\frac{1}{2}\cos\theta - \frac{\sqrt{3}}{2}\sin\theta\right)\\ &= \cos\theta - \sqrt{3}\sin\theta\end{aligned}$$

Therefore, the original equation becomes

$$(\cos\theta + \sqrt{3}\sin\theta) + (\cos\theta - \sqrt{3}\sin\theta) = \sqrt{3}$$

Simplifying and solving give

$$2\cos\theta = \sqrt{3}$$

$$\therefore \quad \cos\theta = \frac{\sqrt{3}}{2}$$

When $\cos\theta = \frac{\sqrt{3}}{2}$,

$\theta = -30°$ and $30°$, in the required range.

The solutions are $\theta = \pm 30°$.

Compound-angle formulae can also be useful when you are asked to prove trigonometric identities.

Example 5

Prove the identity

$$\frac{\sin(A+B)}{\cos(A-B)} + 1 = \frac{(1+\tan B)(1+\cot A)}{\cot A + \tan B}$$

Expanding the LHS gives

$$\text{LHS} = \frac{\sin A\cos B + \cos A\sin B}{\cos A\cos B + \sin A\sin B} + 1$$

Dividing both the numerator and the denominator of the fraction by $\sin A\cos B$ gives

$$\text{LHS} = \frac{1 + \cot A\tan B}{\cot A + \tan B} + 1$$

$$= \frac{1 + \cot A\tan B + \cot A + \tan B}{\cot A + \tan B}$$

$$= \frac{(1+\tan B)(1+\cot A)}{\cot A + \tan B}$$

$= \text{RHS}$ QED

Remember:
QED stands for 'quod erat demonstrandum', which means 'which was to be proved'.

Exercise 9A

Questions **1** to **9** should be answered without using a calculator.

1 Given that A and B are acute angles and that $\sin A = \frac{3}{5}$ and that $\cos B = \frac{5}{13}$, find the value of each of these.

a) $\sin(A + B)$ b) $\cos(A - B)$ c) $\tan(A - B)$

2 Given that C and D are acute angles and that $\cos C = \frac{12}{13}$ and that $\cos D = \frac{3}{5}$, find the value of each of these.

a) $\cos(C + D)$ b) $\cos(C - D)$ c) $\cot(C + D)$

3 Given that P and Q are acute angles and that $\tan P = \frac{7}{24}$ and that $\tan Q = 1$, find the value of each of these.

a) $\sin(P + Q)$ b) $\cos(P + Q)$ c) $\tan(P - Q)$

4 Given that $\tan(A - B) = \frac{1}{2}$ and $\tan A = 3$, find the value of $\tan B$.

5 Given that $\tan(P + Q) = 5$ and $\tan Q = 2$, find the value of $\tan P$.

6 Given that $\tan(\theta - 45°) = 4$, find the value of $\tan\theta$.

7 Given that $\tan(\theta + 60°) = 2$, find the value of $\cot\theta$.

8 Given that $\cot(30° - \theta) = 3$, find the value of $\cot\theta$.

9 In each part of this question find the value of $\tan\theta$.

a) $\sin(\theta - \frac{\pi}{6}) = \cos\theta$ b) $\sin(\theta + \frac{\pi}{4}) = \cos\theta$

c) $\cos(\theta + \frac{\pi}{3}) = \sin\theta$ d) $\sin(\theta + \frac{\pi}{3}) = \cos(\theta - \frac{\pi}{3})$

e) $\cos(\theta + \frac{\pi}{3}) = 2\cos(\theta + \frac{\pi}{6})$ f) $\sin(\theta + \frac{\pi}{3}) = \cos(\frac{\pi}{4} - \theta)$

10 Prove these results:

a) $\cos 15° = \dfrac{\sqrt{3} + 1}{2\sqrt{2}}$ b) $\sin 75° = \dfrac{\sqrt{3} + 1}{2\sqrt{2}}$

c) $\tan 15° = 2 - \sqrt{3}$ d) $\tan 75° = 2 + \sqrt{3}$

In questions **11** to **22**, prove each of the given identities.

11 $\sin(\theta + 90°) = \cos\theta$ **12** $\cos(\theta + 90°) = -\sin\theta$

13 $\sin(180° - \theta) = \sin\theta$ **14** $\cos(\theta - 180°) = -\cos\theta$

15 $\sin(A + B) + \sin(A - B) = 2\sin A\cos B$

16 $\cos(A - B) - \cos(A + B) = 2\sin A\sin B$

17 $\sin(A + B)\sin(A - B) = \sin^2 A - \sin^2 B$

18 $\cos(A + B)\cos(A - B) = -(\sin A + \cos B)(\sin A - \cos B)$

19 $\tan A - \tan B = \dfrac{\sin(A - B)}{\cos A\cos B}$

20 $\cot A + \cot B = \dfrac{\sin(A + B)}{\sin A\sin B}$

21 $\dfrac{\sin(A - B)}{\sin(A + B)} = \dfrac{\tan A - \tan B}{\tan A + \tan B}$

22 $\dfrac{\cos(A - B)}{\cos(A + B)} = \dfrac{\cot A\cot B + 1}{\cot A\cot B - 1}$

9.2 Double angles

For any angle, x

$$\sin 2x = 2\sin x\cos x$$

You should memorise the formulae in this section. However, the compound angle formulae for $\sin(A \pm B)$ and $\cos(A \pm B)$ are in the formula book, and you can use them to devise the double angle formulae.

Proof
You know that

$$\sin(A+B) = \sin A\cos B + \cos A\sin B$$

Let $A = B = x$, then

$$\sin(x+x) = \sin x\cos x + \cos x\sin x$$
$$\therefore \quad \sin 2x = 2\sin x\cos x$$

as required.

For any angle x,

$$\begin{aligned}\cos 2x &= \cos^2 x - \sin^2 x\\ &= 2\cos^2 x - 1\\ &= 1 - 2\sin^2 x\end{aligned}$$

C4

Proof
You know that

$$\cos(A+B) = \cos A\cos B - \sin A\sin B$$

Let $A = B = x$, then

$$\cos(x+x) = \cos x\cos x - \sin x\sin x$$
$$\therefore \quad \cos 2x = \cos^2 x - \sin^2 x \qquad [1]$$

as required.

You also know that $\sin^2 x + \cos^2 x = 1$. Rearranging this identity gives

$$\sin^2 x = 1 - \cos^2 x \quad \text{and} \quad \cos^2 x = 1 - \sin^2 x$$

Substituting $\sin^2 x = 1 - \cos^2 x$ into [1] gives

$$\begin{aligned}\cos 2x &= \cos^2 x - (1 - \cos^2 x)\\ &= 2\cos^2 x - 1\end{aligned}$$

as required.

Substituting $\cos^2 x = 1 - \sin^2 x$ into [1] gives

$$\begin{aligned}\cos 2x &= (1 - \sin^2 x) - \sin^2 x\\ &= 1 - 2\sin^2 x\end{aligned}$$

as required.

For any angle,

$$\tan 2x = \frac{2\tan x}{1-\tan^2 x}$$

Proof
You know that

$$\tan(A+B) = \frac{\tan A + \tan B}{1-\tan A \tan B}$$

Let $A = B = x$, then

$$\tan(x+x) = \frac{\tan x + \tan x}{1-\tan x \tan x}$$

$$= \frac{2\tan x}{1-\tan^2 x}$$

as required.

Example 6

Given that x is acute and that $\tan x = \frac{1}{2}$, evaluate each of these.

a) $\tan 2x$ b) $\sin 2x$ c) $\sec 2x$

When $\tan x = \frac{1}{2}$, the right-angled triangle shows that

$$h^2 = 2^2 + 1^2 \quad \therefore \quad h = \sqrt{5}$$

Therefore, $\sin x = \dfrac{1}{\sqrt{5}}$ and $\cos x = \dfrac{2}{\sqrt{5}}$.

a) $$\tan 2x = \frac{2\tan x}{1-\tan^2 x} = \frac{2\left(\frac{1}{2}\right)}{1-\left(\frac{1}{2}\right)^2}$$

$$\therefore \quad \tan 2x = \frac{4}{3}$$

b) $$\sin 2x = 2\sin x \cos x = 2\left(\frac{1}{\sqrt{5}}\right)\left(\frac{2}{\sqrt{5}}\right)$$

$$\therefore \quad \sin 2x = \frac{4}{5}$$

c) $$\sec 2x = \frac{1}{\cos 2x} = \frac{1}{\cos^2 x - \sin^2 x} = \frac{1}{\left(\frac{2}{\sqrt{5}}\right)^2 - \left(\frac{1}{\sqrt{5}}\right)^2}$$

$$\therefore \quad \sec 2x = \frac{5}{3}$$

Again, these formulae are useful for solving trigonometric equations.

Example 7

Solve the equation $3\sin 2\theta = \cos\theta$ for $-180° \leqslant \theta \leqslant 180°$.

Substitute $\sin 2\theta = 2\sin\theta\cos\theta$ into the equation:

$$3(2\sin\theta\cos\theta) = \cos\theta$$
$$\therefore\quad 6\sin\theta\cos\theta - \cos\theta = 0$$
$$\therefore\quad \cos\theta(6\sin\theta - 1) = 0$$

When $\cos\theta = 0$ the solutions are $\theta = \pm 90°$.
When $\sin\theta = \frac{1}{6}$, one solution is $\theta = 9.6°$. In the range $-180° \leqslant \theta \leqslant 180°$ the other solution is $\theta = 180° - 9.6° = 170.4°$.
The solutions are $\theta = -90°, 9.6°, 90°$ and $170.4°$.

Remember to set your calculator to work in degrees.

Example 8

Solve the equation $4\cos 2\theta - 2\cos\theta + 3 = 0$ for $0 \leqslant \theta \leqslant 2\pi$.

Substitute $\cos 2\theta = 2\cos^2\theta - 1$ into the equation:

$$4(2\cos^2\theta - 1) - 2\cos\theta + 3 = 0$$
$$8\cos^2\theta - 2\cos\theta - 1 = 0$$
$$\therefore\quad (4\cos\theta + 1)(2\cos\theta - 1) = 0$$

C4

Remember θ is in radians here.

When $\cos\theta = -\frac{1}{4}$, one solution is $\theta = 1.82$. In the range $0 \leqslant \theta \leqslant 2\pi$ the other solution is $\theta = 2\pi - 1.82 = 4.46$.

When $\cos\theta = \frac{1}{2}$, one solution is $\frac{\pi}{3}$. In the range $0 \leqslant \theta \leqslant 2\pi$ the other solution is $\theta = 2\pi - \frac{\pi}{3} = \frac{5\pi}{3}$.

The solutions are $\theta = \frac{\pi}{3}, \frac{5\pi}{3}$, 1.82 and 4.46.

Example 9

Prove that $\dfrac{\sin A + \sin 2A}{1 + \cos A + \cos 2A} = \tan A$.

Consider the left-hand side:

$$\text{LHS} \equiv \frac{\sin A + 2\sin A\cos A}{1 + \cos A + (2\cos^2 A - 1)} = \frac{\sin A + 2\sin A\cos A}{\cos A + 2\cos^2 A}$$
$$= \frac{\sin A(1 + 2\cos A)}{\cos A(1 + 2\cos A)}$$
$$= \frac{\sin A}{\cos A}$$
$$= \tan A = \text{RHS} \qquad \text{QED}$$

Using the double-angle formula for $\cos 2A$.

Exercise 9B

1 Given that x is an acute angle and that $\sin x = \frac{4}{5}$, find the value of each of these.

a) $\sin 2x$ b) $\cos 2x$ c) $\tan 2x$

2 Given that x is an acute angle and that $\cos x = \frac{5}{13}$, find the value of each of these.

a) $\cos 2x$ b) $\sin 2x$ c) $\operatorname{cosec} 2x$

3 Given that x is an acute angle and that $\tan x = 2$, find the value of each of these.

a) $\tan 2x$ b) $\sin 2x$ c) $\sec 2x$

4 Given that x is an acute angle and that $\sin x = \frac{1}{2}$, find the value of each of these.

a) $\sin 2x$ b) $\cos 2x$ c) $\cot 2x$

5 Given that x is an acute angle and that $\cos x = \frac{7}{25}$, find the value of each of these.

a) $\sec 2x$ b) $\sin 2x$ c) $\cot 2x$

6 Given that x is an acute angle and that $\tan x = \frac{3}{2}$, find the value of each of these.

a) $\tan 2x$ b) $\operatorname{cosec} 2x$ c) $\cos 2x$

C4

7 Solve each of these equations for $0 \leqslant x \leqslant 360°$, giving your answers correct to one decimal place.

a) $3 \sin 2x = \sin x$ b) $4 \cos x = 3 \sin 2x$

c) $\sin 2x + \cos x = 0$ d) $3 \cos 2x - \cos x + 2 = 0$

e) $6 \cos 2x - 7 \sin x + 6 = 0$ f) $2 \cos 2x = 1 - 3 \sin x$

g) $2 \cos 2x = 2 + 15 \cos x$ h) $\cos 2x \sin 2x = 0$

8 Solve each of these equations for $0 \leqslant x \leqslant 2\pi$, giving each of your answers to two decimal places.

a) $\sin 2x = \sin x$ b) $\sqrt{3} \cos x = \sin 2x$

c) $\sin 2x + \sin x = 0$ d) $\cos 2x + \sin x = 0$

e) $\cos 2x - 7 \cos x - 3 = 0$ f) $2 + \cos 2x = 3 \cos x$

g) $\cos 2x = \cos x$ h) $\sin x \sin 2x = \cos x$

In Questions **9** to **24**, prove each of the given identities.

9 $2 \cos^2 \theta - \cos 2\theta = 1$

10 $2 \operatorname{cosec} 2\theta = \operatorname{cosec} \theta \sec \theta$

11 $2 \cos^3 \theta + \sin 2\theta \sin \theta = 2 \cos \theta$

12 $\tan \theta + \cot \theta = 2 \operatorname{cosec} 2\theta$

13 $\cos^4 \theta - \sin^4 \theta = \cos 2\theta$

14 $\dfrac{1 - \cos 2\theta}{1 + \cos 2\theta} = \tan^2 \theta$

15 $\cot \theta - \tan \theta = 2 \cot 2\theta$

16 $\cot 2\theta + \operatorname{cosec} 2\theta = \cot \theta$

17 $\dfrac{\cos 2\theta}{\cos \theta + \sin \theta} = \cos \theta - \sin \theta$

18 $\dfrac{\sin 2\theta}{1 - \cos 2\theta} = \cot \theta$

19 $\cos 2\theta = \dfrac{1 - \tan^2\theta}{1 + \tan^2\theta}$

20 $\sin 2\theta = \dfrac{2\tan\theta}{1 + \tan^2\theta}$

21 $\dfrac{1}{\cos\theta + \sin\theta} + \dfrac{1}{\cos\theta - \sin\theta} = \dfrac{2\cos\theta}{\cos 2\theta}$

22 $\dfrac{\sin\theta + \sin 2\theta}{1 + \cos\theta + \cos 2\theta} = \tan\theta$

23 $\dfrac{\sin A}{\sin B} - \dfrac{\cos A}{\cos B} = \dfrac{2\sin(A - B)}{\sin 2B}$

24 $\dfrac{\cos A}{\sin B} - \dfrac{\sin A}{\cos B} = \dfrac{2\cos(A + B)}{\sin 2B}$

9.3 Triple-angle formulae

It is possible to derive expressions for the sine, cosine and tangent of a triple angle, $3x$, in terms of x.

> For any angle x
>
> $$\sin 3x = 3\sin x - 4\sin^3 x$$

Proof

You know that

$$\sin(A + B) = \sin A\cos B + \cos A\sin B$$

This is a compound angle formula (page 179).

Let $A = 2x$ and $B = x$, then

$$\sin(2x + x) = \sin 2x\cos x + \cos 2x\sin x$$

$$\therefore\quad \sin 3x = \sin 2x\cos x + \cos 2x\sin x \qquad [1]$$

You also know that

$$\sin 2x = 2\sin x\cos x \quad \text{and} \quad \cos 2x = 1 - 2\sin^2 x$$

These are double-angle formulae (page 186).

Substituting these in [1] gives

$$\begin{aligned}\sin 3x &= (2\sin x\cos x)\cos x + (1 - 2\sin^2 x)\sin x\\ &= 2\sin x\cos^2 x + \sin x - 2\sin^3 x\\ &= 2\sin x(1 - \sin^2 x) + \sin x - 2\sin^3 x\\ &= 2\sin x - 2\sin^3 x + \sin x - 2\sin^3 x\\ &= 3\sin x - 4\sin^3 x\end{aligned}$$

$\sin^2 x + \cos^2 x = 1$

as required.

The identities for the cosine and tangent of a triple angle are:

> For any angle x
>
> $$\cos 3x = 4\cos^3 x - 3\cos x$$
>
> For any angle x
>
> $$\tan 3x = \dfrac{3\tan x - \tan^3 x}{1 - 3\tan^2 x}$$

You will be able to prove these identities in questions 1 and 2 of Exercise 9C on page 191.

Example 10

Solve the equation

$$\sin^2\theta + \sin\theta = 1 - \sin 3\theta$$

for $0° \leqslant \theta \leqslant 360°$.

You know that

$$\sin 3\theta = 3\sin\theta - 4\sin^3\theta$$

So:

$$\sin^2\theta + \sin\theta = 1 - (3\sin\theta - 4\sin^3\theta)$$
$$\sin^2\theta + \sin\theta = 1 - 3\sin\theta + 4\sin^3\theta$$
$$\therefore\ 4\sin^3\theta - \sin^2\theta - 4\sin\theta + 1 = 0$$

This is a cubic equation in $\sin\theta$: you can see this more clearly if you let $\sin\theta = x$ and write it as

$$4x^3 - x^2 - 4x + 1 = 0$$

Using the Factor Theorem, you can see that $x = 1$ and $x = -1$ are two roots. That is, $(x-1)$ and $(x+1)$ are both factors of the cubic expression. So,

$$4x^3 - x^2 - 4x + 1 = (4x-1)(x+1)(x-1)$$

Expand brackets and compare coefficients to find the third factor.

which means that

$$4\sin^3\theta - \sin^2\theta - 4\sin\theta + 1 = 0$$

becomes

$$(4\sin\theta - 1)(\sin\theta + 1)(\sin\theta - 1) = 0$$

Solving gives $\sin\theta = \frac{1}{4}$, $\sin\theta = -1$ and $\sin\theta = 1$.
When $\sin\theta = \frac{1}{4}$, one solution is $\theta = 14.5°$. In the range $0° \leqslant \theta \leqslant 360°$, the other solution is $\theta = 180° - 14.5° = 165.5°$.
When $\sin\theta = -1$, $\theta = 270°$ in the required range. When $\sin\theta = 1$, $\theta = 90°$ in the required range.
The solutions are $\theta = 14.5°, 90°, 165.5°$ and $270°$.

Exercise 9C

1 Use the identity $\cos(A+B) = \cos A\cos B - \sin A\sin B$ with $A = 2x$ and $B = x$ to prove that

$$\cos 3x = 4\cos^3 x - 3\cos x$$

2 Use the identity $\tan(A+B) = \dfrac{\tan A + \tan B}{1 - \tan A\tan B}$ with $A = 2x$ and $B = x$ to prove that

$$\tan 3x = \frac{3\tan x - \tan^3 x}{1 - 3\tan^2 x}$$

3 a) Factorise the cubic $4y^3 - 4y^2 - y + 1$.

b) Use the triple-angle formula for $\sin 3\theta$ to show that the equation

$1 - \sin 3\theta = 2\sin\theta(2\sin\theta - 1)$ can be written as

$4\sin^3\theta - 4\sin^2\theta - \sin\theta + 1 = 0$

c) Hence solve the equation $1 - \sin 3\theta = 2\sin\theta(2\sin\theta - 1)$ for $0 \leqslant \theta \leqslant 360°$.

4 a) Factorise the cubic $4y^3 - y^2 - 4y + 1$.

b) Use the triple-angle formula for $\cos 3\theta$ to show that the equation

$1 + \cos 3\theta = \cos\theta(1 + \cos\theta)$ can be written as

$4\cos^3\theta - \cos^2\theta - 4\cos\theta + 1 = 0$

c) Hence solve the equation $1 + \cos 3\theta = \cos\theta(1 + \cos\theta)$ for $0 \leqslant \theta \leqslant 360°$.

9.4 Applications in integration

The double and triple-angle formulae can help you to integrate powers of $\sin x$ and $\cos x$.

Example 11

C4

Work out each of these integrals.

a) $\int \cos^2 x \, dx$ b) $\int 8\sin^3 x \, dx$

a) From the double-angle formula, $\cos 2x = 2\cos^2 x - 1$.
Hence $\cos^2 x = \frac{1}{2}(1 + \cos 2x)$.
So:

$$\begin{aligned}\int \cos^2 x \, dx &= \int \tfrac{1}{2}(1 + \cos 2x) \, dx \\ &= \int (\tfrac{1}{2} + \tfrac{1}{2}\cos 2x) \, dx \\ &= \tfrac{1}{2}x + \tfrac{1}{4}\sin 2x + c\end{aligned}$$

So $\int \cos^2 x \, dx = \frac{1}{2}x + \frac{1}{4}\sin 2x + c$

b) From the triple-angle formula, $\sin 3x = 3\sin x - 4\sin^3 x$.
Hence $\sin^3 x = \frac{1}{4}(3\sin x - \sin 3x)$.
So:

$$\begin{aligned}\int 8\sin^3 x \, dx &= \int 8 \times \tfrac{1}{4}(3\sin x - \sin 3x) \, dx \\ &= \int 2(3\sin x - \sin 3x) \, dx \\ &= \int (6\sin x - 2\sin 3x) \, dx \\ &= -6\cos x + \tfrac{2}{3}\cos 3x + c\end{aligned}$$

So $\int 8\sin^3 x \, dx = -6\cos x + \frac{2}{3}\cos 3x + c$

The next example is of a definite integral.

Example 12

Evaluate $\int_0^{\frac{\pi}{4}} 4\sin^2 x\,dx$.

From the double-angle formula, $\cos 2x = 1 - 2\sin^2 x$.
Hence $\sin^2 x = \frac{1}{2}(1 - \cos 2x)$.
So:

$$\begin{aligned}
\int_0^{\frac{\pi}{4}} 4\sin^2 x\,dx &= \int_0^{\frac{\pi}{4}} 4 \times \tfrac{1}{2}(1 - \cos 2x)\,dx \\
&= \int_0^{\frac{\pi}{4}} 2(1 - \cos 2x)\,dx \\
&= \int_0^{\frac{\pi}{4}} 2 - 2\cos 2x\,dx \\
&= \left[2x - \sin 2x\right]_0^{\frac{\pi}{4}} \\
&= 2\left(\frac{\pi}{4}\right) - \sin\left(2\left(\frac{\pi}{4}\right)\right) - 0 \\
&= \frac{\pi}{2} - \sin\left(\frac{\pi}{2}\right) \\
&= \frac{\pi}{2} - 1
\end{aligned}$$

So $\int_0^{\frac{\pi}{4}} 4\sin^2 x\,dx = \frac{\pi}{2} - 1$

C4

Exercise 9D

1 Work out each of these integrals.

a) $\int \sin^2 x\,dx$ b) $\int 8\cos^2 dx$ c) $\int 6\cos^2 x\,dx$

d) $\int 5\sin^2 x\,dx$ e) $\int \sin^2 3x\,dx$ f) $\int 16\cos^2 4x\,dx$

2 Work out each of these.

a) $\int \cos^3 x\,dx$ b) $\int 16\sin^3 x\,dx$ c) $\int 6\cos^3 x\,dx$

d) $\int 5\cos^3 x\,dx$ e) $\int \sin^3 2x\,dx$ f) $\int 36\cos^3 3x\,dx$

3 Evaluate each of these definite integrals.

a) $\int_0^{\frac{\pi}{4}} 4\cos^2 x\,dx$ b) $\int_{-\frac{\pi}{4}}^{\frac{\pi}{4}} 8\sin^2 x\,dx$ c) $\int_{\frac{\pi}{2}}^{\pi} 6\sin^2 x\,dx$

d) $\int_{\frac{\pi}{4}}^{\frac{\pi}{2}} 16\cos^2 x\,dx$ e) $\int_{-\frac{\pi}{8}}^{\frac{\pi}{8}} \sin^2 2x\,dx$ f) $\int_{-\pi}^{\pi} 3\cos^2 6x\,dx$

9.5 Harmonic form

Consider the following situation. A large rectangular mirror, $ABCD$, is tilted to fit through a doorway. $AB = 4$ m and $BC = 3$ m. The height of the doorway is h metres. The diagonal, AC, of the mirror is R metres, and this diagonal makes an angle of $\alpha°$ with the side AB. The mirror will just fit when this side, AB, makes an angle of $\theta°$ with the ground.

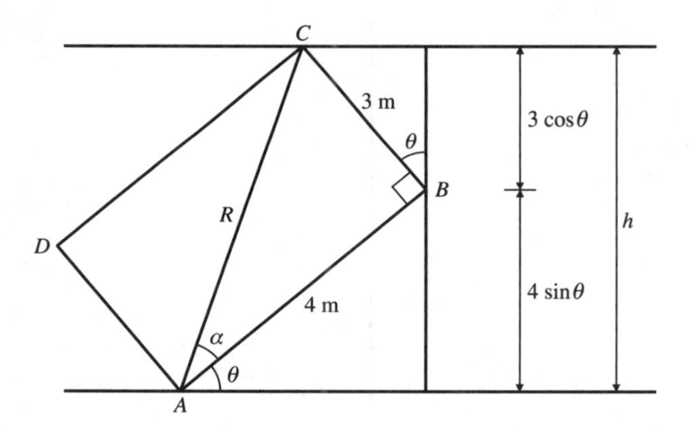

From the diagram you see that

$$h = 4\sin\theta + 3\cos\theta$$

and also

$$h = R\sin(\theta + \alpha)$$

So

$$R\sin(\theta + \alpha) = 4\sin\theta + 3\cos\theta$$

From the triangle ABC, $R^2 = 4^2 + 3^2$, and hence $R = 5$. Also, $\tan\alpha = \frac{3}{4}$, and hence $\alpha = 36.87°$, correct to two decimal places.

So:

$$4\sin\theta + 3\cos\theta = 5\sin(\theta + 36.87°)$$

This is shown graphically.

You can use a graphics calculator to see that the graph of $y = 5\sin(\theta + 36.87°)$ is identical to the graph of $y = 4\sin\theta + 3\cos\theta$.

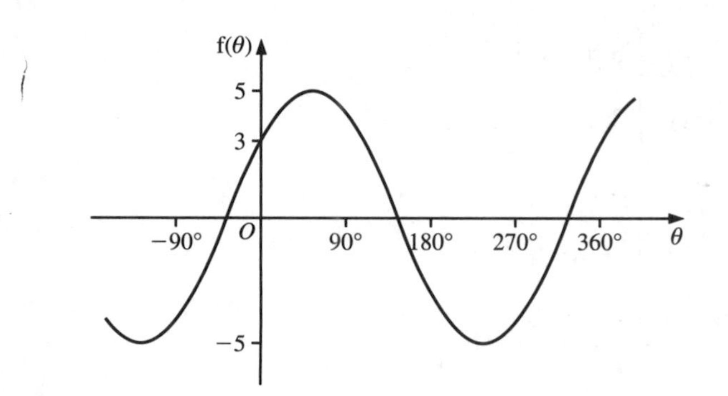

It is also possible to derive the result $4\sin\theta + 3\cos\theta = 5\sin(\theta + 36.87°)$ using the compound angle formulae.

Let $4\sin\theta + 3\cos\theta = R\sin(\theta + \alpha)$

You know that $\sin(A + B) = \sin A\cos B + \cos A\sin B$

So:

$$\begin{aligned} 4\sin\theta + 3\cos\theta &= R(\sin\theta\cos\alpha + \cos\theta\sin\alpha) \\ &= R\sin\theta\cos\alpha + R\cos\theta\sin\alpha \end{aligned}$$

Equate the coefficients of $\sin\theta$ and $\cos\theta$ to obtain

$$4 = R\cos\alpha \quad \text{and} \quad 3 = R\sin\alpha$$

Squaring and adding gives

$$\begin{aligned} R^2\cos^2\alpha + R^2\sin^2\alpha &= 4^2 + 3^2 \\ \therefore\quad R^2(\cos^2\alpha + \sin^2\alpha) &= 25 \\ R^2 &= 25 \text{ (since } \cos^2\alpha + \sin^2\alpha = 1) \\ \therefore\quad R &= 5 \text{ (since } R > 0) \end{aligned}$$

Dividing $R\sin\alpha$ by $R\cos\alpha$ gives

$$\frac{R\sin\alpha}{R\cos\alpha} = \frac{3}{4} = \tan\alpha$$

$$\therefore\quad \alpha = 36.87°$$

Note that in problems like this, it may help to draw a right-angled triangle:

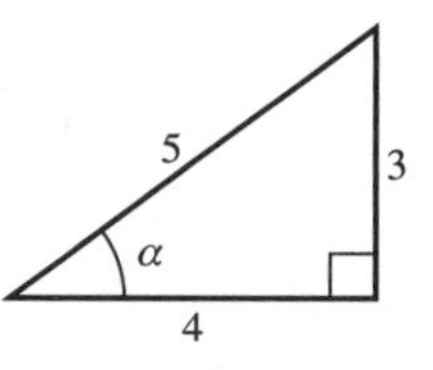

> In general, expressing any function $f(\theta) = a\cos\theta + b\sin\theta$ in the form
>
> $$R\sin(\theta \pm \alpha) \text{ or } R\cos(\theta \pm \alpha)$$
>
> where $R > 0$ and α is acute is known as expressing $f(\theta)$ in **harmonic form.**

This alternative form allows you to solve equations of the form

$$a\cos\theta + b\sin\theta = c$$

and to find the maximum and minimum values of such functions.

C4

Example 13

a) Express $3 \cos \theta - 4 \sin \theta$ in the form $R \cos(\theta + \alpha)$.

b) Solve the equation $3 \cos \theta - 4 \sin \theta = 1$, for $0° \leq \theta \leq 360°$.

a) Let

$$\begin{aligned} 3 \cos \theta - 4 \sin \theta &= R \cos(\theta + \alpha) \\ &= R(\cos \theta \cos \alpha - \sin \theta \sin \alpha) \\ &= R \cos \theta \cos \alpha - R \sin \theta \sin \alpha \end{aligned}$$

and equate the corresponding coefficients of $\cos \theta$ and $\sin \theta$ to obtain

$$3 = R \cos \alpha \quad [1] \qquad \text{and} \qquad 4 = R \sin \alpha \quad [2]$$

Squaring each of [1] and [2] and adding gives

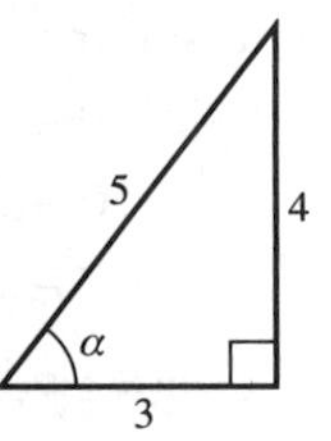

$$R^2 \cos^2 \alpha + R^2 \sin^2 \alpha = 3^2 + 4^2$$

$$R^2(\cos^2 \alpha + \sin^2 \alpha) = 25$$

$$\therefore \quad R^2 = 25 \quad (\text{since } \cos^2 \alpha + \sin^2 \alpha = 1)$$

$$\therefore \quad R = 5 \quad (\text{since } R > 0)$$

Dividing [2] by [1] gives

$$\frac{R \sin \alpha}{R \cos \alpha} = \frac{4}{3}$$

$$\therefore \quad \tan \alpha = \frac{4}{3}$$

$$\therefore \quad \alpha = 53.1°$$

So $3 \cos \theta - 4 \sin \theta = 5 \cos(\theta + 53.1°)$

b) The equation

$$3 \cos \theta - 4 \sin \theta = 1$$

becomes

$$5 \cos(\theta + 53.1°) = 1$$

$$\therefore \quad \cos(\theta + 53.1°) = \frac{1}{5}$$

Changing the range for $\theta + 53.1°$ gives $53.1° \leq \theta + 53.1° \leq 413.1°$. When $\cos(\theta + 53.1°) = \frac{1}{5}$, one solution in the required range is

$$\theta + 53.1° = 78.5° \quad \therefore \quad \theta = 25.4°$$

The other solution is

$$\theta + 53.1° = 360° - 78.5° \quad \therefore \quad \theta = 228.4°$$

The solutions are $\theta = 25.4°$ and $228.4°$.

Example 14

Given $f(\theta) = 8 + 5\cos\theta + 3\sin\theta$ find the greatest and least values of $f(\theta)$, and state, in radians, correct to two decimal places, the smallest non-negative value of θ for which each occurs.

Start by expressing $5\cos\theta + 3\sin\theta$ in the form $R\cos(\theta - \alpha)$.

Using $\cos(A - B) = \cos A\cos B + \sin A\sin B$ gives

$$5\cos\theta + 3\sin\theta = R(\cos\theta\cos\alpha + \sin\theta\sin\alpha)$$
$$= R\cos\theta\cos\alpha + R\sin\theta\sin\alpha$$

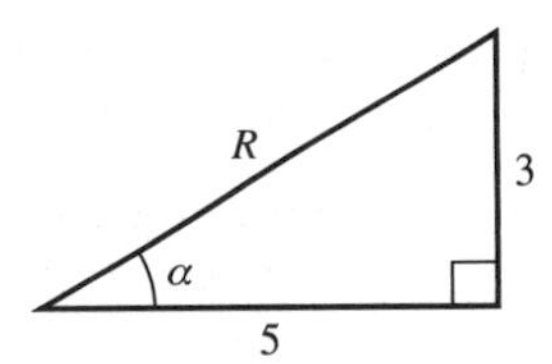

Equate the coefficients of $\sin\theta$ and $\cos\theta$ to obtain

$$5 = R\cos\alpha \quad [1] \qquad \text{and} \qquad 3 = R\sin\alpha \quad [2]$$

Squaring each of [1] and [2] and adding gives

$$R^2\cos^2\alpha + R^2\sin^2\alpha = 5^2 + 3^2$$
$$R^2(\cos^2\alpha + \sin^2\alpha) = 34$$
$$\therefore \quad R^2 = 34 \text{ (since } \cos^2\alpha + \sin^2\alpha = 1)$$
$$\therefore \quad R = \sqrt{34} \text{ (since } R > 0)$$

Dividing [2] by [1] gives

$$\frac{R\sin\alpha}{R\cos\alpha} = \frac{3}{5}$$
$$\therefore \quad \alpha = 0.54 \text{ rad}$$

So $5\cos\theta + 3\sin\theta = \sqrt{34}\cos(\theta - 0.54)$

And $f(\theta) = 8 + 5\cos\theta + 3\sin\theta = 8 + \sqrt{34}\cos(\theta - 0.54)$

The function $\sqrt{34}\cos(\theta - 0.54)$ attains its greatest value when the cosine function attains its maximum value, that is, at $\theta - 0.54 = 0$. Therefore the greatest value of $f(\theta)$ is $8 + \sqrt{34}$ when $\theta = 0.54$.

Similarly the function $\sqrt{34}\cos(\theta - 0.54)$ attains its least value when $\cos(\theta - 0.54)$ equals -1, at $\theta - 0.54 = \pi$. Therefore the least value of $f(\theta)$ is $8 - \sqrt{34}$ when $\theta = \pi + 0.54 = 3.68$.

Exercise 9E

1 Find the value of R and $\tan\alpha$ in each of these identities.

a) $3\sin\theta + 4\cos\theta = R\sin(\theta+\alpha)$
b) $5\sin\theta - 12\cos\theta = R\sin(\theta-\alpha)$
c) $2\cos\theta + 5\sin\theta = R\cos(\theta-\alpha)$
d) $2\cos\theta + 5\sin\theta = R\sin(\theta+\alpha)$
e) $\cos\theta - \sin\theta = R\cos(\theta+\alpha)$
f) $20\sin\theta - 15\cos\theta = R\sin(\theta-\alpha)$
g) $\sqrt{3}\cos\theta + \sin\theta = R\cos(\theta-\alpha)$
h) $2\cos\theta - 4\sin\theta = R\cos(\theta+\alpha)$

2 a) Express $12\sin\theta + 5\cos\theta$ in the form $R\sin(\theta+\alpha)$ where $R > 0$ and α, measured in degrees to 1 decimal place, is acute.

b) Hence solve the equation $12\sin\theta + 5\cos\theta = \frac{13}{2}$ for $0 \leqslant \theta \leqslant 360°$, giving your answer to 1 decimal place.

3 a) Express $4\sin\theta - 3\cos\theta$ in the form $R\sin(\theta-\alpha)$ where $R > 0$ and α, measured in degrees to 1 decimal place, is acute.

b) Hence solve the equation $4\sin\theta - 3\cos\theta = 2$ for $0 \leqslant \theta \leqslant 360°$, giving your answer to 1 decimal place.

4 Solve each of these equations for $0 \leqslant \theta \leqslant 2\pi$ giving your answers correct to two decimal places.

a) $\sin\theta + \sqrt{3}\cos\theta = 1$
b) $\sin\theta + \cos\theta = \dfrac{1}{\sqrt{2}}$
c) $5\sin\theta + 12\cos\theta = 7$
d) $7\sin\theta - 4\cos\theta = 3$
e) $\cos\theta - 3\sin\theta = 2$
f) $5\cos\theta + 2\sin\theta = 4$

5 a) Express $9\cos 2\theta - 4\sin 2\theta$ in the form $R\cos(2\theta+\alpha)$ where $R > 0$ and α, measured in degrees to 1 decimal place, is acute.

b) Hence solve the equation $9\cos 2\theta - 4\sin 2\theta = 6$ for $0 \leqslant \theta \leqslant 180°$.

6 a) Show that the equation $\dfrac{\sqrt{5}}{2}\sec\theta - \tan\theta = 2$ can be written as $\sin\theta + 2\cos\theta = \dfrac{\sqrt{5}}{2}$.

b) Hence solve the equation $\dfrac{\sqrt{5}}{2}\sec\theta - \tan\theta = 2$ for $0 \leqslant \theta \leqslant 2\pi$, giving your answer to 2 decimal places.

7 Solve each of these equations for $0 \leqslant \theta \leqslant 360°$, giving your answers correct to one decimal place.

a) $\sqrt{2}\tan\theta - \sqrt{3}\sec\theta = \sqrt{2}$
b) $1 - \sqrt{13}\sec\theta = 5\tan\theta$
c) $10\operatorname{cosec}\theta - 7\cot\theta = 24$

8 a) Express $f(\theta) = 8\sin\theta + 6\cos\theta$ in the form $R\sin(\theta+\alpha)$ where $R > 0$ and α, measured in degrees to 1 decimal place, is acute.

b) Hence find the greatest and least values of $f(\theta)$, and state, in degrees, correct to one decimal place, the smallest non-negative value of θ for which each occurs.

9 a) Express $f(\theta) = 3 + 2\cos\theta + \sin\theta$ in the form $3 + R\cos(\theta - \alpha)$ where $R > 0$ and α, measured in radians to 2 decimal places, is acute.

b) Hence find the greatest and least values of $f(\theta)$, and state, in radians, correct to two decimal places, the smallest non-negative value of θ for which each occurs.

10 Find the greatest and least values of each of these functions, and state, in degrees, correct to one decimal place, the smallest non-negative value of θ for which each occurs.

a) $12\sin\theta + 5\cos\theta$
b) $2\cos\theta + \sin\theta$
c) $7 + 3\sin\theta - 4\cos\theta$
d) $2 - \sin\theta - \cos\theta$
e) $10 - 2\sin\theta + \cos\theta$
f) $7 - 2\cos\theta - \sqrt{5}\sin\theta$

Summary

You should know how to ...	Check out
1 Use compound angle identities.	**1** a) Show that $\cos(\alpha - \beta) - \cos(\alpha + \beta) = 2\sin\alpha\sin\beta$ b) Simplify $\sin\left(x - \frac{\pi}{2}\right)$
2 Use double-angle identities.	**2** a) Show that $(\cos\theta + \sin\theta)^2 = 1 + \sin 2\theta$ b) Solve the equation $6\cos 2x + 5\sin x = 4$ for $0 \leqslant x \leqslant 360°$
3 Solve equations of the form $a\sin x + b\cos x = c$.	**3** a) Express $5\cos x + 3\sin x$ in the form $R\cos(x - \alpha)\quad 0 < \alpha < \frac{\pi}{2}$ b) i) Find the minimum value of $5\cos x + 3\sin x$ ii) Find the value of x, $0 < x < 2\pi$, at which the minimum occurs. c) Solve the equation $5\cos x + 3\sin x = 4$ for $0 < x < 2\pi$.

Revision exercise 9

1 a) Express $6\cos x - 8\sin x$ in the form $R\cos(x + \alpha)$ where R is a positive constant and $0° < \alpha < 90°$. Give the value of α to the nearest 0.1°.

b) Hence find the general solution, in degrees, of the equation $6\cos x - 8\sin x = 3$. *(AQA, 2004)*

C4

2 a) Find the value of $\tan^{-1} 2.4$ giving your answer in radians to three decimal places.

b) Express $10 \sin \theta + 24 \cos \theta$ in the form $r \sin(\theta + \alpha)$, where $R > 0$ and $0 < \alpha < \frac{\pi}{2}$.

c) Hence
i) write down the maximum value of $10 \sin \theta + 24 \cos \theta$
ii) find a value of θ at which this maximum occurs. *(AQA, 2004)*

3 The diagram shows a rectangle $OABC$ in which B has coordinates $(\sin \theta, 2 \cos \theta)$, where $0 \leqslant \theta \leqslant \frac{\pi}{2}$.

The perimeter of the rectangle is of length L.

a) i) Write down the length L in terms of θ.
ii) Hence obtain an expression for L in the form $R \sin(\theta + \alpha)$ where $R > 0$ and $0 \leqslant \alpha \leqslant \frac{\pi}{2}$.
Give your answer for α to three decimal places.

b) Given that θ varies between 0 and $\frac{\pi}{2}$
i) write down the maximum value of L;
ii) find the value of θ, to two decimal places, for which L is a maximum. *(AQA, 2003)*

C4

4 Part of the curve with equation $y = 3 \sin 2x + \cos 2x$ is sketched in the diagram.

a) Find $\frac{dy}{dx}$.

b) Find the equation of the tangent to the curve at the point where $x = \frac{\pi}{4}$.

c) i) Express $(3 \sin 2x + \cos 2x)^2$ in the form $A \sin 4x + B \cos 4x + C$ where A, B and C are integers.
ii) The region R, shaded in the diagram, is bounded by the curve with equation $y = 3 \sin 2x + \cos 2x$, the coordinate axes and the line $x = \frac{\pi}{4}$.
Find the volume of the solid generated when R is rotated through 2π radians about the x-axis. *(AQA, 2003)*

5 a) Express $4 \sin \theta - 3 \sin \theta$ in the form $R \sin(\theta - \alpha)$ where R is a positive constant and $0° < \alpha < 90°$. Give the value of α to the nearest 0.1°.

b) Hence find the general solution, in degrees, of the equation $4 \sin \theta - 3 \cos \theta = 2$. *(AQA, 2003)*

6 a) Show that $\frac{\cot^2 \theta}{1 + \cot^2 \theta} = \cos^2 \theta$

b) Hence solve $\frac{\cot^2 \theta}{1 + \cot^2 \theta} = 2 \sin 2\theta$ for $0° \leqslant \theta \leqslant 360°$. *(AQA, 2002)*

7 a) Express $6\sin^2\theta$ in the form $a + b\cos 2\theta$ where a and b are constants.

b) Find the exact value of $\int_0^{\frac{\pi}{12}} \sin^2\theta \, d\theta$.

c) Solve the equation $3 - 3\cos 2\theta = 2\operatorname{cosec}\theta$ giving all solutions in radians in the interval $0 < \theta < 2\pi$. *(AQA, 2002)*

8 a) Express $5\cos\theta + 12\sin\theta$ in the form $R\cos(\theta - \alpha)$, where $0 \leqslant \theta \leqslant 2\pi$, R is a positive constant and $0 < \alpha < \frac{\pi}{2}$.

b) Hence find the general solution, in radians, of the equation $5\cos\theta + 12\sin\theta = 6.5$. *(AQA, 2002)*

9 a) Prove that $\frac{2\tan x}{1 + \tan^2 x} = \sin 2x$

b) Hence or otherwise find the exact value of tan 15° in the form $a + b\sqrt{3}$ where a and b are integers. *(AQA, 2002)*

10 a) Given that $\sin\alpha = \frac{12}{13}$ where α is an acute angle, find the exact value of $\cos\alpha$.

b) Given also that $\cos\beta = \frac{4}{5}$, where β is an acute angle, find the exact value of $\sin(\alpha + \beta)$. *(AQA, 2002)*

11 Find the value of $\int_{\frac{\pi}{4}}^{\frac{3\pi}{4}} (\sin\theta + 4\sin^2\theta) \, d\theta$. *(AQA, 2002)*

C4

12 a) Given that $\tan x \neq 1$ show that $\frac{\cos 2x}{\cos x - \sin x} = \cos x + \sin x$.

b) By expressing $\cos x + \sin x$ in the form $R\sin(x + \alpha)$, solve, for $0° \leqslant x \leqslant 360°$, $\frac{\cos 2x}{\cos x - \sin x} = \frac{1}{2}$. *(AQA, 2001)*

13 The graph shows the region R enclosed by the curve $y = x + \sin x$, the x-axis and the line $x = \pi$.

y

R

O

π

x

a) Find the exact value of the area of region R.

b) Show that

i) $\int_0^{\pi} x\sin x \, dx = \pi$

ii) $\int_0^{\pi} \sin^2 x \, dx = \frac{\pi}{2}$.

c) A solid metallic casting is made by rotating the region R through 2π radians about the x-axis. Find the volume of the solid formed. *(AQA, 2001)*

14 Solve the equation $3\cos 2\theta - \cos\theta + 1 = 0$ giving all solutions in degrees to the nearest degree in the interval $0° \leqslant \theta \leqslant 360°$. *(AQA, 2001)*

10 Exponential models

This chapter will show you how to

✦ Solve problems involving exponential growth and decay

Before you start

You should know how to …	Check in
1 Evaluate expressions involving indices.	**1** a) Evaluate these expressions giving answers to three significant figures. i) $\sqrt[4]{25}$ ii) 5×7^{-3} iii) $3e^{-3}$ b) Evaluate each of these expressions for the given value of x. Give your answers to three significant figures. i) 3^{1-2x} at $x = 1$ ii) $1000 \times 5^{\frac{x-3}{10}}$ at $x = 7$ iii) $50 \times e^{-\left(\frac{3x+1}{2}\right)}$ at $x = 2$
2 Solve equations of the form $a^x = b$ and $ax^b = c$.	**2** a) Solve these equations for x. Give your answers to three decimal places. i) $3x^4 = 7$ ii) $10x^{15} = 5000$ iii) $\frac{1}{x^{10}} = 0.1$ b) Solve these equations for x. Give your answer to three decimal places. i) $5^x = 50$ ii) $6^{2x} = 3$ iii) $\frac{7}{3^{2x-1}} = 5$

C4

10.1 Exponential growth and decay

In C2 you learnt about the functions $y = a^x$ and $y = a^{-x}$. Their graphs are shown here for $a > 0$.

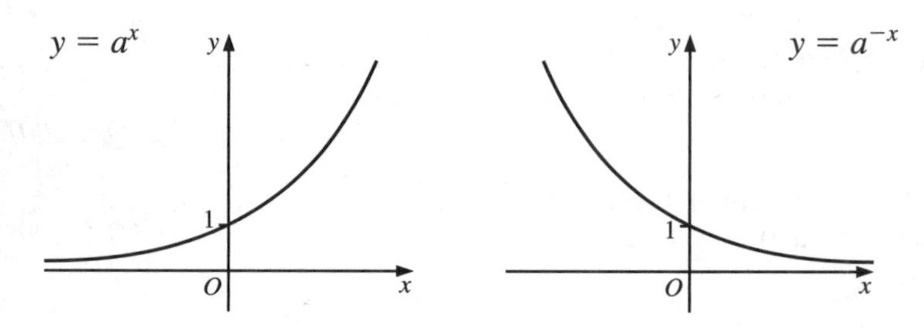

In C4 you will focus on functions of the form $y = A \times a^{bt+c}$ where A, a, b and c are constants, and t is time. These are examples of **exponential growth** and **exponential decay**.

The function $y = a^x$ is an exponential function to the base a.

Example 1

A colony of bacteria grows according to the model $P = A \times 2^t$, where P is the size of the population at time t hours. It is known that at the start of observation the colony has 1000 members.

a) State the value of A.

b) Find the size of the population after 5 hours.

c) Find the time at which the population will exceed 1 000 000 members.

Exponential functions are often used to model the growth of populations of micro-organisms such as bacteria.

a) At $t = 0$, $P = 1000$. Substituting these values into the formula $P = A \times 2^t$ gives

$$1000 = A \times 2^0$$

So $A = 1000$

b) Substituting $t = 5$ in the formula gives

$$P = 1000 \times 2^5$$
$$= 32\,000$$

After 5 hours the size of the population is 32 000.

c) Substituting $P = 1\,000\,000$ into the formula $P = 1000 \times 2^t$ gives

$$1\,000\,000 = 1000 \times 2^t$$
$$\therefore \quad 2^t = \frac{1\,000\,000}{1000}$$
$$= 1000$$

Take logs of both sides to get

$$t \log 2 = \log 1000$$
$$\therefore \quad t = \frac{\log 1000}{\log 2}$$
$$= 9.97$$

So it takes about 10 hours for the population to exceed 1 000 000 members.

C4

Example 2

A type of mould grows according to the formula

$$N = 3(2.7)^{kt}$$

where N is the quantity of mould in grams after t days.

a) What is the initial amount of mould?

After 4 days there are 5.5 grams of mould.

b) Calculate the value of the constant k to 3 decimal places.

c) After what period of time does the quantity of mould reach 10 grams?

Mould is another type of micro-organism that can grow exponentially.

a) Initially $t = 0$, so $N = 3(2.7)^0$
$= 3$

Initially there are 3 grams of mould.

b) When $t = 4$, $N = 5.5$, giving

$$5.5 = 3(2.7)^{4k}$$

$$(2.7)^{4k} = \frac{5.5}{3}$$

$$4k \log 2.7 = \log\left(\frac{5.5}{3}\right)$$

$\therefore \quad k = 0.153$ (to 3 s.f.)

$$4k = \frac{\log 1.833}{\log 2.7}$$
$$4k = 0.61025$$
$$k = 0.153$$

c) $N = 10$, so

$$10 = 3(2.7)^{0.153t}$$

$$(2.7)^{0.153t} = \frac{10}{3}$$

$$0.153t \log 2.7 = \log\left(\frac{10}{3}\right)$$

$\therefore \quad t = 7.92$ (to 3 s.f.)

There is 10 grams of mould after about 8 days.

C4

The decrease in value, or depreciation, of an item after it is purchased can be modelled as exponential decay.

Example 3

Cheryl buys a car for £10 000. The value of her car decreases by 20% every year.

a) How much will her car be worth after 1 year?

b) How much will it be worth after 4 years?

c) Write down an expression for the value of her car after n years.

d) After how many years will the value of Cheryl's car first be less than £2000?

a) Each year the car loses 20% of its value. Hence, each year it is worth 80% of its value the year before.

After 1 year the value, £V, of the car is given by

$$V = \frac{80}{100} \times £10\,000$$
$$= £8000$$

b) After 4 years the value of the car is

$$V = \frac{80}{100} \times \frac{80}{100} \times \frac{80}{100} \times \frac{80}{100} \times £10\,000$$
$$= (0.8)^4 \times £10\,000$$
$$= £4096$$

c) After n years the value of the car is

$$V = (0.8)^n \times 10\,000$$

d) Substituting $V = 2000$ into the formula $V = (0.8)^n \times 10\,000$ gives

$$2000 = (0.8)^n \times 10\,000$$

$$\therefore \quad (0.8)^n = \frac{2000}{10\,000} = 0.2$$

Taking logs:

$$n \log(0.8) = \log(0.2)$$

$$\therefore \quad n = \frac{\log(0.2)}{\log(0.8)} = 7.21$$

The value of the car is first less than £2000 after 7.21 years.

According to the model, the value of Cheryl's car falls below £2000 early in her eighth year of ownership.

Exercise 10A

1 A population of ants grows according to the model $P = A \times 3^t$ where P is the size of the population at time t days. It is known that at the start of observation the size of the population is 500.

a) State the value of A.

b) Find the size of the population after one week.

c) Find, to 2 decimal places, the time at which the size of the population will exceed 1 000 000 ants.

2 A liquid cools in such a way that at a time t minutes its temperature, T°C, is given by the formula $T = 80 \times 2^{-t}$.

a) Write down the initial temperature of the liquid.

b) Find the temperature of the liquid after 8 minutes.

c) Find the time at which the temperature of the liquid first reaches 5 °C.

3 The value of a car depreciates in such a way that t years after it is new its value, £V, is given by the formula $V = 15\,000 \times (1.4)^{-kt}$.

a) What is the initial value of the car?

After 2 years the value of the car is £8000.

b) Calculate the value of the constant k to three decimal places.

c) Calculate the time, to the nearest month, when the value of the car reaches £5000.

4 The number, N, of rabbits on an island grows in such a way that after t months the number of rabbits present is given by the formula $V = 6 \times (5)^{kt}$.

a) What is the initial size of the population?

After 10 months there are 90 rabbits on the island.

b) Calculate the value of the constant k to three decimal places.

c) Calculate the time, to the nearest month, when the size of the population first exceeds one thousand.

5 A man invests 600 euros in a bank account which pays compound interest at the rate of 3% every year.

a) How much will he have after 1 year?

b) How much will he have after 5 years?

c) Write down an expression for the total value of his investment after n years.

d) After how many years will the value of his investment first exceed 1000 euros?

6 The height of a tree increases by 7% every year. When planted the tree is 2 metres high.

a) What will be its height after 1 year?

b) What will be its height after 10 years?

c) Write down an expression for the height of the tree after n years.

d) Assuming that the increase remains steady at 7% per year, after how many years will the height of the tree first exceed 20 metres?

7 The population of a city is 600 000 in 2004 and is increasing by 4% each year.

a) What is the growth factor?

b) What will be the total percentage increase in population after 5 years?

c) Write an expression for the percentage increase after n years.

d) Assuming the increase remains steady at 4% per year, after how many years will the population first exceed 1 million?

8 The formula $P = A(1.05)^t$ models the population, P, of a bacterial culture at time t hours.

a) What does A represent?

b) A culture starts with 100 bacteria. Find the population after 24 hours.

c) What simplifying assumptions are made in using this model?

d) Why might the model not be reliable in giving the population after one week?

10.2 The exponential function

In this section you will study exponential functions to the base e, that is, functions of the form

$$y = Ae^{kx}, \text{ where } A \text{ is a constant.}$$

Example 4

A car had a value of £16 000 when new, and a value of £9000 exactly 3 years later. The value of the car as it depreciates can be modelled by $V = Ae^{-kt}$ where £V is the value of the car t years after it was sold as new, and A and k are constants.

a) State the value of A.

b) Find the value of k correct to three significant figures.

c) Sketch the graph of V against t.

d) Calculate the value of the car after 9 years, giving your answer to the nearest £50.

e) After how many years will the value of the car be less than £1000?

---

a) At $t = 0$, $V = 16\,000$. Substituting these values into the formula $V = Ae^{-kt}$ gives

$$16\,000 = A \times e^0$$

So $A = 16\,000$, and $V = 16\,000e^{-kt}$

$e^0 = 1$

b) At $t = 3$, $V = 9000$. Substituting these values into the formula $V = 16\,000e^{-kt}$ gives

$$9000 = 16\,000 \times e^{-3k}$$

$$\therefore \quad e^{-3k} = \frac{9000}{16\,000} = 0.5625$$

Take logs of both sides to get

$$-3k = \ln(0.5625)$$

$$\therefore \quad k = -\frac{\ln(0.5625)}{3} = 0.192$$

Use natural logarithms, $\log_e x$ or $\ln x$. Remember that $\ln e^x = x$

c) The graph of $V = 16\,000e^{-0.192t}$ is shown.

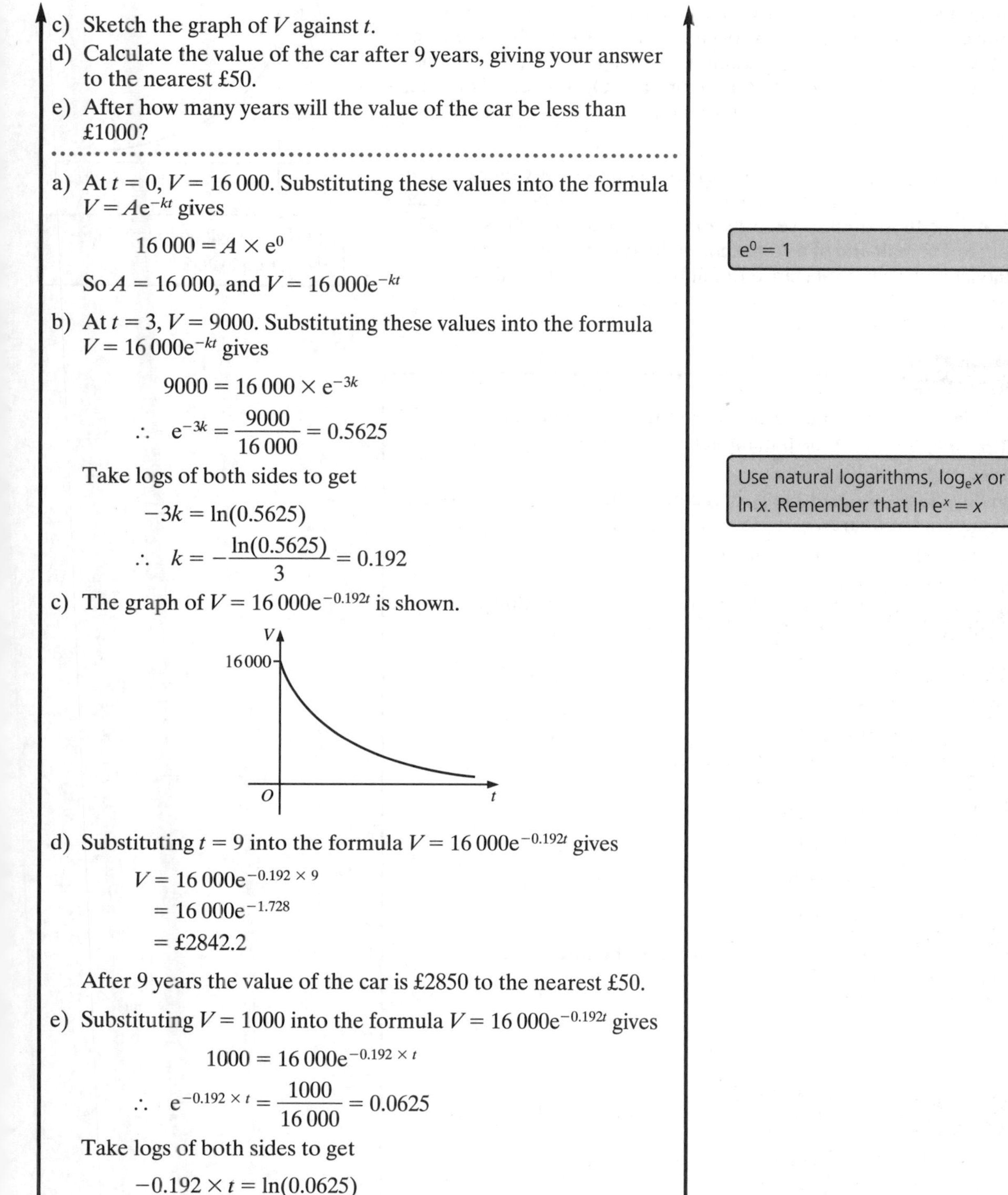

d) Substituting $t = 9$ into the formula $V = 16\,000e^{-0.192t}$ gives

$$V = 16\,000e^{-0.192 \times 9}$$
$$= 16\,000e^{-1.728}$$
$$= £2842.2$$

After 9 years the value of the car is £2850 to the nearest £50.

e) Substituting $V = 1000$ into the formula $V = 16\,000e^{-0.192t}$ gives

$$1000 = 16\,000e^{-0.192 \times t}$$

$$\therefore \quad e^{-0.192 \times t} = \frac{1000}{16\,000} = 0.0625$$

Take logs of both sides to get

$$-0.192 \times t = \ln(0.0625)$$

$$\therefore \quad t = -\frac{\ln(0.0625)}{0.192} = 14.4$$

So the value of the car first reaches £1000 after about $14\frac{1}{2}$ years.

C4

One application of the exponential function is to the decay of a radioactive substance. The radioactivity of a particular substance may be monitored using a Geiger counter – this 'counts' the number of radioactive particles produced per unit time. Over a period of time, the reading on the Geiger counter meter decreases – this is an example of exponential decay.

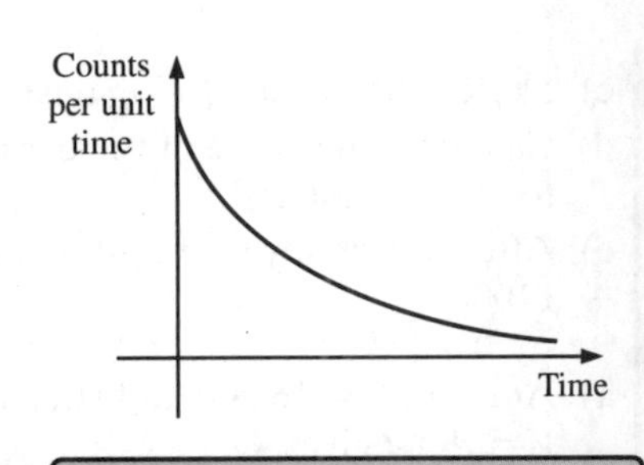

The radioactivity from some substances decreases very quickly, and there may be little left after a few microseconds. For other substances it may take thousands of years before you notice even the smallest change. The **half-life** of a substance is the time taken for the radioactivity of that substance to fall by half of its original value.

Specific knowledge of half-life will not be tested.

Example 5

A radioactive substance decays according to the formula $m = m_0 e^{-kt}$ where m_0 is the initial mass, k is the decay constant, and t is the time in seconds.

a) Given that the half-life is 20 seconds, find the value of k to four significant figures.

b) If $k = 0.8\ \text{s}^{-1}$, calculate the half-life.

C4

a) The half-life is 20 seconds, so at $t = 20$, $m = \frac{1}{2}m_0$. Substituting these values into the formula $m = m_0 e^{-kt}$ gives

$$\tfrac{1}{2}m_0 = m_0 e^{-20k}$$
$$\therefore \quad \tfrac{1}{2} = e^{-20k}$$

Take logs of both sides to give

$$\ln\left(\tfrac{1}{2}\right) = -20k$$
$$\therefore \quad k = -\frac{\ln\left(\frac{1}{2}\right)}{20}$$
$$= 0.034\,66$$

The value of k is $0.034\,66\ \text{s}^{-1}$.

b) Substituting $m = \frac{1}{2}m_0$ and $k = 0.8$ into the formula $m = m_0 e^{-kt}$ gives

$$\tfrac{1}{2}m_0 = m_0 e^{-0.8t}$$
$$\therefore \quad \tfrac{1}{2} = e^{-0.8t}$$

Take logs of both sides to give

$$\ln\left(\tfrac{1}{2}\right) = -0.8t$$
$$\therefore \quad t = -\frac{\ln\left(\frac{1}{2}\right)}{0.8}$$
$$= 0.866$$

The half-life is 0.866 seconds.

Exercise 10B

1 The value of a car depreciates in such a way that when it is t years old, its value $\$V$ is given by the formula

$$V = 12\,000e^{-0.6t}$$

a) What is the purchase price of the car?

b) What is the value of the car after 3 years?

c) The owner decides to sell the car when its value has halved. Find, to the nearest month, the age of the car when it is sold.

2 Rabbits are introduced to a deserted island and the population grows in such a way that at a time t months, the size of the population, N, is given by formula

$$N = 20e^{0.1t}$$

a) What is the initial size of the population?

b) How many rabbits are on the island after one year?

c) How many months does it take for the population to double?

3 A pan of water is heated and then allowed to cool. Its temperature, T °C, t minutes after cooling has begun, is given by the formula

$$T = 80e^{-0.2t}$$

a) What is the temperature of the water when it starts to cool?

b) What is the temperature after 5 minutes?

c) Find, to the nearest minute, the time taken for the water to reach a temperature of 12 °C.

4 A pan of soup is heated. It is then removed from the heat and allowed to cool. Its temperature, T °C, after t minutes is given by $T = 100e^{-0.1t}$

a) What is its temperature when first removed from the heat?

b) What is its temperature after 3 minutes?

c) Find the time for it to cool to 50 °C.

5 A radioactive substance decays according to the formula $m = m_0e^{-kt}$ where m_0 is the initial mass, k is the decay constant and t is the time in years. The half-life is the time taken for half the material to decay.

a) If the half-life is 100 years find the value of k to three significant figures.

b) If $k = 0.2$ find the half-life.

Summary

You should know how to ...	Check out
1 Solve problems involving an exponential model.	**1** An investment bond grows in value according to the formula $P = A.k^n$. where A is the initial value of the bond, k is a constant, and P is the value of the bond after n years. a) If it is assumed an initial investment of £1000 will grow to £3000 in 15 years, find the value of k, giving your answer to three decimal places. b) Show that the growth in a) is approximately an annual percentage increase of 7.6%. c) Find the value of the investment in a) after 30 years. Give your answer to the nearest £10. d) How long will it take for the investment in a) to be worth more than £90 000?
2 Solve problems involving the exponential constant.	**2** A certain species of bacteria causes a disease. A microbiologist is conducting an experiment to see how quickly the bacteria can be destroyed. She uses the model $N = 5000e^{\frac{10-t}{4}}$, where N is the number of bacteria remaining t minutes after a destructive agent starts working. It takes 10 minutes before the bacteria start to be destroyed. a) How many bacteria will there be i) after 10 minutes ii) after 15 minutes? b) Find when the number of bacteria first falls below 1000. c) Find the number of bacteria remaining after 1 hour. Interpret your result.

C4

Revision exercise 10

1 The amount of money, £P, in a special savings account at time t years after 1 January 2000 is given by $P = 100 \times 1.05^t$.

a) State the amount of money in the account on January 2000.

b) Calculate, to the nearest penny, the amount of money in the account on 1 January 2004.

c) Find the value of t when $P = 150$, giving your answer to three significant figures.

(*AQA, 2004*)

2 A microbiologist is studying the growth of simple organisms.
For one such organism, the model proposed is $P = 100 - 50e^{-\frac{1}{4}t}$
where P is the population after t minutes.

a) Write down
   i) the value of the initial population
   ii) the value which the population approaches as t becomes large.

b) Find the time at which the population will have a value of 75, giving your answer to two significant figures. *(AQA, 2004)*

3 The population, P, of insects in a colony is given by $P = Ae^{kt}$, where A and k are constants and the time t is measured in months.

a) Given that $P = 500$ when $t = 0$ and that $P = 750$ when $t = 10$, find the value of k.

b) Find the value of t when $P = 1500$, giving your answers to 3 significant figures. *(AQA, 2004)*

4 Observations were made of the number of bacteria in a certain specimen.
The number N present after t minutes is modelled by the formula

$N = Ac^t$ where A and c are constants.

Initially there are 1000 bacteria in the colony.

a) Write down the value of A.

b) Given that there are 12 000 bacteria after 60 minutes, show that the value of c is 1.0423 to four decimal places.

c) i) Express t in terms of N.
   ii) Calculate, to the nearest minute, the time taken for the number of bacteria to increase from one thousand to one million. *(AQA, 2003)*

C4

5 The decay of a radioactive substance can be modelled by the equation $m = m_0e^{-kt}$, where m grams is the mass at time t years and m_0 grams is the initial mass, and k is a constant.

a) The time taken for a sample of the radioactive substance strontium-90 to decay to half of its mass is 28 years. Show that the value of k is approximately 0.024 755.

b) A sample of strontium-90 has a mass of 1 gram. Assuming the mass has resulted from radioactive decay, use the model to find the mass this sample would have had 100 years ago, giving your answer to three significant figures. *(AQA, 2002)*

6 a) A car has a value of £15 000 when new and a value of £11 000 exactly two years later. The value of the car as it depreciates can be modelled by $V = Pe^{-kt}$ where £V is the value of the car t years after it is sold as new, and P and k are constants.
   i) State the value of P.
   ii) Find the value of k, giving your answer to three decimal places.

b) Another car depreciates in value according to the model $W = 18\,000e^{-0.175t}$ where £W is the value t years after it is sold as new.

Assuming that both cars were sold as new on 1 January 2000, calculate the year during which they will have depreciated to the same value. *(AQA, 2002)*

11 Further calculus

This chapter will show you how to

- ✦ Recognise and use implicit functions
- ✦ Recognise and use parametric equations
- ✦ Use partial fractions to integrate rational functions
- ✦ Solve problems involving differential equations

Before you start

You should know how to ...	Check in
1 Differentiate and integrate functions involving x^n, $\sin ax$ and e^{ax}.	**1** a) Find $\frac{dy}{dx}$ when: i) $y = 3x^4 - \frac{4}{x^3}$ ii) $y = 3\cos 6x$ iii) $y = 2e^{-4x}$ b) Integrate with respect to x: i) $\frac{1}{2x} - \frac{4}{\sqrt{x}}$ ii) $4\sin\frac{1}{2}x$ iii) $\frac{6}{e^{\frac{3}{2}x}}$
2 Differentiate using the chain rule, product rule and quotient rule.	**2** Differentiate with respect to x: a) $y = (4 - x^2)^5$ b) $y = e^{-x}\cos 2x$ c) $\frac{x^2}{2e^{3x}}$
3 Integrate using substitution and integration by parts.	**3** Integrate with respect to x: a) $\frac{x}{\sqrt{4 - x^2}}$ b) $x\cos 2x$ c) x^2e^{3x}
4 Express a rational function in partial fractions.	**4** Express $\frac{x + 11}{(x - 3)(2x + 1)}$ in the form $\frac{A}{x - 3} + \frac{B}{2x + 1}$.

C4

11.1 Implicit functions

So far, most of the curves you have dealt with had equations in the form $y = f(x)$. In other words, the variable y is given explicitly in terms of x. For example, $y = x^3 + 3x + 2$ is an explicit function.

Some curves are defined by **implicit functions**. That is, functions which are *not* expressed in the form $y = f(x)$. For example, $x^2 + 3xy = 4$ is an implicit function. Notice that this implicit function can be rearranged to give an explicit function of x:

$$y = \frac{4 - x^2}{3x}$$

In section 2.5 you learnt to differentiate simple implicit functions such as $x = y^3$.

However, many implicit functions *cannot* be expressed in the form $y = f(x)$. For example, the implicit function

$$x^2 + 3xy - 4y^3 = 7$$

cannot be expressed in the form $y = f(x)$. So you need a technique for differentiating implicit functions.

Implicit differentiation

Consider the implicit function

$$x^2 + y^2 = 2$$

Differentiating each term with respect to x gives

$$\frac{d}{dx}(x^2) + \frac{d}{dx}(y^2) = \frac{d}{dx}(2) \qquad [1]$$

By the chain rule,

$$\frac{d}{dx}(y^2) = \frac{d}{dy}(y^2)\frac{dy}{dx} = 2y\frac{dy}{dx}$$

So [1] becomes

$$2x + 2y\frac{dy}{dx} = 0$$

Rearranging for $\frac{dy}{dx}$ gives

$$\frac{dy}{dx} = -\frac{2x}{2y} = -\frac{x}{y}$$

C4

Example 1

Find $\frac{dy}{dx}$ for each of these functions.

a) $x^2 - 6y^3 + y = 0$

b) $x^2y = 5x + 2$

c) $(x + y)^5 - 7x^2 = 0$

d) $\frac{x^3}{x + y} = 2$

a) Differentiating each term of $x^2 - 6y^3 + y = 0$ with respect to x gives

$$2x - 18y^2\frac{dy}{dx} + \frac{dy}{dx} = 0$$

Rearranging for $\frac{dy}{dx}$ gives

$$\frac{dy}{dx}(1 - 18y^2) = -2x$$

$$\therefore \quad \frac{dy}{dx} = \frac{-2x}{1 - 18y^2} = \frac{2x}{18y^2 - 1}$$

b) To differentiate each term of $x^2y = 5x + 2$ with respect to x you need to use the product rule. This gives

$$x^2\frac{dy}{dx} + y(2x) = 5$$

$$\therefore \quad x^2\frac{dy}{dx} = 5 - 2xy$$

$$\therefore \quad \frac{dy}{dx} = \frac{5 - 2xy}{x^2}$$

c) Differentiating each term of $(x + y)^5 - 7x^2 = 0$ with respect to x gives

$$5(x + y)^4\left(1 + \frac{dy}{dx}\right) - 14x = 0$$

$$\therefore \quad 1 + \frac{dy}{dx} = \frac{14x}{5(x + y)^4}$$

$$\therefore \quad \frac{dy}{dx} = \frac{14x}{5(x + y)^4} - 1$$

Hint for (c)

You can use the chain rule to differentiate $(x + y)^5$.

Let $u = x + y$

$$\frac{d}{dx}(u^5) = 5u^4\frac{du}{dx}$$

but $\frac{du}{dx} = \frac{d}{dx}(x + y)$

$$= 1 + \frac{dy}{dx}$$

So $\frac{d}{dx}(x + y)^5 = 5(x + y)^4\left(1 + \frac{dy}{dx}\right)$

d) $\frac{x^3}{x + y} = 2$. Multiplying throughout by $(x + y)$ gives

$$x^3 = 2x + 2y$$

Differentiating each term with respect to x gives

$$3x^2 = 2 + 2\frac{dy}{dx}$$

$$\therefore \quad 2\frac{dy}{dx} = 3x^2 - 2$$

$$\therefore \quad \frac{dy}{dx} = \frac{3x^2 - 2}{2}$$

You may need to use implicit differentiation to find the equations of tangents and normals to a curve at a given point.

Example 2

Given $(1 + y)^2 + (1 + x^2)^2 = 13$, find an equation of the tangent to the curve at the point $(-1, 2)$.

Since $(1 + y)^2 + (1 + x^2)^2 = 13$, differentiating implicitly gives

$$2(1 + y)\frac{dy}{dx} + 4x(1 + x^2) = 0$$

At $x = -1, y = 2$, so

$$2(1 + 2)\frac{dy}{dx} + 4(-1)(1 + (-1)^2) = 0$$

$$6\frac{dy}{dx} - 8 = 0$$

$$\therefore \quad \frac{dy}{dx} = \frac{4}{3}$$

$(1 + y)^2$ and $(1 + x^2)^2$ can both be differentiated using the chain rule. There is no need to multiply out the brackets.

So the gradient at the point $(-1, 2)$ is $\frac{4}{3}$ and the tangent to the curve at that point has equation of the form $y = \frac{4}{3}x + c$.
Since the tangent passes through $(-1, 2)$,

$$2 = \frac{4}{3}(-1) + c$$

$$\therefore \quad c = \frac{10}{3}$$

The equation of the tangent at $(-1, 2)$ is $y = \frac{4}{3}x + \frac{10}{3}$ or $3y - 4x = 10$.

Example 3

Given $x^2 + 3xy + 2y^2 = 10$

a) find an expression for $\frac{dy}{dx}$ in terms of x and y

b) find the equations of the normals to the curve at the points where $x = -1$.

a) Since $x^2 + 3xy + 2y^2 = 10$, differentiating implicitly gives

$$2x + \left(3x\frac{dy}{dx} + 3y\right) + 4y\frac{dy}{dx} = 0$$

$$\therefore \quad \frac{dy}{dx}(3x + 4y) = -2x - 3y$$

$$\therefore \quad \frac{dy}{dx} = -\frac{2x + 3y}{3x + 4y}$$

b) To find the y-coordinates of the points on the curve where $x = -1$, substitute $x = -1$ into $x^2 + 3xy + 2y^2 = 10$. That is,

$$(-1)^2 + 3(-1)y + 2y^2 = 10$$

$$1 - 3y + 2y^2 = 10$$

$$2y^2 - 3y - 9 = 0$$

$$\therefore \quad (2y + 3)(y - 3) = 0$$

Solving gives $y = -\frac{3}{2}$ and $y = 3$.

The points on the curve at which $x = -1$ are $P\left(-1, -\frac{3}{2}\right)$ and $Q(-1, 3)$. To find the equation of the normal at each point, you need the gradient of the curve at each point.

At the point $P(-1, -\frac{3}{2})$,

$$\left.\frac{dy}{dx}\right|_{x=-1,\,y=-\frac{3}{2}} = -\frac{-2(-1) + 3\left(-\frac{3}{2}\right)}{3(-1) + 4\left(-\frac{3}{2}\right)} = -\frac{13}{18}$$

This is the gradient of the tangent at P.

At the point $Q(-1, 3)$,

$$\left.\frac{dy}{dx}\right|_{x=-1,\,y=3} = -\frac{-2(-1) + 3(3)}{3(-1) + 4(3)} = -\frac{7}{9}$$

C4

The normal to the curve at point P has a gradient of

$$\frac{-1}{\left(-\frac{13}{18}\right)} = \frac{18}{13}$$

So, its equation has the form $y = \frac{18}{13}x + c_1$.
Since the normal passes through $P\left(-1, -\frac{3}{2}\right)$,

$$-\tfrac{3}{2} = \tfrac{18}{13}(-1) + c_1$$

$$\therefore \quad c_1 = -\tfrac{3}{26}$$

The equation of the normal to the curve at P is

$$y = \tfrac{18}{13}x - \tfrac{3}{26} \quad \text{or} \quad 26y = 36x - 3$$

The normal to the curve at Q has a gradient of

$$\frac{-1}{\left(-\frac{7}{9}\right)} = \frac{9}{7}$$

So, its equation has the form $y = \frac{9}{7}x + c_2$.
Since the normal passes through the point $Q(-1, 3)$,

$$3 = \tfrac{9}{7}(-1) + c_2$$

$$\therefore \quad c_2 = \tfrac{30}{7}$$

The equation of the normal to the curve at Q is

$$y = \tfrac{9}{7}x + \tfrac{30}{7} \quad \text{or} \quad 7y = 9x + 30$$

Remember:
If the gradient of the tangent at a point is m, then the gradient of the normal is $-\frac{1}{m}$.

C4

Exercise 11A

1 For each of these curves express $\frac{dy}{dx}$ in terms of x and y.

a) $x^2 - y^3 = 4$ b) $3xy - y^2 = 7$ c) $x^2y + xy^2 = 2$
d) $2x - y^3 = 3xy$ e) $x^4 - xy^2 = 6x$ f) $x^6 - 5xy^3 = 9xy$
g) $x^2(x - 3y) = 4$ h) $\frac{x^2}{x + y} = 2$ i) $\frac{y}{x^2 - 7y^3} = x^5$

2 In each part of this question find the gradient of the stated curve at the point specified.

a) $xy^2 - 6y = 8$ at $(2, -1)$
b) $x^4 - y^3 = 2$ at $(1, -1)$
c) $3y^4 - 7xy^2 - 12y = 5$ at $(-2, 1)$
d) $xy^3 - x^2y = 6$ at $(3, 2)$
e) $(x + y)^2 - 4x + y + 10 = 0$ at $(2, -3)$
f) $\frac{x^2}{x - y} = 8$ at $(4, 2)$
g) $\frac{2}{x} + \frac{5}{y} = 2xy$ at $\left(\frac{1}{2}, 5\right)$
h) $(x + 2y)^4 = 1$ at $(5, -2)$

3 Find the equation of the tangent to the curve $xy^2 + x^2y = 6$ at the point $(1, -3)$.

4 Find the equations of the tangent and the normal to the curve $xy^2 + 3x - 2y = 6$ at the point $(2, 1)$.

5 Find the equations of the tangent and the normal to the curve $xy = 2$ at the point $\left(6, \frac{1}{3}\right)$.

6 Find the equations of the tangents to the curve $x^2y - xy^2 = 12$ at the points where $y = 3$.

7 At what points are the tangents to the circle $x^2 + y^2 - 4x - 6y + 9 = 0$ parallel to the x-axis?

8 Find the equation of the tangents to the curve $x^2 + 3x - 2y^2 = 4$ at the points where the curve crosses the x-axis.

11.2 Parametric equations

In some cases, y is defined as a function of x by expressing both y and x in terms of a third variable known as a parameter. Such equations are called **parametric equations**.

The reason for using parametric equations is that it often makes solving problems in coordinate geometry easier and quicker.

For example, the equations

$$x = t + 1 \quad [1] \qquad y = t^2 \quad [2]$$

are parametric equations, with the parameter being t. In fact, these parametric equations define the parabola with equation

$$y = x^2 - 2x + 1$$

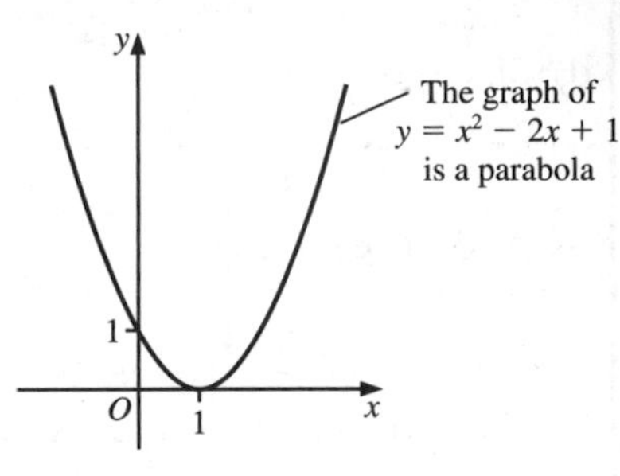

C4

You can check this by eliminating the parameter t between [1] and [2]. From [1], $t = x - 1$. Substituting into [2] gives

$$y = (x - 1)^2$$
$$\therefore \quad y = x^2 - 2x + 1$$

Example 4

Find a cartesian equation for each of these parametric forms.

a) $x = \sqrt{t}, \quad y = 2t^2 - 1$

b) $x = \dfrac{1}{t}, \quad y = 3t - 1$

c) $x = t + \dfrac{1}{t}, \quad y = t - \dfrac{1}{t}$

d) $x = 3\sin t, \quad y = 2\cos t$

a) When $x = \sqrt{t}$, then $x^2 = t$ and $x^4 = t^2$. Substituting into $y = 2t^2 - 1$ gives

$$y = 2x^4 - 1$$

The cartesian equation is $y = 2x^4 - 1$.

b) When $x = \frac{1}{t}$, then $t = \frac{1}{x}$. Substituting into $y = 3t - 1$ gives

$$y = 3\left(\frac{1}{x}\right) - 1 = \frac{3}{x} - 1$$

The cartesian equation is $y = \frac{3}{x} - 1$ or, multiplying throughout by x, gives $xy + x = 3$.

c) When $x = t + \frac{1}{t}$ (1)

and $y = t - \frac{1}{t}$ (2)

(1) + (2) gives $x + y = 2t$

(1) − (2) gives $x - y = \frac{2}{t}$

Multiplying these two results gives

$$(x + y)(x - y) = 2t \times \frac{2}{t}$$

$$\therefore \quad x^2 - y^2 = 4$$

The cartesian equation is $x^2 - y^2 = 4$.

C4

d) Since $x = 3 \sin t$,

$$\sin t = \frac{x}{3}$$

And since $y = 2 \cos t$,

$$\cos t = \frac{y}{2}$$

Using the identity $\sin^2 t + \cos^2 t = 1$ gives

$$\left(\frac{x}{3}\right)^2 + \left(\frac{y}{2}\right)^2 = 1$$

or $\quad 4x^2 + 9y^2 = 36$

This is an example of an ellipse, as shown in the graph.

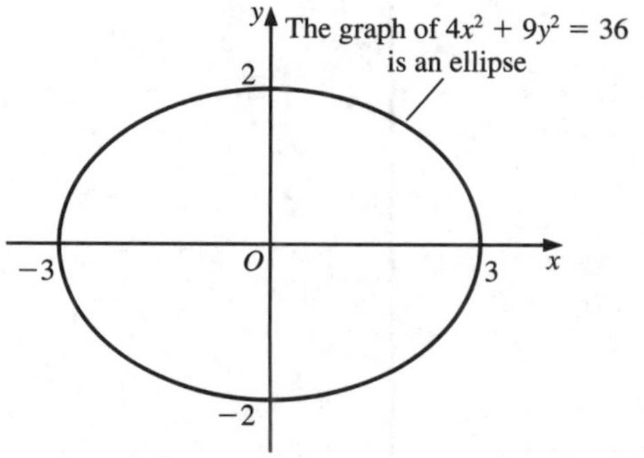

As Example 4 shows, it is often possible to eliminate the parameter t and get the equation in cartesian form. You can then differentiate the cartesian expression. However, it is not always possible to eliminate the parameter and so you need a technique for differentiating parametric equations.

Parametric differentiation

Consider the parametric equations

$$x = t + 1 \quad \text{and} \quad y = t^2$$

You can differentiate x with respect to t, and you can differentiate y with respect to t. This gives

$$\frac{dx}{dt} = 1 \quad \text{and} \quad \frac{dy}{dt} = 2t$$

By the chain rule,

$$\frac{dy}{dx} = \frac{dy}{dt}\frac{dt}{dx} = \frac{\left(\frac{dy}{dt}\right)}{\left(\frac{dx}{dt}\right)}$$

$$\therefore \quad \frac{dy}{dx} = 2t(1) = 2t$$

In this particular case, you know that $t = x - 1$. Therefore,

$$\frac{dy}{dx} = 2(x - 1) = 2x - 2$$

Notice that the derivative of $y = x^2 - 2x + 1$ would be $2x - 2$.

Example 5

Find $\frac{dy}{dx}$ in terms of the parameter, t, for each of these.

a) $y = 3t^2 + 2t$, $x = 1 - 2t$ b) $y = (1 + 2t)^3$, $x = t^3$

a) When $y = 3t^2 + 2t$: $\frac{dy}{dt} = 6t + 2$

When $x = 1 - 2t$: $\frac{dx}{dt} = -2$

By the chain rule,

$$\frac{dy}{dx} = \frac{dy}{dt}\frac{dt}{dx}$$
$$= (6t + 2)\left(-\frac{1}{2}\right)$$
$$\therefore \quad \frac{dy}{dx} = -3t - 1$$

b) When $y = (1 + 2t)^3$: $\frac{dy}{dt} = 3(1 + 2t)^2(2)$

$$\therefore \quad \frac{dy}{dt} = 6(1 + 2t)^2$$

When $x = t^3$: $\frac{dx}{dt} = 3t^2$

By the chain rule,

$$\frac{dy}{dx} = \frac{dy}{dt}\frac{dt}{dx} = 6(1 + 2t)^2 \times \frac{1}{3t^2}$$

$$\therefore \quad \frac{dy}{dx} = \frac{2(1 + 2t)^2}{t^2}$$

Again, you can apply this technique to finding equations of tangents and normals to a curve.

Example 6

Find the equation of the tangent to the curve given parametrically by $x = \frac{2}{t}$ and $y = 3t^2 - 1$, at the point where $t = 1$.

When $t = 1$, $x = \frac{2}{1} = 2$

$$y = 3 \times 1^2 - 1 = 2$$

So $t = 1$ at the point (2, 2) on the curve.
Differentiating parametrically gives

$$\frac{dy}{dt} = 6t \quad \text{and} \quad \frac{dx}{dt} = -\frac{2}{t^2}$$

By the chain rule

$$\frac{dy}{dx} = \frac{dy}{dt} \cdot \frac{dt}{dx} = 6t\left(-\frac{t^2}{2}\right) \quad \therefore \quad \frac{dy}{dx} = -3t^3$$

When $t = 1$,

$$\left.\frac{dy}{dx}\right|_{t=1} = -3$$

The gradient of the tangent is -3.
Hence, the equation of the tangent is of the form

$$y = -3x + c$$

Since the tangent line passes through the point (2, 2),

$$2 = -3(2) + c \quad \therefore \quad c = 8$$

The equation of the tangent is $y = -3x + 8$.

In the next example a trigonometric equation is given in parametric form.

C4

Example 7

Find the equation of the normal to the curve given parametrically by $x = 2\sin t$, $y = 5 - 4\cos 2t$, at the point, P, where $t = \frac{\pi}{6}$.

At $t = \frac{\pi}{6}$, $x = 2\sin\left(\frac{\pi}{6}\right) = 2\left(\frac{1}{2}\right) = 1$

and $y = 5 - 4\cos\left(2\left(\frac{\pi}{6}\right)\right) = 5 - 4\cos\left(\frac{\pi}{3}\right) = 5 - 4\left(\frac{1}{2}\right) = 3$

So P is the point $(1, 3)$.

Differentiating parametrically gives

$$\frac{dy}{dt} = 8\sin 2t \quad \text{and} \quad \frac{dx}{dt} = 2\cos t$$

By the chain rule

$$\frac{dy}{dx} = \frac{8\sin 2t}{2\cos t}$$

When $t = \frac{\pi}{6}$, $\left.\frac{dy}{dx}\right|_{t=\frac{\pi}{6}} = \frac{8\sin\left(\frac{\pi}{3}\right)}{2\cos\left(\frac{\pi}{6}\right)} = \frac{8\left(\frac{\sqrt{3}}{2}\right)}{2\left(\frac{\sqrt{3}}{2}\right)} = 4$

So the gradient of the tangent to the curve at P is 4, and the gradient of the normal is $-\frac{1}{4}$.

Hence the equation of the normal is of the form $y = -\frac{1}{4}x + c$

Since the normal passes through the point $P(1, 3)$,

$$3 = -\tfrac{1}{4}(1) + c \quad \therefore \quad c = \tfrac{13}{4}$$

The equation of the normal is $y = -\frac{1}{4}x + \frac{13}{4}$ or $x + 4y = 13$.

C4

Exercise 11B

1 Find a cartesian equation for each of these parametric forms.

a) $x = t + 3, y = t^2$
b) $x = t^2, y = 2t$
c) $x = 2t - 1, y = 12t^2 - 5$
d) $x = t, y = \frac{4}{t}$
e) $x = \frac{1}{t}, y = 3t$
f) $x = \frac{2}{t}, y = 6t - 1$

2 Find an implicit equation for each of these parametric forms.

a) $x = 2\sin t, y = 3\cos t$
b) $x = \cos t, y = 2\sin t$
c) $x = 5\sin t, y = 4\cos t$
d) $x = t + \frac{3}{t}, y = t - \frac{3}{t}$
e) $x = 2t - \frac{1}{t}, y = 2t + \frac{1}{t}$
f) $x = 3t - \frac{2}{t}, y = 3t + \frac{2}{t}$

3 For these curves, each of which is given in terms of a parameter t, find an expression for $\frac{dy}{dx}$ in terms of t.

a) $x = t^2, y = 4t - 1$
b) $x = 3t^4, y = 2t^2 - 3$
c) $x = 2\sqrt{t}, y = 5t - 4$
d) $x = 6t - 5, y = (2t - 1)^3$
e) $x = 4t(t - 2), y = (t - 1)^3$
f) $x = 4\sqrt{t} - t, y = t^2 - 2\sqrt{t}$
g) $x = \frac{1}{t}, y = t^2 + 4t - 3$
h) $x = \frac{2}{3 + \sqrt{t}}, y = \sqrt{t}$
i) $x = \sin t, y = 2 + \cos t$
j) $x = 3 + \cos t, y = 4 - 2\sin t$
k) $x = \sin 2t, y = \sin 6t$
l) $x = t + \cos t, y = t^2 - \cos 2t$

4 In each part of this question find the gradient of the stated curve at the point defined by the given value of the parameter t.

a) $x = t + 5, y = t^2 - 2t$, where $t = 2$
b) $x = t^6, y = 6t^3 - 5$, where $t = -3$
c) $x = (t - 2)^2, y = (3t + 4)^3$, where $t = 0$
d) $x = (2t^3 - 5), y = (1 - 3t)^3$, where $t = 1$
e) $x = \sqrt{t - 1}, y = \frac{1}{t}$, where $t = 10$
f) $x = 5 + \sin t, y = 3 - \cos t$, at $t = \frac{\pi}{4}$
g) $x = 2 + \cos 3t, y = \sin 2t$, at $t = \frac{\pi}{6}$
h) $x = t^2 - \cos 3t, y = 8t - \cos 4t$, at $t = \frac{\pi}{2}$

C4

5 Find the equation of the tangent to the curve $x = 3t^2, y = 7 + 12t$, at the point where $t = 2$.

6 Find the equations of the tangents to the curve $x = t^2, y = 6t - 7$, at the points where $x = 1$.

7 Find the equation of the tangent to the curve $x = 3 - 2\sin t$, $y = 1 + \cos 2t$ at the point where $t = \frac{\pi}{6}$.

8 Find the equation of the normal to the curve $x = \cos t, y = \cos 2t$, at the point where $t = \frac{\pi}{3}$.

9 Find the equation of the tangent and the normal to the curve $x = 6t^2, y = t^3 - 4t$ at the point where $t = -1$.

10 Find the equations of the tangents to the curve $x = \frac{4}{t}$, $y = t^2 - 3t + 2$, at the points where the curve crosses the x-axis.

11 At which points are the tangents to the curve $x = 2t^2 - 3$, $y = t^3 - 6t^2 + 9t - 4$, parallel to the x-axis?

12 Find the equations of the normals to the curve $x = \frac{16}{t^2}$, $y = 2t - 3$, at the points where the curve crosses the line $x = 1$.

13 A curve is given by $x = t^2, y = 4t$. The tangent at the point where $t = 2$, meets the tangent at the point where $t = -1$, at the point P. Find the coordinates of P.

14 a) Find the equation of the tangents to the curve $x = 8t + 1$, $y = 2t^2$, at the points $(9, 2)$ and $(-31, 32)$.

b) Show that these tangents are perpendicular, and find the coordinates of the point of intersection of the tangents.

15 a) Given the curve $x = \frac{t}{1+t}, y = \frac{t^2}{1+t}$, show that $\frac{dy}{dx} = t(t+2)$.

b) Hence find the points on the curve where the gradient is 15.

16 a) Given the curve $x = t^2(1 - 3t^2), y = 5t^3(4 - t)$, show that

$$\frac{dy}{dx} = \frac{10t(t-3)}{6t^2 - 1}$$

b) Hence find the values of the parameter t at the points on the curve where the gradient is 8.

17 The curve C is given by the parametric equations $x = 3t + \sin t$ and $y = t + 2\cos t$, for $0 \leqslant t \leqslant 2\pi$.

a) Show that $\frac{dy}{dx} = \frac{1 - 2\sin t}{3 + \cos t}$

b) Hence find the points on C where the gradient is zero.

18 The curve C is given by the parametric equations $x = 1 + 2\sin t$ and $y = \sin t + \cos t$, for $0 \leqslant t \leqslant 2\pi$.

a) Show that $\frac{dy}{dx} = \frac{1 - \tan t}{2}$

b) Hence find the points on C where the gradient is zero.

11.3 Using partial fractions

In Chapter 7 you learnt about partial fractions. For example, you learnt how to express

$$\frac{7x+8}{(x+4)(x-6)} \quad \text{as} \quad \frac{2}{x+4} + \frac{5}{x-6}$$

$$\frac{2x-5}{(x-4)^2} \quad \text{as} \quad \frac{2}{x-4} + \frac{3}{(x-4)^2}$$

$$\frac{2x^2+29x-11}{(2x+1)(x-2)^2} \quad \text{as} \quad \frac{3}{x-2} + \frac{11}{(x-2)^2} - \frac{4}{2x+1}$$

and $$\frac{5x^2-71}{(x+5)(x-4)} \quad \text{as} \quad 5 - \frac{6}{x+5} + \frac{1}{x-4}$$

Expressing functions in partial fractions can often enable you to integrate them.

The integrals of these examples are:

- $\int \frac{7x+8}{(x+4)(x-6)}\,dx = \int \frac{2}{x+4} + \frac{5}{x-6}\,dx$
  $= 2\ln(x+4) + 5\ln(x-6) + c$
- $\int \frac{2x-5}{(x-4)^2}\,dx = \int \frac{2}{x-4} + \frac{3}{(x-4)^2}\,dx = 2\ln(x-4) - \frac{3}{(x-4)} + c$
- $\int \frac{2x^2+29x-11}{(2x+1)(x-2)^2}\,dx = \int \frac{3}{x-2} + \frac{11}{(x-2)^2} - \frac{4}{2x+1}\,dx$
  $= 3\ln(x-2) - \frac{11}{(x-2)} - 2\ln(2x+1) + c$
- $\int \frac{5x^2-71}{(x+5)(x-4)}\,dx = \int 5 - \frac{6}{x+5} + \frac{1}{x-4}\,dx$
  $= 5x - 6\ln(x+5) + \ln(x-4) + c$

Example 8

Find $\int \frac{2x^2+x-5}{x-4}\,dx$.

First, express the integral in partial fractions:

$$\frac{2x^2+x-5}{x-4} \equiv Ax + B + \frac{C}{x-4}$$

This is an improper fraction.

C4

Multiplying throughout by $x-4$ gives

$$2x^2+x-5 = (Ax+B)(x-4) + C$$

Let $x = 4$: $2(4)^2 + 4 - 5 = C$

$\therefore\ C = 31$

$0(x^2)$: $2 = A$

$0(1)$: $-5 = -4B + C$

$-5 = -4B + 31$

$\therefore\ B = 9$

$$\int \frac{2x^2+x-5}{x-4}\,dx = \int \left(2x + 9 + \frac{31}{x-4}\right) dx$$
$$= x^2 + 9x + 31\ln(x-4) + c$$

Example 9

Find $\int \frac{2x-5}{(4x-1)(x+2)}\,dx$.

To resolve into partial fractions, let

$$\frac{2x-5}{(4x-1)(x+2)} \equiv \frac{A}{4x-1} + \frac{B}{x+2}$$

This is a fraction with linear factors in the denominator.

Multiplying throughout by $(4x-1)(x+2)$ gives

$$2x-5 = A(x+2) + B(4x-1)$$

Let $x = -2$: $\quad -9 = -9B \quad \therefore \quad B = 1$

Let $x = \frac{1}{4}$: $\quad -\dfrac{9}{2} = \dfrac{9}{4}A \quad \therefore \quad A = -2$

So:

$$\int \frac{2x-5}{(4x-1)(x+2)}\,dx = \int \frac{-2}{4x-1}\,dx + \int \frac{1}{x+2}\,dx$$
$$= \tfrac{1}{2}\int \frac{-4}{4x-1}\,dx + \int \frac{1}{x+2}\,dx$$
$$= -\tfrac{1}{2}\ln(4x-1) + \ln(x+2) + c$$

Example 10

a) Express $\dfrac{x^2+6x+7}{(x+2)(x+3)}$ in the form $A + \dfrac{B}{x+2} + \dfrac{C}{x+3}$.

b) Hence evaluate $\displaystyle\int_0^2 \frac{x^2+6x+7}{(x+2)(x+3)}\,dx$, giving your answer in the form $a + \ln b$, where a and b are rational numbers.

a) $\dfrac{x^2+6x+7}{(x+2)(x+3)} \equiv A + \dfrac{B}{x+2} + \dfrac{C}{x+3}$

This is an improper fraction.

Multiplying throughout by $(x+2)(x+3)$ gives

$$x^2+6x+7 \equiv A(x+2)(x+3) + B(x+3) + C(x+2)$$

Let $x = -2$: $\quad -1 = B$

Let $x = -3$: $\quad -2 = -C \quad \therefore \quad C = 2$

$0(x^2)$: $\quad 1 = A$

So:

$$\int_0^2 \frac{x^2+6x+7}{(x+2)(x+3)}\,dx = \int_0^2 1 - \frac{1}{x+2} + \frac{2}{x+3}\,dx$$
$$= [x - \ln(x+2) + 2\ln(x+3)]_0^2$$
$$= (2 - \ln 4 + 2\ln 5) - (0 - \ln 2 + 2\ln 3)$$
$$= 2 - \ln 4 + \ln(5^2) + \ln 2 - \ln(3^2)$$
$$= 2 - \ln 4 + \ln 25 + \ln 2 - \ln 9$$
$$= 2 + \ln\left(\frac{25\times 2}{4\times 9}\right)$$
$$= 2 + \ln\left(\frac{25}{18}\right)$$

So $\displaystyle\int_0^2 \frac{x^2+6x+7}{(x+2)(x+3)}\,dx = 2 + \ln\left(\frac{25}{18}\right)$

C4

Exercise 11C

You may find it useful to revise the methods of partial fractions before tackling this exercise. Examples are to be found in Exercise 7F, pages 161–162.

1 Find each of these integrals.

a) $\int \frac{x^2-7}{x-2}\,dx$ b) $\int \frac{x^2+x-11}{x+4}\,dx$

c) $\int \frac{2x^2-x-9}{x-3}\,dx$ d) $\int \frac{4x^2+3}{2x-1}\,dx$

e) $\int \frac{x^3-2x^2-x}{x-2}\,dx$ f) $\int \frac{3x^3+x^2-2x+1}{x+1}\,dx$

g) $\int \frac{3x^3-5x^2-18x+19}{x-3}\,dx$ h) $\int \frac{x^4}{x-1}\,dx$

2 Integrate each of these with respect to x.

a) $\frac{x-3}{(x-2)(x-1)}$ b) $\frac{3x+2}{(x+3)(2x-1)}$

c) $\frac{1}{x^2-4}$ d) $\frac{2x^2-6x+1}{(x-1)(x-4)}$

e) $\frac{8x^2}{(x+5)(x-3)}$ f) $\frac{2x^2-5x+4}{(x-1)(2x-3)}$

g) $\frac{x^2+3x+4}{(x+1)^2}$ h) $\frac{5x-4}{(x-2)^2(x+1)}$

i) $\frac{2x-13}{(5-x)^2(2-x)}$ j) $\frac{4x}{(x-1)^2(x+1)}$

k) $\frac{x+2}{(x+1)^2(2x+1)}$ l) $\frac{13x+12}{(x+2)^2(3x-1)}$

3 Evaluate each of these, giving your answers in the form $a+\ln b$ where a and b are rational numbers.

a) $\int_5^7 \frac{1}{(x-3)(x-4)}\,dx$ b) $\int_{-1}^2 \frac{2x+1}{(2x+3)^2}\,dx$

c) $\int_4^6 \frac{x^3-4x+2}{x(x-2)}\,dx$ d) $\int_4^5 \frac{4}{(x-1)^2(x-3)}\,dx$

e) $\int_0^2 \frac{1}{(x+4)(x+2)}\,dx$ f) $\int_{-3}^3 \frac{x^2+11x+29}{(x+6)(x+5)}\,dx$

g) $\int_0^3 \frac{x+2}{(x+3)^2(x+1)}\,dx$ h) $\int_5^9 \frac{x^3-5x^2+6x-5}{(x-1)(x-4)}\,dx$

4 The following integrals may look similar, but they involve a variety of methods. In each case, select an appropriate method and find the integral.

a) $\int \frac{1}{1+x}\,dx$ b) $\int \frac{x}{1-x}\,dx$ c) $\int \frac{x}{1+x^2}\,dx$

d) $\int \frac{1}{\sqrt{1-x^2}}\,dx$ e) $\int \frac{x}{(1-x)^2}\,dx$ f) $\int \frac{1}{\sqrt{1+x}}\,dx$

g) $\displaystyle\int \frac{1}{1+x^2}\,dx$ h) $\displaystyle\int \frac{1}{(1-x)^2}\,dx$ i) $\displaystyle\int \frac{x}{1+x}\,dx$

j) $\displaystyle\int \frac{1}{1-x^2}\,dx$ k) $\displaystyle\int \frac{x}{\sqrt{1-x^2}}\,dx$ l) $\displaystyle\int \frac{x}{\sqrt{1+x}}\,dx$

5 The diagram shows part of the curve with equation $y = \dfrac{1}{9-x^2}$, together with the line $y = \frac{1}{5}$. The curve and the line meet at the points A and B.

a) Find the coordinates of A and B.

b) Calculate the area of the shaded region.

6 The diagram shows part of the curve with equation $y = \dfrac{x}{(4-x)^2}$, together with the line $y = x$. The curve and the line meet at the origin and at the point P.

a) Find the coordinates of the point P.

b) Calculate the area of the shaded region.

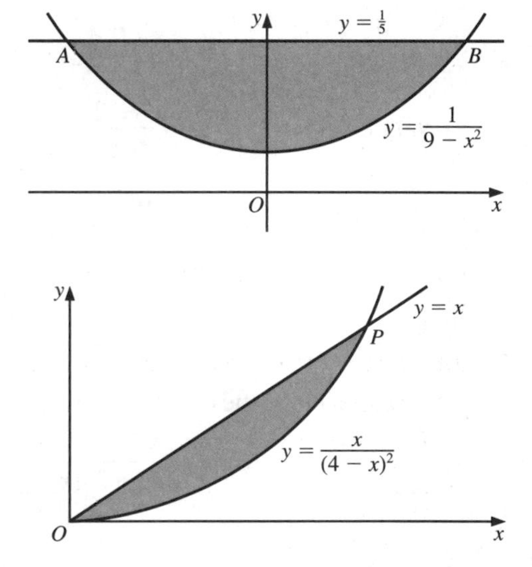

7 a) Sketch the curve with equation $y = \dfrac{4}{(5-x)(x-1)}$.

b) Calculate the area of the finite region bounded by the curve, the x-axis, the line $x = 2$ and the line $x = 4$.

8 The line $y = 4x$ meets the curve $y = \dfrac{9x^2}{25-x^2}$ at the origin and at the points P and Q.

a) Find P and Q.

b) Sketch the line and the curve on the same set of axes.

c) Show that the area of the region, in the positive quadrant, bounded by the line and the curve is given by $68 - 45\ln 3$.

The final sections of this chapter cover differential equations, and in order to be good at these you have to be good at integration. It is all well and good to be able to apply the technique of partial fractions when tackling an exercise which is dedicated to this method, as you have just done. But it is far more tricky when you don't know which method to apply, and you have to decide for yourself. However, as you get more skilled in methods of integration, choosing the right method becomes part of the fun.

The following exercise contains a mix of questions requiring a variety of methods. You don't get any hints, so you have to decide an appropriate method for yourself. In fact, some questions can be done by more than one method.

You may find it useful to spend some time working through some of the questions in Exercise 11D before moving on to the final sections of the chapter.

Exercise 11D

1 Integrate each of these with respect to x.

a) $x(x+3)^2$ b) $\dfrac{x+2}{x(x+1)}$ c) $x\sin x$

d) $\dfrac{4}{(2-x)^2}$ e) $\dfrac{1+x}{\sqrt{1-x^2}}$ f) $\dfrac{x}{x^2-4}$

g) $\sin x\, e^{\cos x}$ h) $\dfrac{x^2+2}{x^3}$ i) $\dfrac{x-5}{x^2-1}$

j) $\ln x$ k) $\dfrac{x}{3-x}$ l) $x\sqrt{x^2+1}$

m) xe^{2x} n) $(x-5)^3$ o) $\sin x\cos^3 x$

p) $\dfrac{x^2+1}{(x+3)(x-2)}$ q) $\dfrac{e^x}{e^x+1}$ r) $x\sqrt{2-x}$

2 Integrate each of these with respect to x.

a) $\dfrac{x-1}{x+2}$ b) $\dfrac{1}{(x+3)(x+4)}$ c) $x^3(x^4-5)^4$

d) $\dfrac{2}{5-x}$ e) xe^{-2x} f) $(x+2)(x-3)$

g) $\dfrac{x^2-9x+5}{(x-1)^2(x+2)}$ h) $\dfrac{x}{x+4}$ i) $\sec^2 3x\tan^2 3x$

j) $\dfrac{x+3}{\sqrt{1-x^2}}$ k) e^{2x} l) $x\cos 2x$

m) $\sqrt{5x-1}$ n) $\dfrac{x^2-x+1}{x(x^2+1)}$ o) $\dfrac{3x^2+5}{x(x^2+5)}$

p) $\dfrac{\sin x}{1-\cos x}$ q) $x\sqrt{x+2}$ r) $x^5\ln x$

3 Integrate each of these with respect to x.

a) $\dfrac{x}{x-5}$ b) $\dfrac{\sec^2 x}{1+\tan x}$ c) $(x^2+1)(x^3+3x-2)^2$

d) $\dfrac{1}{x^2-9}$ e) xe^{5x} f) $\dfrac{x}{(x^2+1)^2}$

g) $\dfrac{4x^2-x-2}{x^2(x+1)}$ h) xe^{x^2} i) $x(2x+3)^3$

j) $\dfrac{x-1}{\sqrt{x}}$ k) $x\sin 3x$ l) $\dfrac{x+1}{x^2+1}$

m) $x\sin(x^2)$ n) $\dfrac{1}{2x+1}$ o) $\dfrac{4x}{(x+1)(x-3)}$

p) $\dfrac{3}{\sqrt{2x+1}}$ q) $\dfrac{x}{(2-x)^2}$ r) $\ln(\sqrt{x})$

11.4 Differential equations

A **differential equation** is an equation that includes not only x and y but also one or more derivatives of y with respect to x.

An equation which involves only a first-order derivative, such as $\frac{dy}{dx}$, is called a **first-order differential equation**.

An equation which involves a second-order derivative, such as $\frac{d^2y}{dx^2}$, is called a **second-order differential equation**.

Suppose $y = x^2 + 1$, then $\frac{dy}{dx} = 2x$. The equation

$$\frac{dy}{dx} = 2x$$

is a first-order differential equation which you can solve by integrating. That is,

$$y = \int 2x\,dx \quad \therefore \quad y = x^2 + c$$

However, solving the differential equation does not give you a unique solution. In fact, you get a whole family of solutions for different values of the constant c, as shown in the diagram.

The solution $y = x^2 + c$ is called the **general solution** of the differential equation.

If you also know that $y = 2$ when $x = 1$, then

$$2 = 1^2 + c \quad \therefore \quad c = 1$$

You now have a **particular solution** of the differential equation, namely $y = x^2 + 1$.

In this example, the differential equation is of the form

$$\frac{dy}{dx} = f(x)$$

In other words, the right-hand side is a function of x only.

However, some differential equations are of the form

$$\frac{dy}{dx} = f(x, y) = h(x)g(y)$$

in which the variables are separable: that is, it is possible to rearrange the equation so that the y terms are on one side and the x terms on the other.

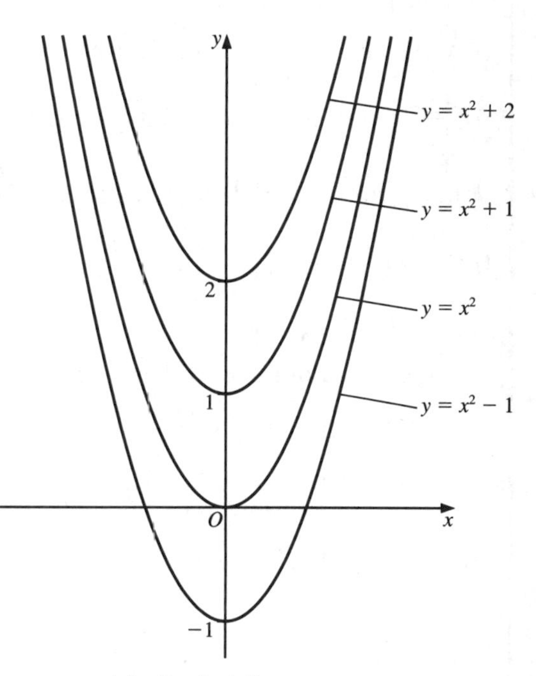

A family of solution curves

Example 11

Find the general solution of the differential equation $\frac{dy}{dx} = y$, for $y > 0$.

You can write the equation as

$$\frac{1}{y}\frac{dy}{dx} = 1$$

This step is called 'separating the variables'.

Integrating both sides with respect to x gives

$$\int \frac{1}{y}\,dy = \int 1\,dx$$

$$\therefore \quad \ln y = x + c$$

Note that you only need one constant, c.

However, you need y in terms of x so:

$$e^{\ln y} = e^{x+c}$$

$$\therefore \quad y = e^x e^c$$

$$\therefore \quad y = Ae^x$$

where $A = e^c$ is a constant.

This is the general solution of the equation, because *A* can take different values.

C4

If you are given the values of x and y at a point on the curve, you can find the particular solution of the differential equation.

Example 12

Find the particular solution of each of these differential equations.

a) $\frac{dy}{dx}(x - 1) = y$, such that $y = 5$ when $x = 2$.

b) $\frac{dy}{dx} = e^{2x - y}$, such that $y = \ln\left(\frac{3}{2}\right)$ when $x = 0$.

c) $\frac{dy}{dx} = 3y^2 \sin x$, such that $y = 1$ when $x = \frac{\pi}{3}$.

a) After rearranging,

$$\frac{1}{y}\frac{dy}{dx} = \frac{1}{x - 1}$$

Separate and integrate.

Integrating both sides with respect to x gives

$$\int \frac{1}{y}\,dy = \int \frac{1}{x - 1}\,dx$$

$$\ln y = \ln(x - 1) + c$$

$$y = e^{\ln(x - 1) + c}$$

$$\therefore \quad y = e^{\ln(x - 1)}e^c$$

Hence

$$y = A(x - 1)$$

where $A = e^c$.

This is the general solution of the differential equation. To find the particular solution, substitute $y = 5$ and $x = 2$:

$$5 = A(2 - 1) \quad \therefore \quad A = 5$$

The particular solution is $y = 5(x - 1)$.

b) After rearranging,

$$\frac{dy}{dx} = e^{2x}e^{-y}$$

$$\therefore \quad e^{y}\frac{dy}{dx} = e^{2x}$$

Integrating both sides with respect to x gives

$$\int e^{y}\,dy = \int e^{2x}\,dx$$

$$\therefore \quad e^{y} = \tfrac{1}{2}e^{2x} + c$$

$$\therefore \quad y = \ln\left(\tfrac{1}{2}e^{2x} + c\right)$$

This is the general solution of the differential equation. To find the particular solution, substitute $y = \ln\left(\frac{3}{2}\right)$ and $x = 0$:

$$\ln\left(\tfrac{3}{2}\right) = \ln\left(\tfrac{1}{2}e^{0} + c\right) = \ln\left(\tfrac{1}{2} + c\right)$$

$$\therefore \quad c = 1$$

The particular solution is $y = \ln\left(\frac{1}{2}e^{2x} + 1\right)$.

c) After rearranging,

$$\frac{1}{3y^2}\frac{dy}{dx} = \sin x$$

Integrating both sides with respect to x gives

$$\int \frac{1}{3y^2}\,dy = \int \sin x\,dx$$

$$\therefore \quad -\frac{1}{3y} = -\cos x + c$$

To find the particular solution, substitute $y = 1$ and $x = \frac{\pi}{3}$:

$$-\frac{1}{3(1)} = -\cos\left(\frac{\pi}{3}\right) + c$$

$$\therefore \quad -\tfrac{1}{3} = -\tfrac{1}{2} + c$$

$$\therefore \quad c = \tfrac{1}{6}$$

The particular solution is given by

$$-\frac{1}{3y} = -\cos x + \tfrac{1}{6}$$

Multiplying by -6 gives

$$\frac{2}{y} = 6\cos x - 1$$

$$\therefore \quad y = \frac{2}{6\cos x - 1}$$

Rearrange the solution into the form $y = f(x)$.

C4

Example 13

Find the equation of a curve given that it passes through the point (1, 0) and that its gradient at any point (x, y) is equal to $x(y - 1)^2$.

If the gradient equals $x(y - 1)^2$ at any point (x, y) then

$$\frac{dy}{dx} = x(y - 1)^2$$

$$\therefore \quad \frac{1}{(y - 1)^2}\frac{dy}{dx} = x$$

Integrating both sides with respect to x gives

$$\int \frac{1}{(y - 1)^2}\,dy = \int x\,dx$$

$$\therefore \quad -\frac{1}{y - 1} = \frac{x^2}{2} + c$$

You know that the curve passes through (1, 0). Therefore,

$$1 = \frac{1}{2} + c \quad \therefore \quad c = \frac{1}{2}$$

The equation of the curve is

$$-\frac{1}{y - 1} = \frac{x^2}{2} + \frac{1}{2} = \frac{x^2 + 1}{2}$$

$$\therefore \quad -\frac{2}{x^2 + 1} = y - 1$$

Hence

$$y = 1 - \frac{2}{x^2 + 1} \quad \text{giving} \quad y = \frac{x^2 - 1}{x^2 + 1}$$

C4

Exercise 11E

1 Find the general solution to each of these differential equations.

a) $\dfrac{dy}{dx} = \dfrac{x + 1}{y}$

b) $\dfrac{dy}{dx} = \dfrac{3x^2 - 2}{2y}$

c) $y^2\dfrac{dy}{dx} = 2x + 3$

d) $\dfrac{dy}{dx} = 3x^2\sqrt{y}$

e) $(x - 1)^2\dfrac{dy}{dx} + 2\sqrt{y} = 0$

f) $\dfrac{dy}{dx} = 2x\sqrt{2y - 1}$

g) $\dfrac{dy}{dx} = 3x^2y$

h) $e^y\dfrac{dy}{dx} = x$

i) $\dfrac{dy}{dx} = \sin x\, e^y$

j) $\cos y\dfrac{dy}{dx} + \sin x = 1$

2 Find the particular solution to each of these differential equations.

a) $\frac{dy}{dx} = 4 - 3x^2, \ y = 5$ at $x = 1$

b) $\frac{dy}{dx} = \frac{x+1}{y}, \ y = 3$ at $x = -2$

c) $(y-3)\frac{dy}{dx} = x + 3, \ y = 4$ at $x = 0$

d) $3y^2\frac{dy}{dx} + 2x = 1, \ y = 2$ at $x = 4$

e) $x^3\frac{dy}{dx} = 2y^2, \ y = -\frac{4}{3}$ at $x = 2$

f) $\frac{dy}{dx} = \sqrt{\frac{y}{x+1}}, \ y = 9$ at $x = 3$

g) $\frac{dy}{dx} = 2xy, \ y = 7$ at $x = 0$

h) $x^2\frac{dy}{dx} = \operatorname{cosec} y, \ y = \frac{\pi}{3}$ at $x = 4$

i) $x(x-1)\frac{dy}{dx} = y, \ y = 1$ at $x = 2$

j) $e^y\frac{dy}{dx} + \sin x = 0, \ y = 0$ at $x = \frac{\pi}{3}$

3 The gradient of a curve at the point (x, y) is given by the expression $2x(y + 1)$. Given that the curve passes through the point $(3, 0)$, find an expression for y in terms of x.

4 The gradient of a curve at the point (x, y) is given by the expression $2\cos x\sqrt{y+3}$. Given that the curve passes through the point $\left(\frac{\pi}{2}, 1\right)$, find an expression for y in terms of x.

11.5 Modelling with differential equations

Now that you have seen how to solve a first-order differential equation you are ready to put what you have learnt into context. There are lots of practical situations which give rise to differential equations, and you will come across some of those in the examples which follow.

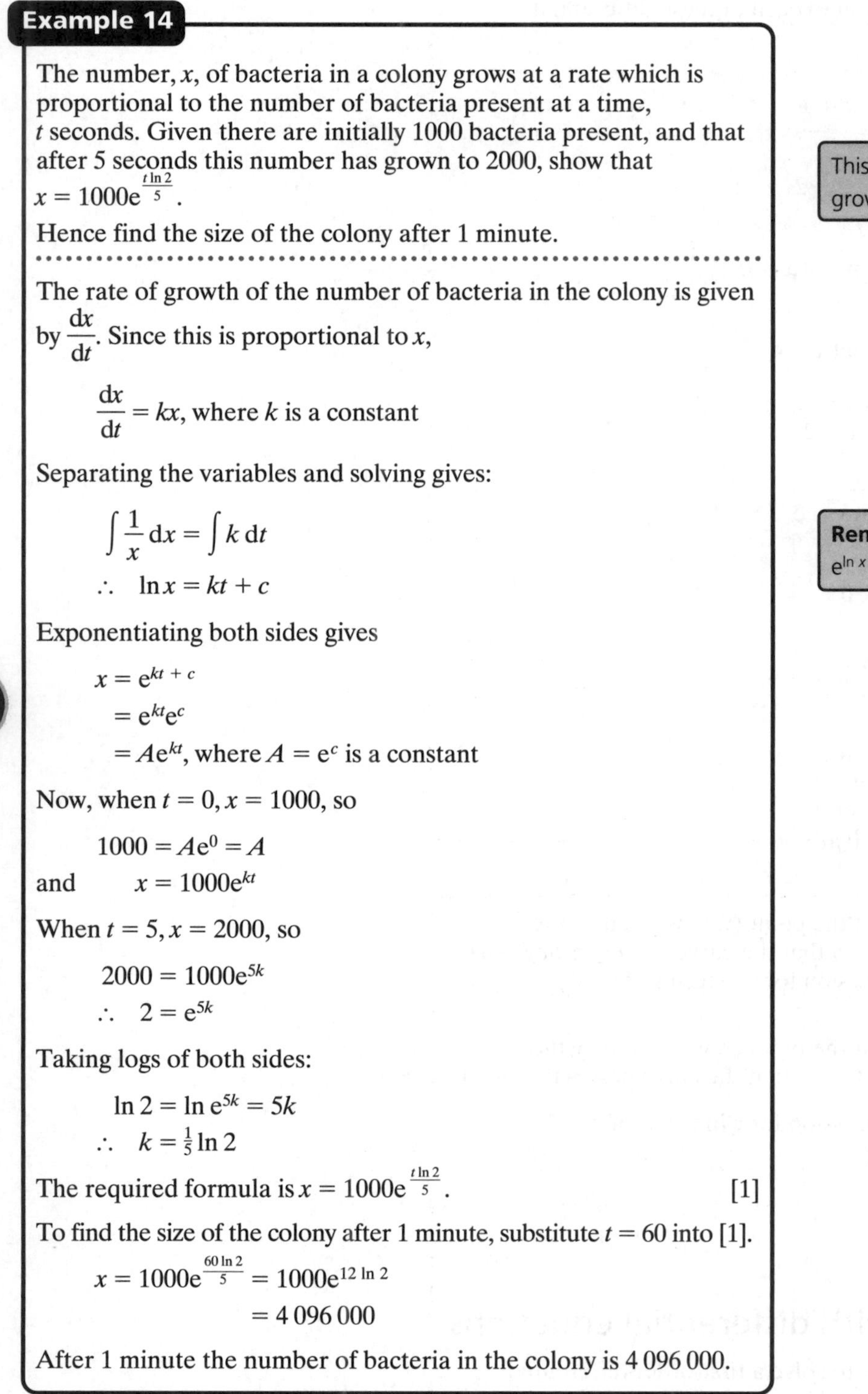

Example 14

The number, x, of bacteria in a colony grows at a rate which is proportional to the number of bacteria present at a time, t seconds. Given there are initially 1000 bacteria present, and that after 5 seconds this number has grown to 2000, show that $x = 1000e^{\frac{t \ln 2}{5}}$.

Hence find the size of the colony after 1 minute.

The rate of growth of the number of bacteria in the colony is given by $\frac{dx}{dt}$. Since this is proportional to x,

$$\frac{dx}{dt} = kx, \text{ where } k \text{ is a constant}$$

Separating the variables and solving gives:

$$\int \frac{1}{x}\,dx = \int k\,dt$$

$$\therefore \quad \ln x = kt + c$$

Exponentiating both sides gives

$$\begin{aligned} x &= e^{kt + c} \\ &= e^{kt}e^{c} \\ &= Ae^{kt}, \text{ where } A = e^{c} \text{ is a constant} \end{aligned}$$

Now, when $t = 0, x = 1000$, so

$$1000 = Ae^{0} = A$$

and $\quad x = 1000e^{kt}$

When $t = 5, x = 2000$, so

$$2000 = 1000e^{5k}$$

$$\therefore \quad 2 = e^{5k}$$

Taking logs of both sides:

$$\ln 2 = \ln e^{5k} = 5k$$

$$\therefore \quad k = \tfrac{1}{5}\ln 2$$

The required formula is $x = 1000e^{\frac{t \ln 2}{5}}$. [1]

To find the size of the colony after 1 minute, substitute $t = 60$ into [1].

$$\begin{aligned} x = 1000e^{\frac{60 \ln 2}{5}} &= 1000e^{12 \ln 2} \\ &= 4\,096\,000 \end{aligned}$$

After 1 minute the number of bacteria in the colony is 4 096 000.

This is an example of exponential growth (see Chapter 10).

Remember: $e^{\ln x} = x$ (see page 79).

A differential equation may be used to model the rate at which something happens.

Example 15

A candle is burning in such a way that when its height is h cm it is burning down at the rate of $2\sqrt{h}$ cm per hour. Given that the initial height of the candle is 64 cm, calculate how long it will take for the candle to burn down completely.

Let t be the time in hours for which the candle is burning.

The height is decreasing, so $\frac{dh}{dt}$ is negative. So

$$\frac{dh}{dt} = -2\sqrt{h}$$

Separating the variables and solving gives

$$\int \frac{1}{\sqrt{h}}\,dh = \int -2\,dt$$

$$\therefore\quad 2\sqrt{h} = -2t + c$$

When $t = 0$, $h = 64$, so

$$2\sqrt{64} = c \qquad \therefore\ c = 16$$

Hence $\quad 2\sqrt{h} = -2t + 16 \qquad [1]$

When the candle has burnt down completely, $h = 0$. Substituting $h = 0$ into [1] gives

$$0 = -2t + 16 \qquad \therefore\ t = 8$$

The candle has burnt down completely after 8 hours.

C4

Example 16

In a population of N people a rumour spreads in such a way that $\frac{dn}{dt} = \frac{n(N-n)}{50}$ where n is the number of people who have heard the rumour at a time t hours.

a) Show that $n = \frac{ANe^t}{Ae^t - 1}$

b) Deduce that, eventually, every member of the population will have heard the rumour.

a) $$\frac{dn}{dt} = \frac{n(N-n)}{50}$$

Separating the variables and integrating give

$$\int \frac{1}{n(N-n)}\,dn = \int \frac{1}{50}\,dt \qquad [1]$$

To work out $\int \frac{1}{n(N-n)}\,dn$, split $\frac{1}{n(N-n)}$ into partial fractions.

Let

$$\frac{1}{n(N-n)} \equiv \frac{A}{n} + \frac{B}{N-n}$$

Multiplying by $n(N-n)$ gives

$$1 \equiv A(N-n) + Bn$$

Let $n = N$: $1 = BN \quad \therefore \quad B = \frac{1}{N}$

Let $n = 0$: $1 = AN \quad \therefore \quad A = \frac{1}{N}$

Returning to [1],

$$\int \frac{\frac{1}{N}}{n} + \frac{\frac{1}{N}}{N-n} \, dn = \int \frac{1}{50} \, dt$$

$$\therefore \quad \frac{1}{N}\ln n - \frac{1}{N}\ln(N-n) = \frac{1}{50}t + c$$

$$\therefore \quad \frac{1}{N}\ln\left(\frac{n}{N-n}\right) = \frac{1}{50}t + c$$

Multiplying throughout by N and exponentiating give

$$\frac{n}{N-n} = e^{N\left(\frac{1}{50}t+c\right)}$$

$$\therefore \quad \frac{n}{N-n} = e^{\frac{Nt}{50}} e^{Nc} = Ae^{\frac{Nt}{50}}, \text{ where } A = e^{Nc} \text{ is a constant.}$$

Rearranging for n gives

$$n = Ae^{\frac{Nt}{50}}(n-N)$$

$$\therefore \quad n = Ane^{\frac{Nt}{50}} - ANe^{\frac{Nt}{50}}$$

$$\therefore ANe^{\frac{Nt}{50}} = n(Ae^{\frac{Nt}{50}} + 1)$$

$$\therefore \quad n = \frac{ANe^{\frac{Nt}{50}}}{Ae^{\frac{Nt}{50}} + 1}$$

b) As $t \to \infty$, the $e^{\frac{Nt}{50}}$ term dominates, so $n \to \dfrac{ANe^{\frac{Nt}{50}}}{Ae^{\frac{Nt}{50}}} = N$. This shows that, after a long time, all N members of the population will have heard the rumour.

Exercise 11F

1 The number, x, of rats in a sewer grows at a rate which is proportional to the number of rats present at a time, t months. Initially, there are 20 rats in the sewer, and after 3 months this number has grown to 100.

a) Show that $x = 20e^{\frac{t\ln 5}{3}}$.

b) Hence find the size of the colony after one year.

c) Why is this model likely to break down over a long period of time?

2 Water is escaping from a tank in such a way that when the depth of water in the tank is h cm the depth of the water in the tank is going down at a rate of $3\sqrt{h}$ cm per hour. Given that the initial depth of the water in the tank is 100 cm,

a) show that $3t = 20 - 2\sqrt{h}$

b) calculate how long it will take for the tank to empty completely.

3 Mould is spreading through a piece of cheese of mass 1 kg in such a way that at a time t days, x kg of the cheese has become mouldy where

$$\frac{dx}{dt} = 2x(1 - x)$$

Given that initially 0.01 kg of the cheese is mouldy

a) show that $x = \dfrac{e^{2t}}{99 + e^{2t}}$

b) calculate after how long 0.99 kg of the cheese will be mouldy, giving your answer to the nearest hour.

4 At an air temperature of $-T$ °C the thickness, x cm, of the ice on a pond at a time t minutes satisfies the equation

$$\frac{dx}{dt} = \frac{T}{14\,000x}, \text{ where } x = 0 \text{ at } t = 0$$

a) Show that $7000x^2 = Tt$.

b) Given $T = 10$ when the ice starts forming, calculate how long it takes for the thickness of the ice to reach 2 cm.

5 A pebble falls through water and at a time t seconds after being released from rest its velocity, v m s^{-1}, satisfies the equation

$$\frac{dv}{dt} = 10 - 5v$$

a) Show that $v = 2(1 - e^{-5t})$.

b) Calculate the time at which the velocity of the pebble is 1.5 m s^{-1}.

c) Sketch the graph of v against t and hence show that the velocity of the pebble can never be more than 2 m s^{-1}.

6 A lake is stocked with 100 fish. Since the fish breed the population has a natural tendency to increase. However, each month some of the fish are removed from the lake. At a time t months after the lake is stocked, the number, x, of fish in the lake satisfies the differential equation

$$\frac{dx}{dt} = x(4 - t)$$

a) Show that $2 \ln x + (4 - t)^2 = 16 + \ln(10\,000)$.

b) Show that after 10 months the population will be extinct.

11.6 Applications to exponential laws of growth and decay

In Chapter 10 you looked at exponential growth and decay, and Example 14 on page 234 showed you how to model the growth of bacteria by a differential equation which gave rise to an equation of exponential growth.

In general, laws of growth and decay can be expressed in the form of differential equations. For example:

- if the rate of growth of x is proportional to x, then
$$\frac{dx}{dt} = kx$$
where k is a positive constant;
- if the rate of decay of x is proportional to x, then
$$\frac{dx}{dt} = -kx$$
where k is a positive constant.

The next example illustrates the solution of a differential equation which has been derived from a 'decay' situation.

C4

Example 17

At time t minutes, the rate of change of temperature of a cooling body is proportional to the temperature T °C of that body at that time. Initially, $T = 72$ °C. Show that

$$T = 72e^{-kt}$$

where k is a constant.

Given also that $T = 32$ °C when $t = 10$, find how much longer it will take the body to cool to 27 °C under these conditions.

The temperature is decreasing, so form the differential equation

$$\frac{dT}{dt} = -kT$$

Separating the variables and solving give

$$\int_{72}^{T} \frac{dT}{T} = -\int_0^t k\,dt$$

$$\therefore \quad [\ln T]_{72}^{T} = [-kt]_0^t$$

$$\ln T - \ln 72 = -kt$$

$$\ln\left(\frac{T}{72}\right) = -kt$$

$$\frac{T}{72} = e^{-kt}$$

$$\therefore \quad T = 72e^{-kt} \qquad [1]$$

as required.

When $t = 10$, $T = 32\,°C$. Substituting into [1] gives the value of the constant k.

$$32 = 72e^{-10k}$$

$$\therefore \quad e^{-10k} = \frac{32}{72} = \frac{4}{9}$$

$$-10k = \ln\left(\tfrac{4}{9}\right)$$

$$\therefore \quad k = -\tfrac{1}{10}\ln\left(\tfrac{4}{9}\right) \qquad [2]$$

Now if t is the time it takes the body to cool to $27\,°C$,

$$27 = 72e^{-kt}$$

$$\therefore \quad e^{-kt} = \frac{27}{72} = \frac{3}{8}$$

Hence

$$-kt = \ln\left(\tfrac{3}{8}\right) \qquad [3]$$

From [2] and [3]:

$$\tfrac{1}{10}\ln\left(\tfrac{4}{9}\right)t = \ln\left(\tfrac{3}{8}\right)$$

$$\therefore \quad t = 10\,\frac{\ln\left(\tfrac{3}{8}\right)}{\ln\left(\tfrac{4}{9}\right)} = 12.1$$

Initially $t = 10$, and $12.1 - 10 = 2.1$

It will take the body just over 2 minutes longer to cool to $27\,°C$.

Don't forget you were asked how much *longer* it would take!

C4

Exercise 11G

1 The rate, in $cm^3\,s^{-1}$, at which air is escaping from a balloon at time t seconds is proportional to the volume of air, $V\,cm^3$, in the balloon at that instant. Initially $V = 1000$.

a) Show that $V = 1000e^{-kt}$, where k is a positive constant.

Given also that $V = 500$ when $t = 6$,

b) show that $k = \frac{1}{6}\ln 2$

c) calculate the value of V when $t = 12$.

2 At time t minutes the rate of change of temperature of a cooling liquid is proportional to the temperature, $T\,°C$, of that liquid at that time. Initially $T = 80$.

a) Show that $T = 80e^{-kt}$, where k is a positive constant.

Given also that $T = 20$ when $t = 6$,

b) show that $k = \frac{1}{3}\ln 2$

c) calculate the time at which the temperature will reach $10\,°C$.

3 The value of a car depreciates in such a way that when it is t years old the rate of decrease in its value is proportional to the value, £V, of the car at that time. The car cost £12 000 when new.

a) Show that $V = 12\,000\mathrm{e}^{-kt}$.

When the car is three years old its value has dropped to £4000.

b) Show that $k = \frac{1}{3}\ln 3$.

The owner decides to sell the car when its value reaches £2000.

c) Calculate, to the nearest month, the age of the car at that time.

4 A lump of a radioactive substance is decaying. At time t hours the rate of decay of the mass of the substance is proportional to the mass M grams of the substance at that time. At $t = 0$, $M = 72$; and at $t = 2$, $M = 50$.

a) Show that $M = 72\mathrm{e}^{-t\ln(6/5)}$.

b) Sketch of graph of M against t.

5 By treatment with certain chemicals a scientist is able to control a killer virus. The rate of decrease in the number of viruses is found to be proportional to n, the number of viruses present. Using these chemicals it is found that the virus population is halved in six days. Show that the virus population is reduced to 1% of its original value after approximately 40 days.

C4

6 A population is growing in such a way that, at time t years, the rate at which the population is increasing is proportional to the size, x, of that population at that time. Initially the size of the population is 2.

a) Show that $x = 2\mathrm{e}^{kt}$, where k is a positive constant.

After 6 years the population size is 100.

b) Show that $k = \frac{1}{6}\ln 50$.

c) Calculate an estimate, to the nearest 1000, of the population size after 20 years.

7 During the initial stages of the spread of a disease in a body, the rate of increase of the number of infected cells in the body is proportional to the number, n, of infected cells present in the body at that time. Initially, n_0 infected cells are introduced to the body, and one day later the number of infected cells has risen to $2n_0$.

a) Show that $n = n_0\mathrm{e}^{t\ln 2}$.

b) Show that five days after the infection was introduced the number of infected cells has risen to $32n_0$.

8 A patient is required to take a course of a certain drug. The rate of decrease in the concentration of the drug in the bloodstream at time t hours is proportional to the amount x mg, of the drug in the bloodstream at that time.

a) Show that $x = x_0\mathrm{e}^{-kt}$, where k is a positive constant and x_0 mg is the size of the dose.

The patient repeats the dose of x_0 mg at regular intervals of T hours.

b) Show that the amount of the drug in the bloodstream will never exceed $\left(\dfrac{x_0}{1 - \mathrm{e}^{-kT}}\right)$ mg.

Summary

You should know how to ...	Check out
1 Solve problems involving implicit functions.	**1** a) Find an expression for the gradient of the function defined implicitly as $$2xy^2 + 3x^2y - 2 = 0$$ b) Find the value of the gradient at the two points where $x = 1$.
2 Solve problems involving parametric equations.	**2** a) Convert the parametric equations $$x = 1 - t^2 \qquad y = \frac{4}{t}$$ to their cartesian form. b) Find the normal to the curve $$x = 1 - t^2 \qquad y = \frac{4}{t}$$ at the point where $t = 2$.
3 Integrate an expression using partial fractions.	**3** Find $\int \frac{x - 1}{(x + 3)(3x - 1)}\,dx$.
4 Solve differential equations.	**4** A curve is given by the differential equation $$\frac{dy}{dx} = \frac{y - 1}{x^2}$$ The point $(-1, 2)$ lies on the curve. Find the equation of the the curve in the form $y = f(x)$.
5 Use differential equations to solve problems in context.	**5** A researcher is investigating the rate at which a species of toadstool grows. a) The researcher proposes the differential equation $$\frac{dx}{dt} = \frac{(5 - x)^2}{4}$$ as a model, for the growth of the toadstools, where x cm is the height after t days. Find an expression for the height of a toadstool after t days. b) Use the model to calculate at what time the toadstool is 2 cm in height. c) Explain how the model shows the maximum height to be 5 cm.

C4

Revision exercise 11

1 A curve is given by the parametric equation $x = 2t - 1, y = \frac{1}{2t}$.

a) Find $\frac{dy}{dx}$ in terms of t.

b) Find the equation of the normal to the curve at the point where $t = 1$. *(AQA, 2004)*

2 a) Express $\frac{30}{(x + 4)(7 - 2x)}$ in the form $\frac{A}{x + 4} + \frac{B}{7 - 2x}$.

b) Hence find $\int_0^3 \frac{30}{(x + 4)(7 - 2x)} dx$, giving your answer in the form $p \ln q$ where p and q are rational numbers. *(AQA, 2004)*

3 A curve is given by the parametric equation $9(y + 2)^2 = 5 + 4(x - 1)^2$.

a) Find the coordinates of the two points on the curve where $x = 2$.

b) Find the gradient of the curve at each of these points. *(AQA, 2004)*

4 a) Express $\frac{13 - 2x}{(x + 4)(2x + 1)}$ in the form $\frac{A}{x + 4} + \frac{B}{2x + 1}$.

b) Hence, prove that $\int_0^4 \frac{13 - 2x}{(x + 4)(2x + 1)} dx = p \ln 3 - q \ln 2$ where p and q are positive integers. *(AQA, 2004)*

C4

5 The gradient of a curve, C, at the point (x, y) is given by $\frac{dy}{dx} = \frac{1}{2y(x + 2)}$ $x > 0, y > 0$. The point $P(1, 1)$ lies on the curve C.

a) i) Write down the gradient of the curve C at the point P.

ii) Show that the equation of the normal at P is $y + 6x = 7$.

b) Find the equation of the curve C in the form $y^2 = f(x)$. *(AQA, 2004)*

6 The speed v m s^{-1} of a pebble falling through still water after t seconds can be modelled by the differential equation $\frac{dv}{dt} = 10 - 5v$.

A pebble is placed carefully on the surface of the water at time $t = 0$ and begins to sink.

a) Show that $t = \frac{1}{5} \ln\left(\frac{2}{2 - v}\right)$.

b) Use the model to find the speed of the pebble after 0.5 seconds, giving your answer to two significant figures. *(AQA, 2004)*

7 Given that $y^3 + 3y = x^3$, use implicit differentiation to show that $\frac{dy}{dx} = \frac{x^2}{y^2 + 1}$. *(AQA, 2004)*

8 A curve is given by the parametric equations $x = 3t - 1,\ y = \frac{1}{t}$.

a) Find $\frac{dy}{dx}$ in terms of t.

b) Hence find the equation of the normal to the curve at the point where $t = 1$. *(AQA, 2003)*

9 Solve the differential equation $\frac{dx}{dt} = \frac{10 - x}{5}$ using integration, given that $x = 1$ when $t = 0$.
Hence find the value of t when x is 2, giving your answer to three decimal places. *(AQA, 2003)*

10 A curve is defined by the parametric equations $x = 3\sin t$, $y = \cos t$.

a) Show that, at the point P where $t = \frac{\pi}{4}$, the gradient of the curve is $-\frac{1}{3}$.

b) Find the equation of the tangent to the curve at the point P, giving your answer in the form $y = mx + c$. *(AQA, 2003)*

11 a) i) Show that $\frac{x^2}{x^2 - 16} = 1 + \frac{16}{x^2 - 16}$.

ii) Express $\frac{16}{x^2 - 16}$ in the form $\frac{A}{x - 4} + \frac{B}{x + 4}$.

b) Hence find $\int_5^8 \frac{x^2}{x^2 - 16}\,dx$ giving your answer in the form $p + q\ln r$. *(AQA, 2003)*

12 A group of students are researching the rate at which ice thickens on a frozen pond. They have experimental evidence that when the air temperature is $-T$°C, the ice thickens at a rate $\frac{T}{14\,000x}$ cm s^{-1} where x cm is the thickness of the ice that has already formed.
On a particular winter day, the temperature is -7°C. At 12.00 noon the students note that the ice is 2 cm thick. Time t seconds later the thickness of the ice is x cm.

a) Show that $\frac{dx}{dt} = \frac{1}{2000x}$.

b) Solve the differential equation and hence find the time when the students predict the ice will be 3 cm thick. *(AQA, 2003)*

13 a) Given that $\frac{25x + 1}{(2x - 1)(x + 1)^2} \equiv \frac{A}{2x - 1} + \frac{B}{x + 1} + \frac{C}{(x + 1)^2}$

i) show that $25x + 1 \equiv A(x + 1)^2 + B(x + 1)(2x - 1) + C(2x - 1)$

ii) find the values of A, B and C.

b) Hence find $\int_1^2 \frac{25x + 1}{(2x - 1)(x + 1)^2}\,dx$, leaving your answer in the form $p + q\ln 2$. *(AQA, 2003)*

14 A curve has implicit equation $y^3 + xy = 4x - 2$.

a) Show that the value of $\frac{dy}{dx}$ at the point (1, 1) is $\frac{3}{4}$.

b) Find the equation of the normal to the curve at the point (1, 1). *(AQA, 2003)*

15 a) Use integration by parts to find $\int xe^x \, dx$.

b) Hence find the solution of the differential equation $\frac{dy}{dx} = yxe^x$ given that $y = e$ when $x = 1$. *(AQA, 2003)*

16 A curve has equation $\frac{x^2}{9} + \frac{y^2}{25} = 1$.

a) Find the y-coordinates of the two points on the curve at which the x-coordinate is 2.

b) Find the values of the gradient of the curve at these two points, giving your answer to two significant figures. *(AQA, 2002)*

17 The function f is defined by $f(x) = \frac{2x}{(1 + 2x)^3} \quad x \neq -\frac{1}{2}$.

Using the substitution $u = 1 + 2x$, or otherwise, find $\int f(x) \, dx$. *(AQA, 2002)*

18 A curve is given by the equation $3(x + 1)^2 - 9(y - 1)^2 = 32$.

a) Find the coordinates of the two points on the curve at which $x = 3$.

b) Find the gradient of the curve at each of these two points. *(AQA, 2002)*

C4

19 A rectangular water tank rests on a horizontal surface. The tank is emptied in such a way that the depth of water decreases at a rate which is proportional to the square root of the depth of the water.

The depth of water is h metres at time t hours. Initially there is depth of 1 metre of water in the tank. The depth of water is 0.5 metres after 2 hours.

a) i) Write down a differential equation for h.

ii) Hence show that $2\sqrt{h} = 2 - kt$ where k is a constant.

iii) Find the value of k giving your answer to three decimal places.

b) Find how long it will take to empty the tank completely, giving your answer to the nearest minute. *(AQA, 2002)*

20 a) Express $\frac{2x^2 - x + 11}{(2x - 3)(x + 2)}$ in the form $A + \frac{B}{2x - 3} + \frac{C}{x + 2}$.

b) Hence find $\int_2^6 \frac{2x^2 - x + 11}{(2x - 3)(x + 2)} \, dx$ giving your answer in the form $p + q \ln 2 + r \ln 3$ where p, q and r are integers. *(AQA, 2002)*

21 A pond covers an area of 300 m^2. A specimen of pondweed grows on the surface of the pond. At time t days after the weed is first discovered, it covers an area of A m^2.

The area of the pond covered by the weed increases at a rate which is proportional to the square root of the area of the pond already covered by the weed.

Initially, the area covered by the weed is 0.25 m^2 and its rate of growth is 1 m^2 per day.

a) Show that $\frac{dA}{dt} = 2A^{\frac{1}{2}}$.

b) Find a relationship between t and A.

c) Deduce, to the nearest day, the time taken for the pond's surface to be completely covered by this weed. *(AQA, 2002)*

22 The rate of increase in a population of bacteria is proportional to the size of the population that exists at any particular time.

a) Explain briefly why this situation can be modelled by a differential equation of the form $\frac{dP}{dt} = kP$ where P is the size of the population, k is a constant and t is the time in minutes measured from a given starting time.

b) i) At time $t = 0$, the population of bacteria of type A is 1000. After 30 minutes, this population is 2000. Solve the differential equation, stating the value of k, in the form $q \ln 2$, where q is a rational number to be found.

ii) As the population of type A increases, the population of another type of bacteria, B, decreases. The population, Q, of the type B bacteria at time t minutes is modelled by $Q = 5000e^{-0.05t}$. Find, to the nearest minute, the value of t when the populations of bacteria of types A and B are the same. *(AQA, 2002)*

23 a) Use the substitution $u = 3 + \cos x$, or otherwise, to determine $\int \frac{\sin x}{(3 + \cos x)^2} dx$.

b) Hence find the solution to the differential equation $(3 + \cos x)^2 \frac{dy}{dx} = y \sin x$ given that $y = e$ when $x = 0$. *(AQA, 2001)*

12 Vectors

This chapter will show you how to

✦ Solve problems with vectors in two and three dimensions
✦ Calculate a scalar product
✦ Find the angle between two lines
✦ Determine and use the vector equation of a line

Before you start

You should know how to ...	Check in
1 Find the gradient of a line joining two points.	**1** Find the gradient of the line AB, where a) A is $(2, 5)$, B is $(-1, -4)$ b) A is $(4, 1)$, B is $(-4, 5)$.
2 Find the distance between two points, using Pythagoras' theorem.	**2** a) Find the distance between points A and B if A is $(3, 2)$ and B is $(-5, -4)$. b) P is the point $(-10, 4)$ and Q is the point $(-3, -3)$. Find the length PQ in the form $p\sqrt{q}$ where p and q are integers.
3 Use the cosine rule.	**3** Find the lengths of the sides and the angles in triangle ABC. $A\ (0, 4)$ $B\ (-3, 1)$ $C\ (5, -5)$

C4

A quantity which is specified by a magnitude and a direction is called a **vector**.

For example, force and velocity are both specified by a magnitude and a direction and are therefore examples of vector quantities.

> A quantity which is specified by just its magnitude is called a scalar.

For example, distance and speed are both fully specified by a magnitude and are therefore examples of scalar quantities.

In two dimensions, a vector is represented by a straight line with an arrowhead. In the diagram, the line OA represents a vector $\overrightarrow{OA}$.

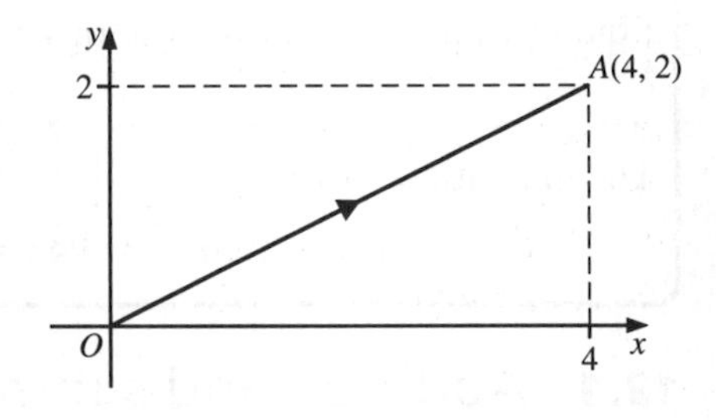

One way of writing this vector is

$$\overrightarrow{OA} = \begin{pmatrix} 4 \\ 2 \end{pmatrix}$$

which means that to go from O to A you move 4 units in the positive x direction and 2 units in the positive y direction. This is called a **column vector**. The numbers 4, 2 in the column are called the **components** of the vector.

The magnitude or modulus of the vector $\overrightarrow{OA}$ is represented by the length OA and is denoted by $|\overrightarrow{OA}|$. In this example,

$$|\overrightarrow{OA}|^2 = 4^2 + 2^2 = 20$$

$$\therefore \quad |\overrightarrow{OA}| = \sqrt{20} = 2\sqrt{5}$$

A **unit vector** is a vector of length 1. The standard unit vectors in two dimensions are

$$\mathbf{i} = \begin{pmatrix} 1 \\ 0 \end{pmatrix} \text{ and } \mathbf{j} = \begin{pmatrix} 0 \\ 1 \end{pmatrix}$$

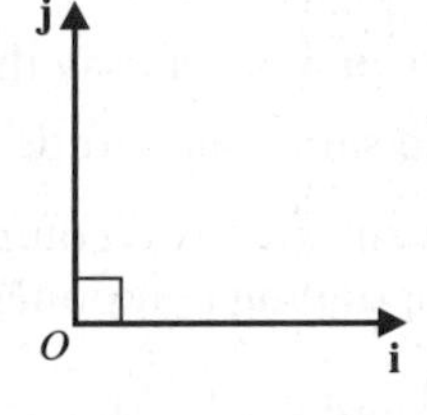

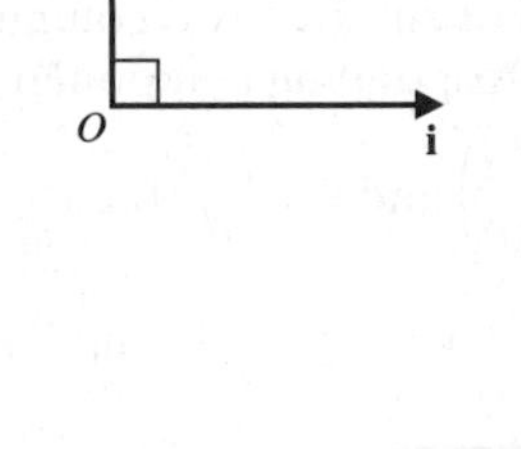

They can be represented diagramatically as shown.

Notice that **i** and **j** are in bold face. This is to indicate that they are vectors. When you write vectors by hand, draw a line beneath the letter representing it: for example, $\underset{\sim}{i}$ and $\underset{\sim}{j}$.

The vector $\overrightarrow{OA}$ can be written as

$$\overrightarrow{OA} = 4\mathbf{i} + 2\mathbf{j} = \begin{pmatrix} 4 \\ 2 \end{pmatrix}$$

In three dimensions, the standard unit vectors are

$$\mathbf{i} = \begin{pmatrix} 1 \\ 0 \\ 0 \end{pmatrix} \quad \mathbf{j} = \begin{pmatrix} 0 \\ 1 \\ 0 \end{pmatrix} \quad \mathbf{k} = \begin{pmatrix} 0 \\ 0 \\ 1 \end{pmatrix}$$

For example, the column vector $\mathbf{v} = \begin{pmatrix} -4 \\ 2 \\ 3 \end{pmatrix}$ can be written as $\mathbf{v} = -4\mathbf{i} + 2\mathbf{j} + 3\mathbf{k}$. The magnitude of $\mathbf{v}$ is given by

$$|\mathbf{v}| = \sqrt{(-4)^2 + 2^2 + 3^2} = \sqrt{29}$$

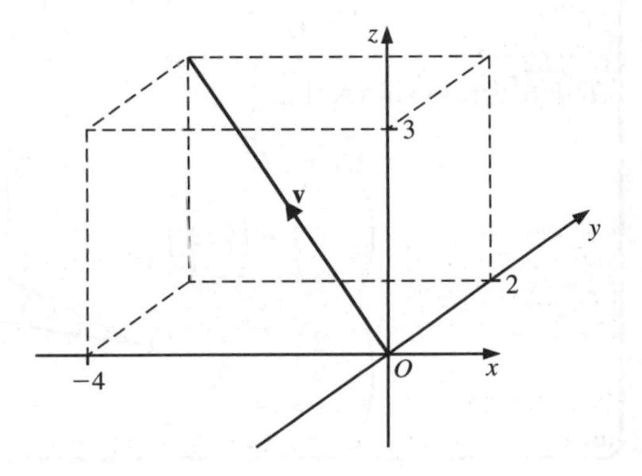

C4

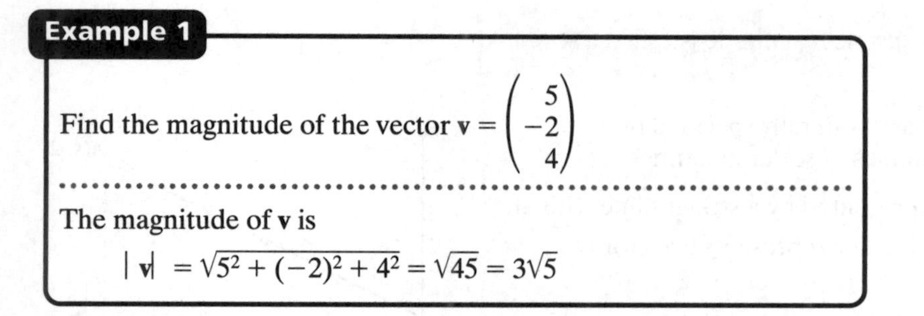

Example 1

Find the magnitude of the vector $\mathbf{v} = \begin{pmatrix} 5 \\ -2 \\ 4 \end{pmatrix}$

The magnitude of **v** is

$$|\mathbf{v}| = \sqrt{5^2 + (-2)^2 + 4^2} = \sqrt{45} = 3\sqrt{5}$$

12.1 Addition and subtraction of vectors

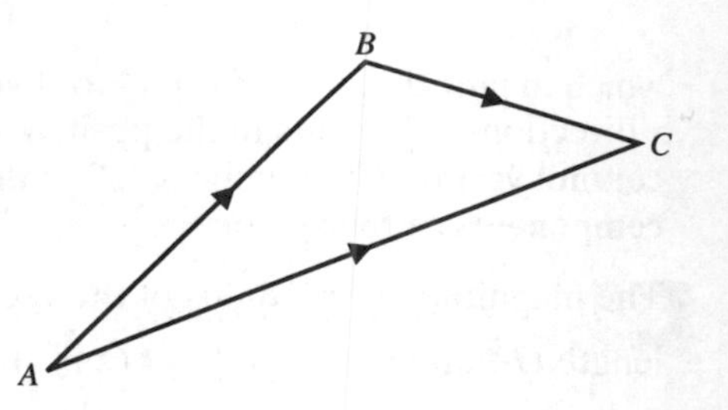

The diagram shows two possible paths you could take to travel from A to C.

One route is A to B then B to C. The other route is to go directly from A to C. You can write this as a vector equation:

$$\overrightarrow{AC} = \overrightarrow{AB} + \overrightarrow{BC}$$

The vector $\overrightarrow{AC}$ is called the **resultant** of vectors $\overrightarrow{AB}$ and $\overrightarrow{BC}$.

Since vectors may also be written as single letters in bold type, if you let $\mathbf{u} = \overrightarrow{AB}$, $\mathbf{v} = \overrightarrow{BC}$ and $\mathbf{w} = \overrightarrow{BC}$, then

$$\mathbf{w} = \mathbf{u} + \mathbf{v}$$

C4

Since $\overrightarrow{AB} = \mathbf{u}$, you can say that $\overrightarrow{BA} = -\mathbf{u}$. In other words, the vector $\overrightarrow{AB}$ has the same magnitude as $\overrightarrow{BA}$ but is in the opposite direction.

To add or subtract two column vectors, you simply add or subtract the corresponding components:

If $\mathbf{u} = \begin{pmatrix} a \\ b \end{pmatrix}$ and $\mathbf{v} = \begin{pmatrix} c \\ d \end{pmatrix}$ then

$$\mathbf{u} + \mathbf{v} = \begin{pmatrix} a + c \\ b + d \end{pmatrix} \quad \text{and} \quad \mathbf{u} - \mathbf{v} = \begin{pmatrix} a - c \\ b - d \end{pmatrix}$$

Example 2

Given that $\overrightarrow{AB} = \begin{pmatrix} 3 \\ 5 \\ -4 \end{pmatrix}$ and $\overrightarrow{BC} = \begin{pmatrix} -1 \\ 4 \\ -1 \end{pmatrix}$, find $\overrightarrow{AC}$.

The diagram shows that

$$\overrightarrow{AC} = \overrightarrow{AB} + \overrightarrow{BC}$$

$$= \begin{pmatrix} 3 \\ 5 \\ -4 \end{pmatrix} + \begin{pmatrix} -1 \\ 4 \\ -1 \end{pmatrix}$$

$$\therefore \quad \overrightarrow{AC} = \begin{pmatrix} 2 \\ 9 \\ -5 \end{pmatrix}$$

$\begin{pmatrix} 3 \\ 5 \\ -4 \end{pmatrix}$ B $\begin{pmatrix} -1 \\ 4 \\ -1 \end{pmatrix}$ C A ?

Example 3

Given that $\overrightarrow{BC} = \begin{pmatrix} 7 \\ -2 \\ 1 \end{pmatrix}$ and $\overrightarrow{AC} = \begin{pmatrix} 1 \\ 0 \\ -6 \end{pmatrix}$, find $\overrightarrow{BA}$.

$$\overrightarrow{BA} = \overrightarrow{BC} + \overrightarrow{CA}$$
$$= \overrightarrow{BC} - \overrightarrow{AC}$$
$$= \begin{pmatrix} 7 \\ -2 \\ 1 \end{pmatrix} - \begin{pmatrix} 1 \\ 0 \\ -6 \end{pmatrix}$$
$$\therefore \quad \overrightarrow{BA} = \begin{pmatrix} 6 \\ -2 \\ 7 \end{pmatrix}$$

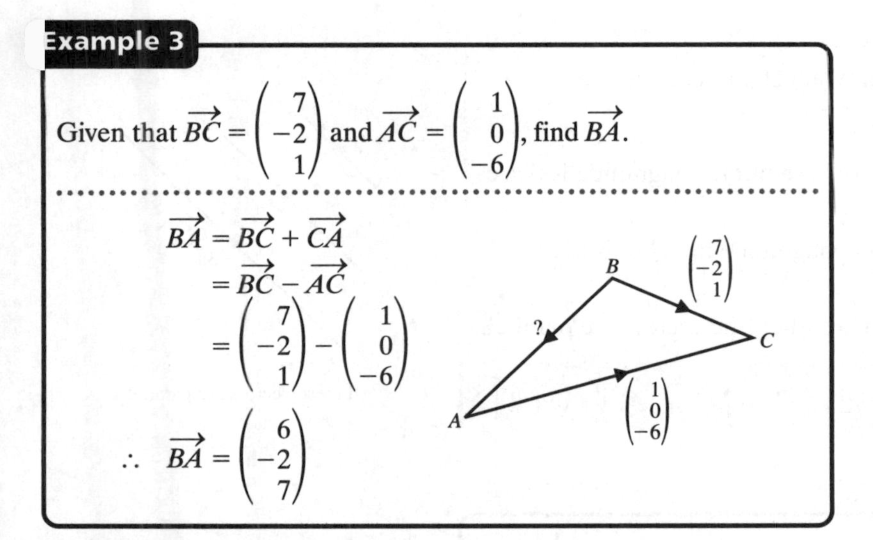

Example 4

Two vectors are given by $\mathbf{a} = \begin{pmatrix} 2 \\ -1 \\ -1 \end{pmatrix}$ and $\mathbf{b} = \begin{pmatrix} -1 \\ 3 \\ 4 \end{pmatrix}$.

a) Find $\mathbf{a} + \mathbf{b}$ and $\mathbf{a} - \mathbf{b}$.

b) Draw a diagram showing $\mathbf{a} + \mathbf{b}$ and another showing $\mathbf{a} - \mathbf{b}$.

a) Adding the two vectors gives

$$\mathbf{a} + \mathbf{b} = \begin{pmatrix} 2 \\ -1 \\ -1 \end{pmatrix} + \begin{pmatrix} -1 \\ 3 \\ 4 \end{pmatrix} = \begin{pmatrix} 1 \\ 2 \\ 3 \end{pmatrix}$$

Subtracting the two vectors gives

$$\mathbf{a} - \mathbf{b} = \begin{pmatrix} 2 \\ -1 \\ -1 \end{pmatrix} - \begin{pmatrix} -1 \\ 3 \\ 4 \end{pmatrix} = \begin{pmatrix} 3 \\ -4 \\ -5 \end{pmatrix}$$

b) Adding $\mathbf{a}$ to $\mathbf{b}$ gives

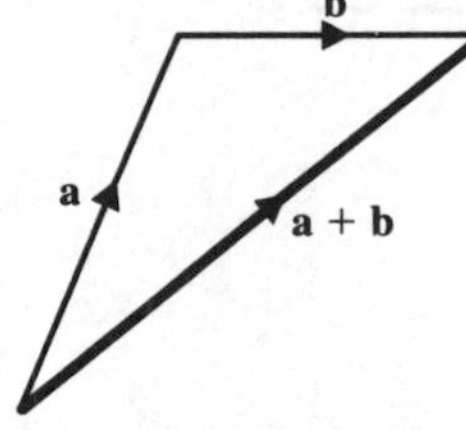

To illustrate $\mathbf{a} - \mathbf{b}$, add $-\mathbf{b}$ to $\mathbf{a}$ giving

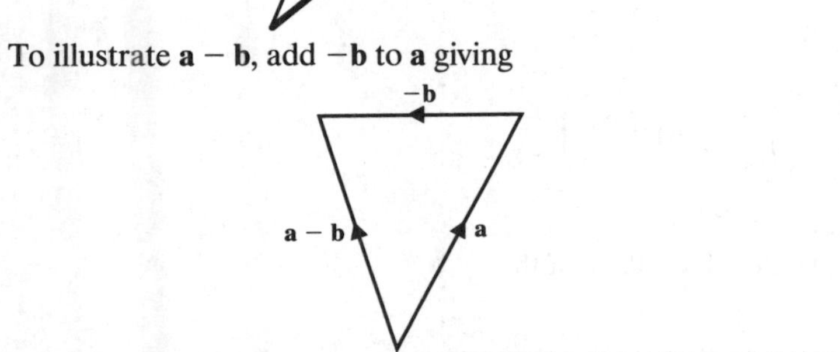

Scalar multiplication

If you multiply a vector by a scalar, you get another vector.

For example, $2 \times \mathbf{a} = 2\mathbf{a}$.

The vector $2\mathbf{a}$ has the same direction as $\mathbf{a}$ but its magnitude is twice as great.

The vector $-3\mathbf{a}$ has three times the magnitude of $\mathbf{a}$, but is in the opposite direction.

If one vector is a scalar multiple of another, the vectors are parallel.

> If $\mathbf{u} = k\mathbf{v}$ then $\mathbf{u}$ is parallel to $\mathbf{v}$.

a
2a
−3a

All these vectors are parallel.

Example 5

Given $\mathbf{u} = \begin{pmatrix} 4 \\ -1 \\ 2 \end{pmatrix}$ and $\mathbf{v} = \begin{pmatrix} -2 \\ 3 \\ 0 \end{pmatrix}$

a) find $3\mathbf{u} + 5\mathbf{v}$ b) find also $|3\mathbf{u} + 5\mathbf{v}|$.

a) $3\mathbf{u} + 5\mathbf{v} = 3\begin{pmatrix} 4 \\ -1 \\ 2 \end{pmatrix} + 5\begin{pmatrix} -2 \\ 3 \\ 0 \end{pmatrix}$

$= \begin{pmatrix} 12 \\ -3 \\ 6 \end{pmatrix} + \begin{pmatrix} -10 \\ 15 \\ 0 \end{pmatrix}$

$= \begin{pmatrix} 2 \\ 12 \\ 6 \end{pmatrix}$

b) $|3\mathbf{u} + 5\mathbf{v}| = \sqrt{2^2 + 12^2 + 6^2} = \sqrt{184} = 2\sqrt{46}$

> To multiply a vector by a scalar, multiply each of its components by the scalar.

C4

Exercise 12A

1 Find the magnitude of each of these vectors.

a) $\begin{pmatrix} 4 \\ 3 \end{pmatrix}$ b) $\begin{pmatrix} 5 \\ -7 \end{pmatrix}$ c) $\begin{pmatrix} 12 \\ 5 \end{pmatrix}$ d) $\begin{pmatrix} 2 \\ -4 \end{pmatrix}$

e) $\begin{pmatrix} 2 \\ -2 \\ 1 \end{pmatrix}$ f) $\begin{pmatrix} 6 \\ -3 \\ 4 \end{pmatrix}$ g) $\begin{pmatrix} -9 \\ 7 \\ 5 \end{pmatrix}$ h) $\begin{pmatrix} 5 \\ -7 \\ 3 \end{pmatrix}$

2 Given $\mathbf{v} = \begin{pmatrix} x \\ 5 \\ -\sqrt{7} \end{pmatrix}$ and $|\mathbf{v}| = 9$, find the possible values of the constant x.

3 Given that $\mathbf{a} = \begin{pmatrix} 2 \\ y \\ -4 \end{pmatrix}$ and $|\mathbf{a}| = 6$, find the possible values of the constant y.

4 If $\mathbf{b} = \begin{pmatrix} x \\ 4x \\ 4 \end{pmatrix}$, find the possible values of the constant x such that $|\mathbf{b}| = 13$.

5 Given that $\overrightarrow{AB} = \begin{pmatrix} 2 \\ -4 \\ 5 \end{pmatrix}$ and $\overrightarrow{BC} = \begin{pmatrix} 3 \\ 6 \\ -2 \end{pmatrix}$ find $\overrightarrow{AC}$.

6 Given that $\overrightarrow{PQ} = \begin{pmatrix} -2 \\ 3 \\ -6 \end{pmatrix}$ and $\overrightarrow{QR} = \begin{pmatrix} 5 \\ 0 \\ -7 \end{pmatrix}$ find $\overrightarrow{PR}$.

7 Given that $\overrightarrow{RS} = \begin{pmatrix} 3 \\ 6 \\ -8 \end{pmatrix}$ and $\overrightarrow{ST} = \begin{pmatrix} 5 \\ -5 \\ 0 \end{pmatrix}$ find $\overrightarrow{RT}$.

8 Given that $\overrightarrow{AB} = \begin{pmatrix} 5 \\ -7 \\ -2 \end{pmatrix}$ and $\overrightarrow{AC} = \begin{pmatrix} 2 \\ 3 \\ -2 \end{pmatrix}$ find $\overrightarrow{BC}$.

9 Given that $\overrightarrow{ML} = \begin{pmatrix} 10 \\ -4 \\ -6 \end{pmatrix}$ and $\overrightarrow{MN} = \begin{pmatrix} 0 \\ 7 \\ -5 \end{pmatrix}$ find $\overrightarrow{NL}$.

10 Given that $\overrightarrow{PQ} = \begin{pmatrix} 5 \\ 2 \\ -8 \end{pmatrix}$ and $\overrightarrow{PR} = \begin{pmatrix} -2 \\ 5 \\ -6 \end{pmatrix}$ find $\overrightarrow{QR}$.

11 Given $\mathbf{u} = \begin{pmatrix} 6 \\ 2 \\ 5 \end{pmatrix}$ and $\mathbf{v} = \begin{pmatrix} 4 \\ 3 \\ 1 \end{pmatrix}$

a) find $5\mathbf{u} - 7\mathbf{v}$

b) find also $|5\mathbf{u} - 7\mathbf{v}|$.

12 Given $\mathbf{u} = \begin{pmatrix} 0 \\ -4 \\ 1 \end{pmatrix}$ and $\mathbf{v} = \begin{pmatrix} 1 \\ 5 \\ -2 \end{pmatrix}$

a) find $2\mathbf{u} + 5\mathbf{v}$

b) find also $|2\mathbf{u} + 5\mathbf{v}|$.

13 Given $\mathbf{p} = \begin{pmatrix} 4 \\ 2 \\ -2 \end{pmatrix}$, $\mathbf{q} = \begin{pmatrix} 4 \\ 5 \\ -1 \end{pmatrix}$, $\mathbf{r} = \begin{pmatrix} 3 \\ 4 \\ 4 \end{pmatrix}$ and $\mathbf{s} = \begin{pmatrix} 7 \\ 4 \\ 0 \end{pmatrix}$

a) find $3\mathbf{p} + 2\mathbf{q}$

b) find $5\mathbf{r} - \mathbf{s}$

c) show that $|3\mathbf{p} + 2\mathbf{q}| = |5\mathbf{r} - \mathbf{s}|$.

12.2 Position vectors and direction vectors

The **position vector** of a point P with respect to a fixed origin O is the vector $\overrightarrow{OP}$.

The usual notation is $\overrightarrow{OP} = \mathbf{p}$.

The cartesian coordinates of the point are the components of its position vector.

For example, if P has coordinates (3, 4, 5) then $\overrightarrow{OP} = \mathbf{p} = \begin{pmatrix} 3 \\ 4 \\ 5 \end{pmatrix}$.

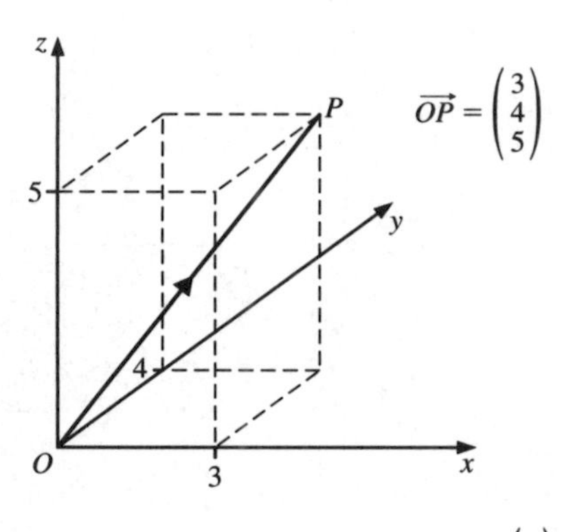

Suppose Q is the point (5, 7, 6). Then $\overrightarrow{OQ} = \mathbf{q} = \begin{pmatrix} 5 \\ 7 \\ 6 \end{pmatrix}$.

To get from P to Q you need to go in the direction $\overrightarrow{PQ}$ specified by $\begin{pmatrix} 2 \\ 3 \\ 1 \end{pmatrix}$: 2 units along Ox, 3 units along Oy and 1 unit along Oz.

Note that:

$$\overrightarrow{PQ} = \overrightarrow{OQ} - \overrightarrow{OP} = \mathbf{q} - \mathbf{p} = \begin{pmatrix} 5 \\ 7 \\ 6 \end{pmatrix} - \begin{pmatrix} 3 \\ 4 \\ 5 \end{pmatrix} = \begin{pmatrix} 2 \\ 3 \\ 1 \end{pmatrix}$$

$\overrightarrow{PQ}$ is the **direction vector** from P to Q.

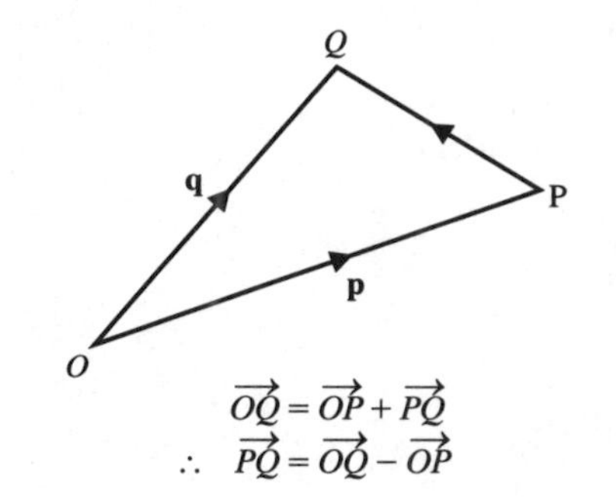

$$\overrightarrow{OQ} = \overrightarrow{OP} + \overrightarrow{PQ}$$
$$\therefore \quad \overrightarrow{PQ} = \overrightarrow{OQ} - \overrightarrow{OP}$$

Example 6

Given $\overrightarrow{OP} = \begin{pmatrix} 3 \\ -1 \\ 2 \end{pmatrix}$, $\overrightarrow{OQ} = \begin{pmatrix} 5 \\ 2 \\ -2 \end{pmatrix}$, $\overrightarrow{OR} = \begin{pmatrix} -4 \\ 5 \\ 7 \end{pmatrix}$ and $\overrightarrow{OS} = \begin{pmatrix} 2 \\ 14 \\ -5 \end{pmatrix}$

a) find $\overrightarrow{PQ}$

b) find $\overrightarrow{RS}$

c) deduce that $\overrightarrow{PQ}$ is parallel to $\overrightarrow{RS}$.

a) $\overrightarrow{PQ} = \overrightarrow{OQ} - \overrightarrow{OP} = \begin{pmatrix} 5 \\ 2 \\ -2 \end{pmatrix} - \begin{pmatrix} 3 \\ -1 \\ 2 \end{pmatrix} = \begin{pmatrix} 2 \\ 3 \\ -4 \end{pmatrix}$

b) $\overrightarrow{RS} = \overrightarrow{OS} - \overrightarrow{OR} = \begin{pmatrix} 2 \\ 14 \\ -5 \end{pmatrix} - \begin{pmatrix} -4 \\ 5 \\ 7 \end{pmatrix} = \begin{pmatrix} 6 \\ 9 \\ -12 \end{pmatrix}$

c) $\overrightarrow{RS} = \begin{pmatrix} 6 \\ 9 \\ -12 \end{pmatrix} = 3\begin{pmatrix} 2 \\ 3 \\ -4 \end{pmatrix} = 3\overrightarrow{PQ}$

This means that $\overrightarrow{RS}$ is in the same direction as $\overrightarrow{PQ}$, but three times as long. Hence $\overrightarrow{RS}$ and $\overrightarrow{PQ}$ are parallel vectors.

Remember: Two vectors are parallel if one is a scalar multiple of the other.

C4

Mid-point of a line

In C2 you learnt how to find the mid-point of a line using coordinate geometry. You can also do it using vectors.

If points A and B have position vectors $\mathbf{a}$ and $\mathbf{b}$ respectively, then the position vector of the mid-point, M, of AB is given by $\overrightarrow{OM} = \frac{1}{2}(\mathbf{a} + \mathbf{b})$.

In the triangle OAB,

$$\begin{aligned} \overrightarrow{OM} &= \overrightarrow{OA} + \tfrac{1}{2}\overrightarrow{AB} \\ &= \mathbf{a} + \tfrac{1}{2}(\mathbf{b} - \mathbf{a}) \\ &= \tfrac{1}{2}(\mathbf{a} + \mathbf{b}) \end{aligned}$$

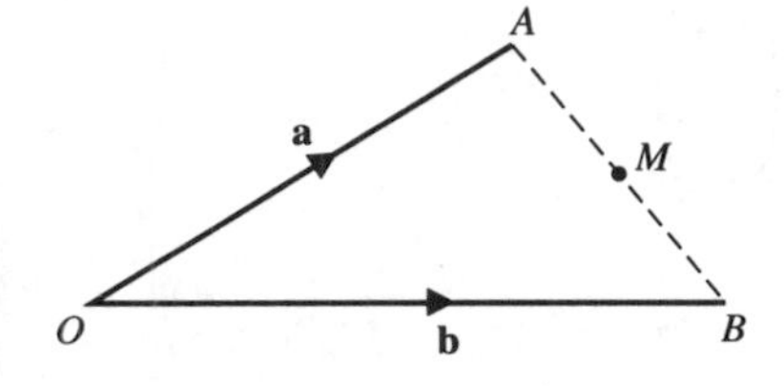

Example 7

$\overrightarrow{OA} = \begin{pmatrix} 5 \\ -4 \\ 7 \end{pmatrix}$ and $\overrightarrow{OB} = \begin{pmatrix} 3 \\ -2 \\ 9 \end{pmatrix}$

a) Find the distance between the points A and B.
b) Find the coordinates of the mid-point, M, of the line AB.

a) $\overrightarrow{AB} = \overrightarrow{OB} - \overrightarrow{OA}$

$$= \begin{pmatrix} 3 \\ -2 \\ 9 \end{pmatrix} - \begin{pmatrix} 5 \\ -4 \\ 7 \end{pmatrix}$$

$$= \begin{pmatrix} -2 \\ 2 \\ 2 \end{pmatrix}$$

The distance AB is given by $|\overrightarrow{AB}| = \sqrt{(-2)^2 + 2^2 + 2^2}$
$= \sqrt{12}$
$= 2\sqrt{3}$

b) $\overrightarrow{OM} = \frac{1}{2}(\mathbf{a} + \mathbf{b})$

$$= \frac{1}{2}\left[\begin{pmatrix} 5 \\ -4 \\ 7 \end{pmatrix} + \begin{pmatrix} 3 \\ -2 \\ 9 \end{pmatrix}\right]$$

$$= \frac{1}{2}\begin{pmatrix} 8 \\ -6 \\ 16 \end{pmatrix}$$

$$= \begin{pmatrix} 4 \\ -3 \\ 8 \end{pmatrix}$$

The coordinates of M are $(4, -3, 8)$.

$\overrightarrow{AB}$ is the direction vector from A to B.

$\overrightarrow{OM}$ is the position vector of M.

The components of the position vector $\overrightarrow{OM}$ are the x, y and z coordinates of M.

C4

Exercise 12B

1 Given $\overrightarrow{OP} = \begin{pmatrix} -1 \\ 2 \\ 5 \end{pmatrix}$, $\overrightarrow{OQ} = \begin{pmatrix} 2 \\ 0 \\ 7 \end{pmatrix}$, $\overrightarrow{OR} = \begin{pmatrix} -3 \\ -5 \\ 4 \end{pmatrix}$ and $\overrightarrow{OS} = \begin{pmatrix} 3 \\ -9 \\ 8 \end{pmatrix}$

a) find $\overrightarrow{PQ}$ b) find $\overrightarrow{RS}$.

Deduce that $\overrightarrow{PQ}$ is parallel to $\overrightarrow{RS}$.

2 $\overrightarrow{OA} = \begin{pmatrix} -4 \\ 2 \\ 6 \end{pmatrix}$, $\overrightarrow{OB} = \begin{pmatrix} -5 \\ 4 \\ 3 \end{pmatrix}$, $\overrightarrow{OC} = \begin{pmatrix} 8 \\ -2 \\ 5 \end{pmatrix}$ and $\overrightarrow{OD} = \begin{pmatrix} 5 \\ 4 \\ -4 \end{pmatrix}$

a) Find $\overrightarrow{AB}$ b) find $\overrightarrow{CD}$.

Deduce that $\overrightarrow{AB}$ is parallel to $\overrightarrow{CD}$.

3 $\overrightarrow{OA} = \begin{pmatrix} 6 \\ 7 \\ -5 \end{pmatrix}$, $\overrightarrow{OB} = \begin{pmatrix} 8 \\ 4 \\ -4 \end{pmatrix}$, $\overrightarrow{OC} = \begin{pmatrix} 12 \\ -2 \\ 7 \end{pmatrix}$ and $\overrightarrow{OD} = \begin{pmatrix} 4 \\ 10 \\ 3 \end{pmatrix}$

a) Find $\overrightarrow{AB}$.

b) Find $\overrightarrow{CD}$.

Deduce that $\overrightarrow{\mathrm{AB}}$ is parallel to $\overrightarrow{CD}$.

4 $\overrightarrow{OA} = \begin{pmatrix} -6 \\ 3 \\ 5 \end{pmatrix}$ and $\overrightarrow{OB} = \begin{pmatrix} 2 \\ -1 \\ -1 \end{pmatrix}$

a) Find the distance between the points A and B.

b) Find the coordinates of the mid-point, M, of the line AB.

5 $\overrightarrow{OP} = \begin{pmatrix} 0 \\ 3 \\ -2 \end{pmatrix}$ and $\overrightarrow{OQ} = \begin{pmatrix} 5 \\ 3 \\ 3 \end{pmatrix}$

a) Find the distance between the points P and Q.

b) Find the coordinates of the mid-point, M, of the line PQ.

C4

6 $\overrightarrow{OA} = \begin{pmatrix} 3 \\ -6 \\ -5 \end{pmatrix}$, $\overrightarrow{OB} = \begin{pmatrix} 5 \\ 4 \\ 1 \end{pmatrix}$, $\overrightarrow{OC} = \begin{pmatrix} 0 \\ -5 \\ 9 \end{pmatrix}$ and $\overrightarrow{OD} = \begin{pmatrix} -2 \\ -1 \\ 3 \end{pmatrix}$

a) Find the coordinates of the mid-point, M, of AB.

b) Find the coordinates of the mid-point, N, of CD.

c) Calculate the distance MN.

7 $\overrightarrow{OA} = \begin{pmatrix} -2 \\ 1 \\ 0 \end{pmatrix}$, $\overrightarrow{OB} = \begin{pmatrix} 6 \\ 3 \\ -2 \end{pmatrix}$, $\overrightarrow{OC} = \begin{pmatrix} -5 \\ 0 \\ 5 \end{pmatrix}$ and $\overrightarrow{OD} = \begin{pmatrix} -5 \\ -8 \\ 3 \end{pmatrix}$

a) Find the coordinates of the mid-point, M, of AB.

b) Find the coordinates of the mid-point, N, of CD.

c) Calculate the distance MN.

8 The points A, B, C and D have coordinates (2, 3, 4), (8, −5, 2), (3, 1, 5) and (−1, 2, −3) respectively. P is the mid-point of AB, Q is the mid-point of BC, R is the mid-point of CD and S is the mid-point of DA.

a) Find $\overrightarrow{OP}$, $\overrightarrow{OQ}$, $\overrightarrow{OR}$ and $\overrightarrow{OS}$.

b) Show that $\overrightarrow{PQ} = \overrightarrow{SR}$.

c) Show that $\overrightarrow{QR} = \overrightarrow{PS}$.

Collinearity

You can use vectors to show that three points lie in the same straight line, that is, they are **collinear**.

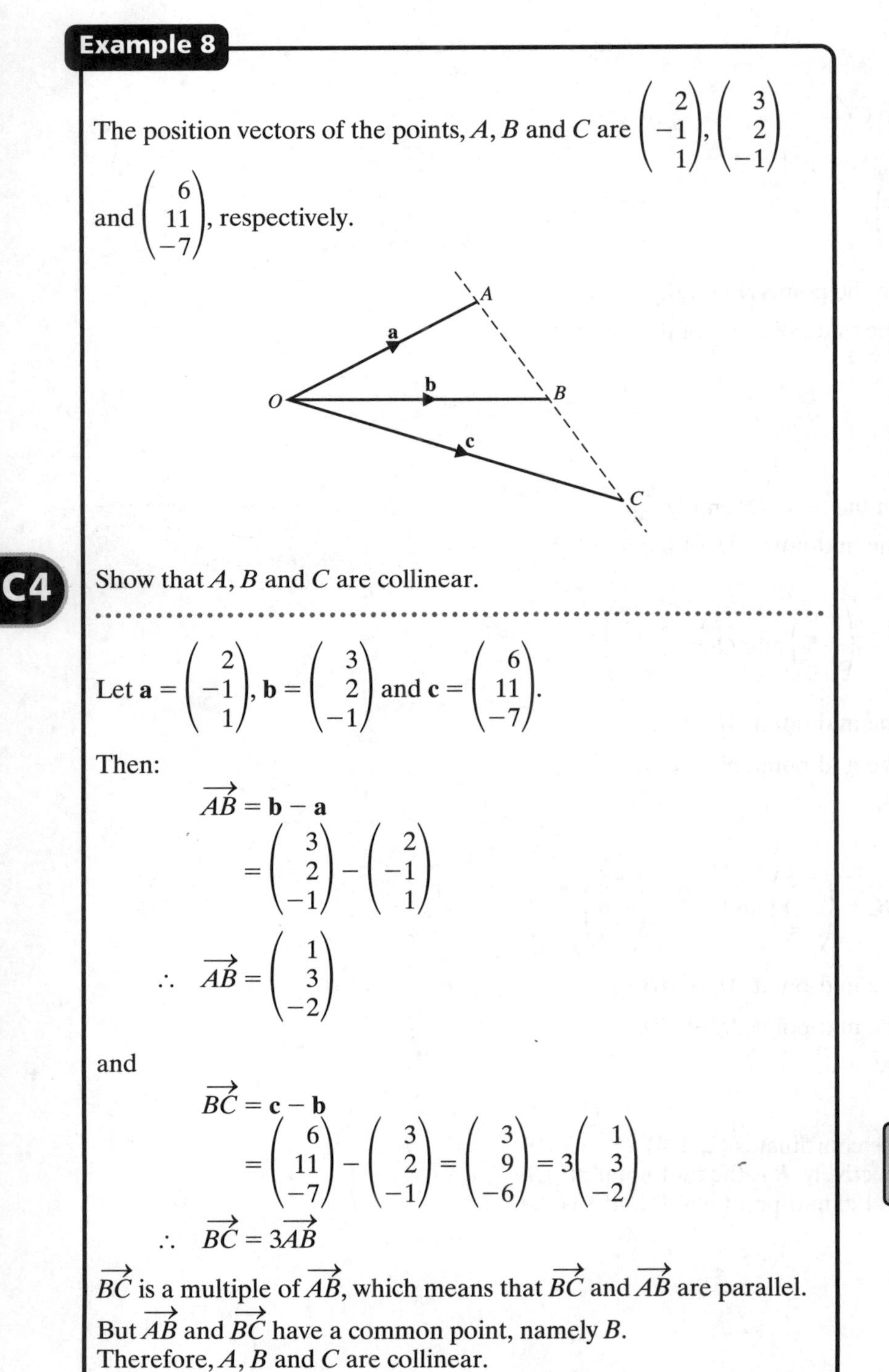

Example 8

The position vectors of the points, A, B and C are $\begin{pmatrix} 2 \\ -1 \\ 1 \end{pmatrix}$, $\begin{pmatrix} 3 \\ 2 \\ -1 \end{pmatrix}$ and $\begin{pmatrix} 6 \\ 11 \\ -7 \end{pmatrix}$, respectively.

Show that A, B and C are collinear.

Let $\mathbf{a} = \begin{pmatrix} 2 \\ -1 \\ 1 \end{pmatrix}$, $\mathbf{b} = \begin{pmatrix} 3 \\ 2 \\ -1 \end{pmatrix}$ and $\mathbf{c} = \begin{pmatrix} 6 \\ 11 \\ -7 \end{pmatrix}$.

Then:

$$\overrightarrow{AB} = \mathbf{b} - \mathbf{a}$$
$$= \begin{pmatrix} 3 \\ 2 \\ -1 \end{pmatrix} - \begin{pmatrix} 2 \\ -1 \\ 1 \end{pmatrix}$$
$$\therefore \quad \overrightarrow{AB} = \begin{pmatrix} 1 \\ 3 \\ -2 \end{pmatrix}$$

and

$$\overrightarrow{BC} = \mathbf{c} - \mathbf{b}$$
$$= \begin{pmatrix} 6 \\ 11 \\ -7 \end{pmatrix} - \begin{pmatrix} 3 \\ 2 \\ -1 \end{pmatrix} = \begin{pmatrix} 3 \\ 9 \\ -6 \end{pmatrix} = 3\begin{pmatrix} 1 \\ 3 \\ -2 \end{pmatrix}$$
$$\therefore \quad \overrightarrow{BC} = 3\overrightarrow{AB}$$

Reduce the final vector to its simplest form.

$\overrightarrow{BC}$ is a multiple of $\overrightarrow{AB}$, which means that $\overrightarrow{BC}$ and $\overrightarrow{AB}$ are parallel.
But $\overrightarrow{AB}$ and $\overrightarrow{BC}$ have a common point, namely B.
Therefore, A, B and C are collinear.

Exercise 12C

1 Points P, Q and R have position vectors $\begin{pmatrix}5\\4\\1\end{pmatrix}$, $\begin{pmatrix}7\\5\\4\end{pmatrix}$ and $\begin{pmatrix}11\\7\\10\end{pmatrix}$, respectively.

a) Find $\overrightarrow{PQ}$ and $\overrightarrow{QR}$.

b) Deduce that P, Q and R are collinear and find the ratio $PQ : QR$.

2 The points A, B and C have coordinates $(1, -5, 6)$, $(3, -2, 10)$ and $(7, 4, 18)$ respectively. Show that A, B and C are collinear.

3 Show that the points $P(5, 4, -3)$, $Q(3, 8, -1)$ and $R(0, 14, 2)$ are collinear.

4 Given that $A(2, 13, -5)$, $B(3, y, -3)$ and $C(6, -7, z)$ are collinear, find the values of the constants y and z.

12.3 Scalar product

You can also multiply vectors together. One form of vector multiplication is the **scalar product**.

The scalar product is sometimes called the dot product.

The scalar product **a.b** of two vectors **a** and **b** is defined by

$$\mathbf{a}.\mathbf{b} = |\mathbf{a}|\,|\mathbf{b}| \cos \theta$$

where θ is the angle between the vectors.

It is called the scalar product because the answer is a scalar, not a vector.

- When the two vectors **a** and **b** are perpendicular, $\theta = 90°$ and $\cos 90° = 0$. Therefore $\mathbf{a}.\mathbf{b} = 0$.

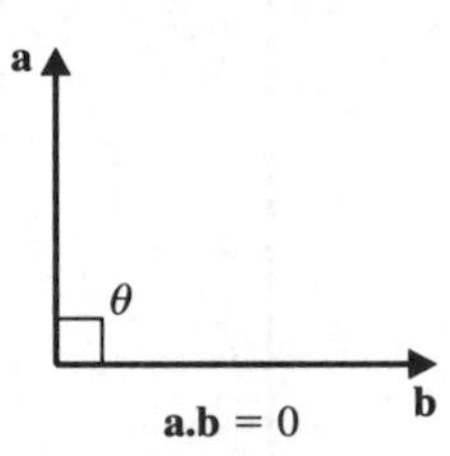

- When the angle between the vectors **a** and **b** is acute, $\cos \theta > 0$ and therefore $\mathbf{a}.\mathbf{b} > 0$.

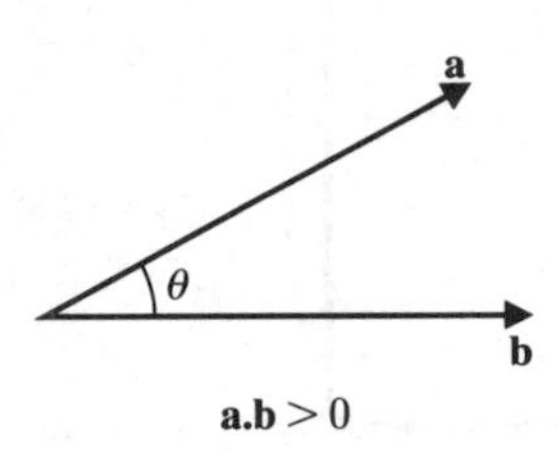

- When the angle between the vectors **a** and **b** is between 90° and 180°, $\cos\theta < 0$ and therefore $\mathbf{a.b} < 0$.

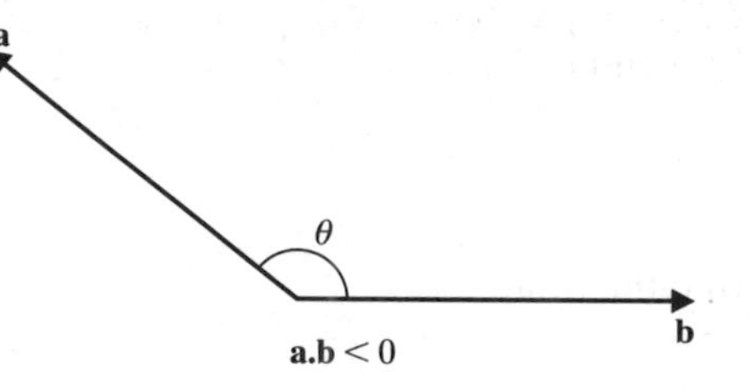

$\mathbf{a.b} < 0$

If the vectors are in column vector (component) form, it is easy to calculate the scalar product.

If $\mathbf{a} = \begin{pmatrix} a_1 \\ a_2 \end{pmatrix}$ and $\mathbf{b} = \begin{pmatrix} b_1 \\ b_2 \end{pmatrix}$, then

$$\mathbf{a.b} = a_1b_1 + a_2b_2$$

In three dimensions, if $\mathbf{a} = \begin{pmatrix} a_1 \\ a_2 \\ a_3 \end{pmatrix}$ and $\mathbf{b} = \begin{pmatrix} b_1 \\ b_2 \\ b_3 \end{pmatrix}$, then

$$\mathbf{a.b} = a_1b_1 + a_2b_2 + a_3b_3$$

C4

Example 9

Find the scalar product of each of these pairs of vectors

a) $\begin{pmatrix} 2 \\ 3 \end{pmatrix}$ and $\begin{pmatrix} 1 \\ -6 \end{pmatrix}$

b) $\begin{pmatrix} 4 \\ -2 \\ 1 \end{pmatrix}$ and $\begin{pmatrix} 2 \\ 1 \\ -3 \end{pmatrix}$

c) $\begin{pmatrix} 2 \\ -1 \\ 4 \end{pmatrix}$ and $\begin{pmatrix} -3 \\ 1 \\ -5 \end{pmatrix}$

a) $\begin{pmatrix} 2 \\ 3 \end{pmatrix} . \begin{pmatrix} 1 \\ -6 \end{pmatrix} = (2 \times 1) + (3 \times -6) = -16$

b) $\begin{pmatrix} 4 \\ -2 \\ 1 \end{pmatrix} . \begin{pmatrix} 2 \\ 1 \\ -3 \end{pmatrix} = (4 \times 2) + (-2 \times 1) + (1 \times -3) = 3$

c) $\begin{pmatrix} 2 \\ -1 \\ 4 \end{pmatrix} . \begin{pmatrix} -3 \\ 1 \\ -5 \end{pmatrix} = (2 \times -3) + (-1 \times 1) + (4 \times -5) = -27$

You can use the scalar product to find the angle between two vectors.

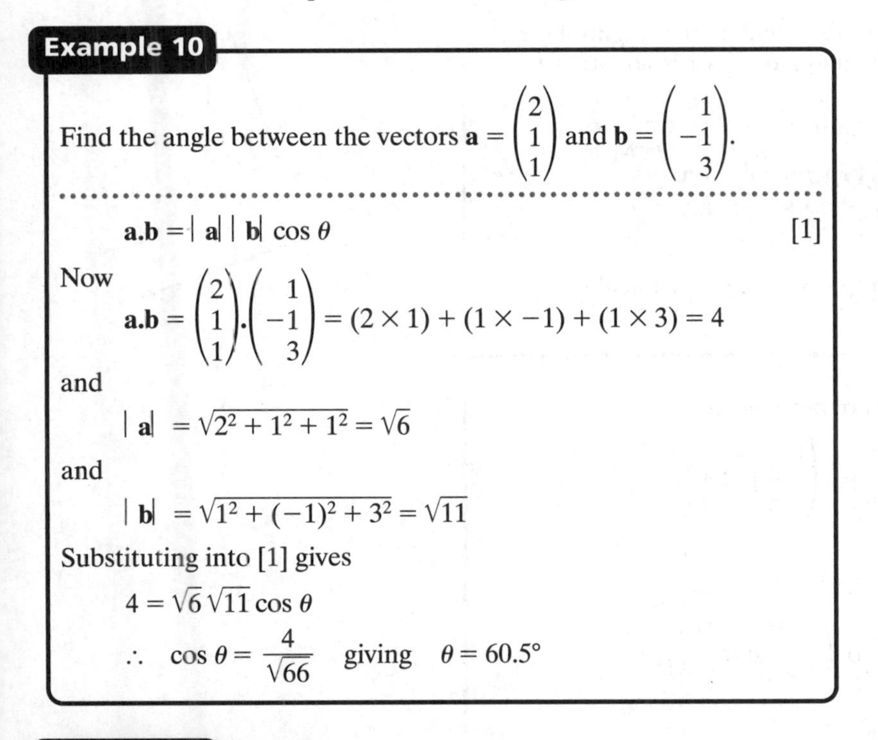

Example 10

Find the angle between the vectors $\mathbf{a} = \begin{pmatrix} 2 \\ 1 \\ 1 \end{pmatrix}$ and $\mathbf{b} = \begin{pmatrix} 1 \\ -1 \\ 3 \end{pmatrix}$.

$$\mathbf{a.b} = |\mathbf{a}|\,|\mathbf{b}| \cos\theta \qquad [1]$$

Now

$$\mathbf{a.b} = \begin{pmatrix} 2 \\ 1 \\ 1 \end{pmatrix} . \begin{pmatrix} 1 \\ -1 \\ 3 \end{pmatrix} = (2 \times 1) + (1 \times -1) + (1 \times 3) = 4$$

and

$$|\mathbf{a}| = \sqrt{2^2 + 1^2 + 1^2} = \sqrt{6}$$

and

$$|\mathbf{b}| = \sqrt{1^2 + (-1)^2 + 3^2} = \sqrt{11}$$

Substituting into [1] gives

$$4 = \sqrt{6}\sqrt{11} \cos\theta$$

$$\therefore \quad \cos\theta = \frac{4}{\sqrt{66}} \quad \text{giving} \quad \theta = 60.5°$$

Example 11

Given the points $P(5, 2, -1)$, $Q(4, -2, 3)$ and $R(0, 3, 9)$, find the angle $P\hat{Q}R$.

$$\overrightarrow{QP} = \overrightarrow{OP} - \overrightarrow{OQ} = \begin{pmatrix} 5 \\ 2 \\ -1 \end{pmatrix} - \begin{pmatrix} 4 \\ -2 \\ 3 \end{pmatrix} = \begin{pmatrix} 1 \\ 4 \\ -4 \end{pmatrix}$$

$$\text{and } \overrightarrow{QR} = \overrightarrow{OR} - \overrightarrow{OQ} = \begin{pmatrix} 0 \\ 3 \\ 9 \end{pmatrix} - \begin{pmatrix} 4 \\ -2 \\ 3 \end{pmatrix} = \begin{pmatrix} -4 \\ 5 \\ 6 \end{pmatrix}$$

P (5, 2, −1)
θ
Q (4, −2, 3)
R (0, 3, 9)

You know that $\overrightarrow{QP}.\overrightarrow{QR} = |\overrightarrow{QP}|\,|\overrightarrow{QR}| \cos\theta$ [1]

where θ is the angle $P\hat{Q}R$.

$$\text{Now } \overrightarrow{QP}.\overrightarrow{QR} = \begin{pmatrix} 1 \\ 4 \\ -4 \end{pmatrix} . \begin{pmatrix} -4 \\ 5 \\ 6 \end{pmatrix} = (1 \times -4) + (4 \times 5) + (-4 \times 6) = -8$$

$$|\overrightarrow{QP}| = \sqrt{1^2 + 4^2 + (-4)^2} = \sqrt{33}$$

$$\text{and} \quad |\overrightarrow{QR}| = \sqrt{(-4)^2 + 5^2 + 6^2} = \sqrt{77}$$

Substituting into [1] gives

$$-8 = \sqrt{33}\sqrt{77} \cos\theta$$

$$\therefore \quad \cos\theta = -\frac{8}{11\sqrt{21}}, \quad \text{giving} \quad \theta = 99.1°$$

Perpendicular vectors

As you saw on page 257, if two vectors, **a** and **b** are at right-angles, then the angle between them is 90° and $\mathbf{a.b} = |\mathbf{a}|\,|\mathbf{b}| \cos 90° = 0$.

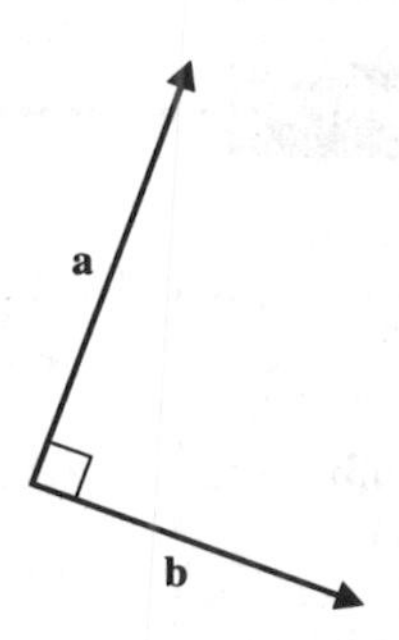

> So, for two non-zero vectors, **a** and **b**
> - if $\mathbf{a.b} = 0$, then **a** and **b** are perpendicular vectors
> - if **a** and **b** are perpendicular vectors, then $\mathbf{a.b} = 0$.

You can use this fact to show that vectors are perpendicular.

Example 12

Show that these pairs of vectors are perpendicular.

a) $\begin{pmatrix} 2 \\ -5 \end{pmatrix}$ and $\begin{pmatrix} 10 \\ 4 \end{pmatrix}$ b) $\begin{pmatrix} 2 \\ 4 \\ -3 \end{pmatrix}$ and $\begin{pmatrix} -1 \\ 5 \\ 6 \end{pmatrix}$

a) $\begin{pmatrix} 2 \\ -5 \end{pmatrix} \cdot \begin{pmatrix} 10 \\ 4 \end{pmatrix} = (2 \times 10) + (-5 \times 4) = 0$

Since the scalar product is zero, $\begin{pmatrix} 2 \\ -5 \end{pmatrix}$ and $\begin{pmatrix} 10 \\ 4 \end{pmatrix}$ are perpendicular.

b) $\begin{pmatrix} 2 \\ 4 \\ -3 \end{pmatrix} \cdot \begin{pmatrix} -1 \\ 5 \\ 6 \end{pmatrix} = (2 \times -1) + (4 \times 5) + (-3 \times 6) = 0$

Since the scalar product is zero, $\begin{pmatrix} 2 \\ 4 \\ -3 \end{pmatrix}$ and $\begin{pmatrix} -1 \\ 5 \\ 6 \end{pmatrix}$ are perpendicular.

Example 13

Given that the two vectors $\mathbf{a} = \begin{pmatrix} 3t+1 \\ 1 \\ -1 \end{pmatrix}$ and $\mathbf{b} = \begin{pmatrix} t+3 \\ 3 \\ -2 \end{pmatrix}$ are perpendicular, find the possible values of the constant t.

a and **b** are perpendicular, so $\mathbf{a.b} = 0$. That is,

$$\begin{pmatrix} 3t+1 \\ 1 \\ -1 \end{pmatrix} \cdot \begin{pmatrix} t+3 \\ 3 \\ -2 \end{pmatrix} = 0$$

$$(3t+1)(t+3) + (1 \times 3) + (-1 \times -2) = 0$$

$$3t^2 + 10t + 8 = 0$$

$$(3t+4)(t+2) = 0$$

Solving gives $t = -\frac{4}{3}$ and $t = -2$, so the values of t are $-\frac{4}{3}$ and -2.

C4

Exercise 12D

1 Given $\mathbf{a} = \begin{pmatrix} 3 \\ 4 \end{pmatrix}$, $\mathbf{b} = \begin{pmatrix} 1 \\ -3 \end{pmatrix}$ and $\mathbf{c} = \begin{pmatrix} 2 \\ 5 \end{pmatrix}$, evaluate each of these scalar products.

a) **a.b** b) **b.a** c) **a.c**
d) **c.b** e) **a.a** f) **c.(a + b)**

2 Given $\mathbf{x} = \begin{pmatrix} 2 \\ -3 \\ 1 \end{pmatrix}$, $\mathbf{y} = \begin{pmatrix} 5 \\ 2 \\ -7 \end{pmatrix}$ and $\mathbf{z} = \begin{pmatrix} 1 \\ -4 \\ -2 \end{pmatrix}$, evaluate each of these.

a) **x.y** b) **y.x** c) **x.z**
d) **z.z** e) **x.(y + z)** f) **y.(z − x)**

3 Given

$$\mathbf{p} = \begin{pmatrix} -2 \\ 3 \end{pmatrix} \quad \mathbf{q} = \begin{pmatrix} 1 \\ 1 \end{pmatrix} \quad \text{and} \quad \mathbf{r} = \begin{pmatrix} 5 \\ -2 \end{pmatrix}$$

evaluate each of these.

a) **p.q** b) **q.r** c) **r.q**
d) **q.q** e) **r.(q + p)** f) **p.(q − r)**

4 Given

$$\mathbf{c} = \begin{pmatrix} 3 \\ 1 \\ -4 \end{pmatrix} \quad \mathbf{d} = \begin{pmatrix} -5 \\ -2 \\ 7 \end{pmatrix} \quad \text{and} \quad \mathbf{e} = \begin{pmatrix} 0 \\ 4 \\ -5 \end{pmatrix}$$

evaluate each of these.

a) **c.d** b) **d.e** c) **c.e**
d) **d.(e − c)** e) **c.(c + d)** f) **e.(2c − d)**

5 Decide which of these pairs of vectors are perpendicular, which are parallel, and which are neither perpendicular nor parallel.

a) $\begin{pmatrix} 2 \\ 8 \end{pmatrix}$ and $\begin{pmatrix} 4 \\ -1 \end{pmatrix}$ b) $\begin{pmatrix} 3 \\ 5 \end{pmatrix}$ and $\begin{pmatrix} 6 \\ 10 \end{pmatrix}$

c) $\begin{pmatrix} -3 \\ 1 \end{pmatrix}$ and $\begin{pmatrix} 6 \\ -2 \end{pmatrix}$ d) $\begin{pmatrix} 12 \\ 6 \end{pmatrix}$ and $\begin{pmatrix} 1 \\ -2 \end{pmatrix}$

e) $\begin{pmatrix} 6 \\ -8 \\ 2 \end{pmatrix}$ and $\begin{pmatrix} 9 \\ -12 \\ 3 \end{pmatrix}$ f) $\begin{pmatrix} 5 \\ -6 \\ -3 \end{pmatrix}$ and $\begin{pmatrix} 3 \\ 2 \\ 1 \end{pmatrix}$

g) $\begin{pmatrix} 3 \\ -1 \\ 4 \end{pmatrix}$ and $\begin{pmatrix} 9 \\ -3 \\ 12 \end{pmatrix}$ h) $\begin{pmatrix} 1 \\ 2 \\ 3 \end{pmatrix}$ and $\begin{pmatrix} 3 \\ 2 \\ 1 \end{pmatrix}$

6 Find the angle between each of these pairs of vectors, giving your answers correct to one decimal place.

a) $\begin{pmatrix} 3 \\ -4 \end{pmatrix}$ and $\begin{pmatrix} 12 \\ 5 \end{pmatrix}$ b) $\begin{pmatrix} 1 \\ 0 \end{pmatrix}$ and $\begin{pmatrix} 1 \\ 1 \end{pmatrix}$

c) $\begin{pmatrix} 2 \\ -1 \end{pmatrix}$ and $\begin{pmatrix} 6 \\ 3 \end{pmatrix}$ d) $\begin{pmatrix} 3 \\ 5 \end{pmatrix}$ and $\begin{pmatrix} 2 \\ -3 \end{pmatrix}$

e) $\begin{pmatrix} 2 \\ 1 \\ -2 \end{pmatrix}$ and $\begin{pmatrix} 4 \\ -3 \\ 12 \end{pmatrix}$ f) $\begin{pmatrix} 3 \\ -5 \\ -2 \end{pmatrix}$ and $\begin{pmatrix} 1 \\ 0 \\ -6 \end{pmatrix}$

g) $\begin{pmatrix} -2 \\ 1 \\ 3 \end{pmatrix}$ and $\begin{pmatrix} 4 \\ -3 \\ 3 \end{pmatrix}$ h) $\begin{pmatrix} 3 \\ 0 \\ -1 \end{pmatrix}$ and $\begin{pmatrix} 2 \\ 5 \\ 0 \end{pmatrix}$

7 $\mathbf{a} = \begin{pmatrix} 4 \\ 5 \end{pmatrix}$, $\mathbf{b} = \begin{pmatrix} \lambda \\ -8 \end{pmatrix}$ and $\mathbf{c} = \begin{pmatrix} 1 \\ \mu \end{pmatrix}$.

a) Find the value of the constant λ given that **a** and **b** are perpendicular.

b) Find the value of the constant μ given that **a** and **c** are parallel.

8 $\mathbf{p} = \begin{pmatrix} 6 \\ -1 \end{pmatrix}$, $\mathbf{q} = \begin{pmatrix} \lambda \\ 2 \end{pmatrix}$ and $\mathbf{r} = \begin{pmatrix} 2 \\ \mu \end{pmatrix}$.

a) Find the value of the constant λ given that **p** and **q** are parallel.

b) Find the value of the constant μ given that **p** and **r** are perpendicular.

9 Given that the vectors $\begin{pmatrix} 2 \\ t \\ -4 \end{pmatrix}$ and $\begin{pmatrix} 1 \\ -3 \\ t-4 \end{pmatrix}$ are perpendicular, find the value of the constant t.

C4

10 Given that $\begin{pmatrix} \lambda \\ 2+\lambda \\ 3 \end{pmatrix}$ and $\begin{pmatrix} -1 \\ 3 \\ 4-\lambda \end{pmatrix}$ are perpendicular vectors, find the value of constant λ.

11 Find the possible values of the constant t, given that the vectors

$$\begin{pmatrix} t \\ 8 \\ 3t+1 \end{pmatrix} \text{ and } \begin{pmatrix} t+1 \\ t-1 \\ -2 \end{pmatrix}$$

are perpendicular.

12 Given that the vectors $\begin{pmatrix} t \\ 4 \\ 2t+1 \end{pmatrix}$ and $\begin{pmatrix} t+2 \\ 1-t \\ -1 \end{pmatrix}$ are perpendicular, find the possible values of the constant t.

12.4 Vector equation of a line

Let **a** and **b** be the position vectors of two points A and B with respect to an origin O. Let **r** be the position vector of a point P on the line AB.

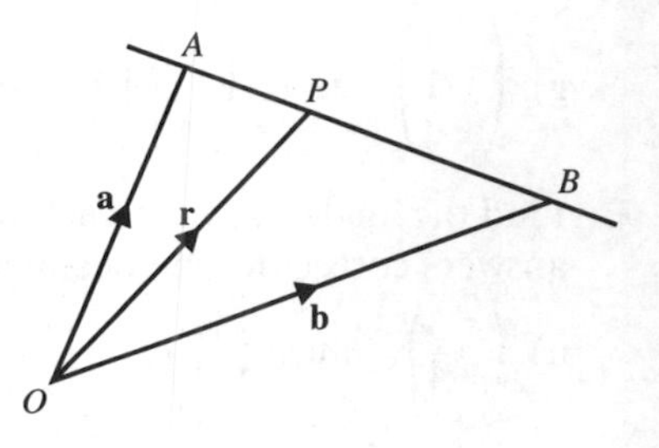

Then:

$$\begin{aligned} \overrightarrow{OP} &= \overrightarrow{OA} + \overrightarrow{AP} \\ &= \overrightarrow{OA} + t\overrightarrow{AB} \end{aligned}$$

where t is a scalar. Therefore,

$$\mathbf{r} = \mathbf{a} + t(\mathbf{b} - \mathbf{a}) \quad \text{or} \quad \mathbf{r} = (1-t)\mathbf{a} + t\mathbf{b}$$

This is the vector equation of the line AB. The vector $(\mathbf{b} - \mathbf{a})$ is the direction vector of the line.

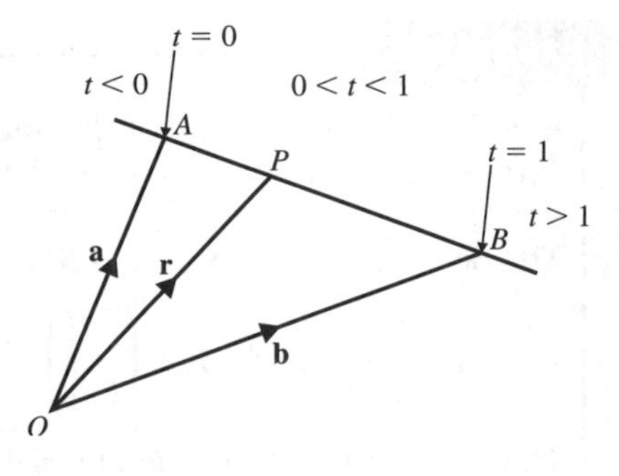

Each value of the parameter t corresponds to a point on the line AB.

- When $t < 0$, point P is on the line BA produced.
- When $t = 0$, $\mathbf{r} = \mathbf{a}$, i.e. $P = A$
- When $0 < t < 1$, point P is between A and B.
- When $t = 1$, $\mathbf{r} = \mathbf{b}$, i.e. $P = B$
- When $t > 1$, point P is on the line AB produced.

Example 14

a) Find the vector equation of the line passing through $A(-1, 2, 0)$ and $B(5, 11, 12)$.

b) Show that the point $Q(1, 5, 4)$ lies on the line AB.

a) Let A be the 'jumping on' point for the line, and $\overrightarrow{AB}$ be the direction.

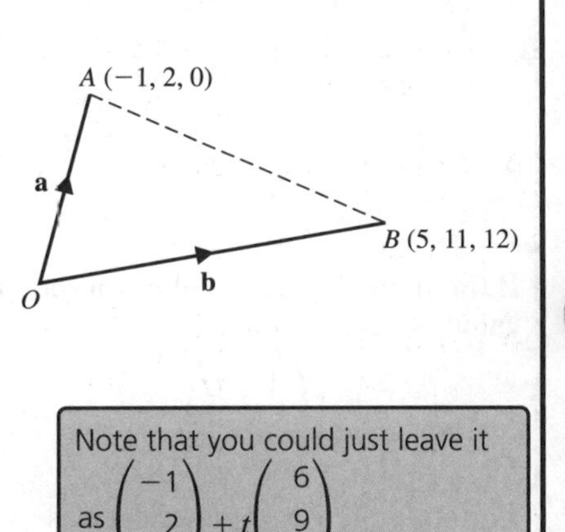

If $P(x, y, z)$ is any point on the line, then $\overrightarrow{OP} = \overrightarrow{OA} + t\overrightarrow{AB}$. That is:

$$\overrightarrow{OP} = \mathbf{a} + t(\mathbf{b} - \mathbf{a})$$
$$= \begin{pmatrix} -1 \\ 2 \\ 0 \end{pmatrix} + t\left[\begin{pmatrix} 5 \\ 11 \\ 12 \end{pmatrix} - \begin{pmatrix} -1 \\ 2 \\ 0 \end{pmatrix}\right]$$
$$\therefore \quad \begin{pmatrix} x \\ y \\ z \end{pmatrix} = \begin{pmatrix} -1 \\ 2 \\ 0 \end{pmatrix} + t\begin{pmatrix} 6 \\ 9 \\ 12 \end{pmatrix} = \begin{pmatrix} -1 + 6t \\ 2 + 9t \\ 12t \end{pmatrix}$$

So the vector equation of the line is

$$\begin{pmatrix} x \\ y \\ z \end{pmatrix} = \begin{pmatrix} -1 + 6t \\ 2 + 9t \\ 12t \end{pmatrix}$$

Note that you could just leave it as $\begin{pmatrix} -1 \\ 2 \\ 0 \end{pmatrix} + t\begin{pmatrix} 6 \\ 9 \\ 12 \end{pmatrix}$

b) If $Q(1, 5, 4)$ lies on the line AB, then there must be a value of t for which

$$\begin{pmatrix} 1 \\ 5 \\ 4 \end{pmatrix} = \begin{pmatrix} -1 + 6t \\ 2 + 9t \\ 12t \end{pmatrix}$$

Equating the x components gives

$$x: \quad 1 = -1 + 6t$$
$$\therefore \quad t = \tfrac{1}{3}$$

Checking this value of t in the y and z components gives

$$y: \quad 2 + 9t = 2 + 9\left(\tfrac{1}{3}\right) = 5$$
$$z: \quad 12t = 12\left(\tfrac{1}{3}\right) = 4$$

So $(x, y, z) = (1, 5, 4)$ as required, and Q does lie on AB.

C4

Example 15

Find the vector equation of the line passing through $A(1, 3, 2)$ and $B(0, -1, 4)$. Does the point $P(-2, 9, 1)$ lie on the line AB?

The position vectors of A and B are given by

$$\mathbf{a} = \begin{pmatrix} 1 \\ 3 \\ 2 \end{pmatrix} \quad \text{and} \quad \mathbf{b} = \begin{pmatrix} 0 \\ -1 \\ 4 \end{pmatrix}$$

The vector equation of the line is given by

$$\mathbf{r} = \mathbf{a} + t(\mathbf{b} - \mathbf{a})$$

$$= \begin{pmatrix} 1 \\ 3 \\ 2 \end{pmatrix} + t\left[\begin{pmatrix} 0 \\ -1 \\ 4 \end{pmatrix} - \begin{pmatrix} 1 \\ 3 \\ 2 \end{pmatrix}\right]$$

$$\therefore \quad \mathbf{r} = \begin{pmatrix} 1 \\ 3 \\ 2 \end{pmatrix} + t\begin{pmatrix} -1 \\ -4 \\ 2 \end{pmatrix}$$

i.e. $\begin{pmatrix} x \\ y \\ z \end{pmatrix} = \begin{pmatrix} 1 - t \\ 3 - 4t \\ 2 + 2t \end{pmatrix}$

C4

If the point $P(-2, 9, 1)$ lies on the line AB, there will exist a unique value of t for which

$$\begin{pmatrix} -2 \\ 9 \\ 1 \end{pmatrix} = \begin{pmatrix} 1 - t \\ 3 - 4t \\ 2 + 2t \end{pmatrix}$$

Equating the x components gives

$$x: \quad -2 = 1 - t \quad \therefore \quad t = 3$$

However, $t = 3$ does not satisfy $9 = 3 - 4t$. Therefore, the point $P(-2, 9, 1)$ does not lie on the line AB.

If two lines intersect, the position vector of their point of intersection must satisfy the vector equations of both lines, as the point lies on both lines.

Example 16

Two lines l and m have vector equations

$$\mathbf{r}_l = \begin{pmatrix} 2 \\ 1 \\ 0 \end{pmatrix} + \lambda\begin{pmatrix} -3 \\ 1 \\ 4 \end{pmatrix} \quad \text{and} \quad \mathbf{r}_m = \begin{pmatrix} -1 \\ 3 \\ 7 \end{pmatrix} + \mu\begin{pmatrix} 3 \\ 0 \\ -1 \end{pmatrix}$$

respectively.

a) Show that the lines l and m intersect.

b) Find the coordinates of their point of intersection.

c) Find the angle between the lines.

This example uses the Greek letters λ and μ in place of t.

a) You can write the vector equations of the two lines as

$$\mathbf{r}_l = \begin{pmatrix} 2-3\lambda \\ 1+\lambda \\ 4\lambda \end{pmatrix} \quad \text{and} \quad \mathbf{r}_m = \begin{pmatrix} -1+3\mu \\ 3 \\ 7-\mu \end{pmatrix}$$

If the lines intersect, they share a common point. At the common point:

$$\mathbf{r}_l = \mathbf{r}_m$$

$$\therefore \quad \begin{pmatrix} 2-3\lambda \\ 1+\lambda \\ 4\lambda \end{pmatrix} = \begin{pmatrix} -1+3\mu \\ 3 \\ 7-\mu \end{pmatrix}$$

Equating x, y and z components:

$$x: \quad 2-3\lambda = -1+3\mu \qquad [1]$$

$$y: \quad 1+\lambda = 3 \qquad [2]$$

$$z: \quad 4\lambda = 7-\mu \qquad [3]$$

This is a set of simultaneous equations.

From [2], $\lambda = 2$. Substituting into [1] gives

$$2-3(2) = -1+3\mu$$

$$\therefore \quad \mu = -1$$

If the lines intersect, $\lambda = 2$, $\mu = -1$ must also satisfy equation [3].

Check: $4 \times 2 = 8 \qquad 7-(-1) = 8$

So $\lambda = 2$, $\mu = -1$ satisfy all three component equations. Therefore the lines l and m intersect.

Alternatively, you can substitute these values of λ and μ in the vector equations of the two lines:

$$\mathbf{r}_l = \begin{pmatrix} 2-3(2) \\ 1+2 \\ 4(2) \end{pmatrix} = \begin{pmatrix} -4 \\ 3 \\ 8 \end{pmatrix}, \quad \mathbf{r}_m = \begin{pmatrix} -1+3(-1) \\ 3 \\ 7-(-1) \end{pmatrix} = \begin{pmatrix} -4 \\ 3 \\ 8 \end{pmatrix}$$

b) The coordinates of the common point are $(-4, 3, 8)$.

c) You know that

$$\mathbf{r}_l = \begin{pmatrix} 2-3\lambda \\ 1+\lambda \\ 4\lambda \end{pmatrix} = \begin{pmatrix} 2 \\ 1 \\ 0 \end{pmatrix} + \lambda \begin{pmatrix} -3 \\ 1 \\ 4 \end{pmatrix}$$

and

$$\mathbf{r}_m = \begin{pmatrix} -1+3\mu \\ 3 \\ 7-\mu \end{pmatrix} = \begin{pmatrix} -1 \\ 3 \\ 7 \end{pmatrix} + \mu \begin{pmatrix} 3 \\ 0 \\ -1 \end{pmatrix}$$

To find the angle between the lines you need to find the angle between the direction vectors $\begin{pmatrix} -3 \\ 1 \\ 4 \end{pmatrix}$ and $\begin{pmatrix} 3 \\ 0 \\ -1 \end{pmatrix}$.

Remember
If the vector equation of a line is $\mathbf{r} = \mathbf{a} + t(\mathbf{b} - \mathbf{a})$, $(\mathbf{b} - \mathbf{a})$ is the direction vector of the line.

Let $\mathbf{a} = \begin{pmatrix} -3 \\ 1 \\ 4 \end{pmatrix}$ and $\mathbf{b} = \begin{pmatrix} 3 \\ 0 \\ -1 \end{pmatrix}$.

The scalar product of **a** and **b** is

$$\mathbf{a.b} = |\mathbf{a}|\,|\mathbf{b}| \cos\theta$$

$$\mathbf{a.b} = \begin{pmatrix} -3 \\ 1 \\ 4 \end{pmatrix}.\begin{pmatrix} 3 \\ 0 \\ -1 \end{pmatrix} = (-3 \times 3) + (1 \times 0) + (4 \times -1)$$

$$= -13$$

Also

$$|\mathbf{a}| = \sqrt{(-3)^2 + 1^2 + 4^2} = \sqrt{26}$$

and

$$|\mathbf{b}| = \sqrt{3^2 + (-1)^2} = \sqrt{10}$$

So:

$$\therefore \quad -13 = \sqrt{26}\sqrt{10}\cos\theta$$

$$\therefore \quad \cos\theta = -\frac{13}{\sqrt{260}}$$

$$\therefore \quad \theta = 143.7°$$

The angle between the lines l and m is 143.7°.

θ is the **obtuse** angle between the lines l and m.

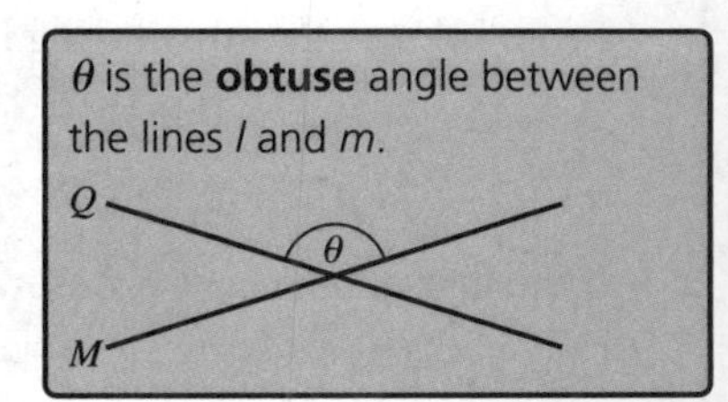

C4

Exercise 12E

1 Find a vector equation for the line passing through the point (4, 3) and parallel to the vector $\begin{pmatrix} 1 \\ -2 \end{pmatrix}$.

2 Find a vector equation for the line passing through the point (5, −2, 3) and parallel to the vector $\begin{pmatrix} 4 \\ -3 \\ 1 \end{pmatrix}$.

3 Find a vector equation for the line passing through the point (5, −1) and perpendicular to the vector $\begin{pmatrix} 1 \\ 1 \end{pmatrix}$.

4 Find a vector equation for the line joining the points (2, 6) and (5, −2).

5 Find a vector equation for the line joining the points (−1, 2, −3) and (6, 3, 0).

6 a) Find the vector equation of the line passing through $A(-4, 1, -3)$ and $B(6, -4, 12)$.

b) Show that the point $Q(-2, 0, 0)$ lies on the line AB.

7 a) Find the vector equation of the line passing through $M(3, 6, -1)$ and $N(0, 12, 5)$.

b) Show that the point $C(2, 8, 1)$ lies on the line MN.

8 a) Find the vector equation of the line passing through $L(12, 19, 15)$ and $M(2, 4, -10)$.

b) Show that the point $R(10, 16, 10)$ lies on the line LM.

9 a) Find the vector equation of the line passing through $A(-3, 2, 6)$ and $B(2, 5, 0)$.

b) Show that the point $Q(1, 2, -3)$ does not lie on the line AB.

12.5 Parallel and skew lines

Parallel lines

Suppose a line has vector equation $\mathbf{r} = \mathbf{a} + \lambda(\mathbf{b} - \mathbf{a})$, where $(\mathbf{b} - \mathbf{a})$ is the direction vector of the line, and another line has equation $\mathbf{s} = \mathbf{c} + \mu(\mathbf{d} - \mathbf{c})$ with direction vector $(\mathbf{d} - \mathbf{c})$.

You know that if one vector is a scalar multiple of another the vectors are parallel. So if

See page 250.

$$\mathbf{b} - \mathbf{a} = k(\mathbf{d} - \mathbf{c})$$

the direction vectors of the two lines are parallel. This means that the lines are parallel.

C4

Example 17

The line l_1 has equation $\mathbf{r} = \begin{pmatrix} 5 \\ 0 \\ -2 \end{pmatrix} + \lambda \begin{pmatrix} -1 \\ 3 \\ -5 \end{pmatrix}$.

The line l_2 passes through the points $A(2, 5, -1)$ and $B(4, -1, 9)$.

a) Find the equation of the line l_2.

b) Show that l_1 and l_2 are parallel lines.

a) $\mathbf{r} = \overrightarrow{OA} + \mu(\overrightarrow{AB})$

$$\therefore \quad \mathbf{r} = \begin{pmatrix} 2 \\ 5 \\ -1 \end{pmatrix} + \mu \left[\begin{pmatrix} 4 \\ -1 \\ 9 \end{pmatrix} - \begin{pmatrix} 2 \\ 5 \\ -1 \end{pmatrix} \right]$$

$$= \begin{pmatrix} 2 \\ 5 \\ -1 \end{pmatrix} + \mu \begin{pmatrix} 2 \\ -6 \\ 10 \end{pmatrix}$$

b) The direction of l_1 is given by $\begin{pmatrix} -1 \\ 3 \\ -5 \end{pmatrix}$, and the direction of l_2 is given by $\begin{pmatrix} 2 \\ -6 \\ 10 \end{pmatrix}$. Since $\begin{pmatrix} 2 \\ -6 \\ 10 \end{pmatrix} = -2 \begin{pmatrix} -1 \\ 3 \\ -5 \end{pmatrix}$, l_1 and l_2 are parallel.

Skew lines

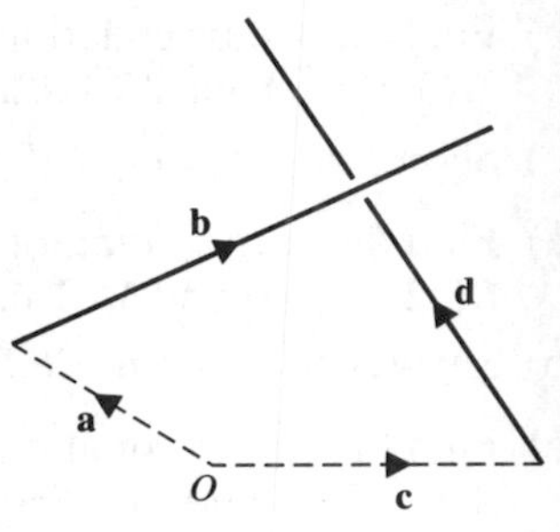

In a two-dimensional plane two lines are either parallel or they intersect. However, in three dimensions it is possible to have two lines which are not parallel but do not intersect. Such lines are called **skew**. It is still possible to find the angle between two skew lines. You just need to find the angle between their direction vectors.

If you have two skew lines, $\mathbf{r} = \mathbf{a} + \lambda\mathbf{b}$ and $\mathbf{s} = \mathbf{c} + \mu\mathbf{d}$, the angle between the two lines is simply the angle between the direction vectors **b** and **d**.

Example 18

a) Show that these two lines are skew:

$$\begin{pmatrix}3\\-2\\0\end{pmatrix} + \lambda\begin{pmatrix}-4\\1\\-2\end{pmatrix} \quad \text{and} \quad \begin{pmatrix}-6\\-3\\4\end{pmatrix} + \mu\begin{pmatrix}1\\3\\-2\end{pmatrix}$$

b) Show that the angle between the lines is given by $\cos^{-1}\left(\frac{\sqrt{6}}{14}\right)$.

a) To show that two lines are skew you must show that they do not intersect. So you need to find the values of λ and μ so that two of the components are equal, and show that these values do *not* satisfy the third component.

Equating the x and y components gives:

$$x: \quad 3 - 4\lambda = -6 + \mu \qquad [1]$$

$$y: \quad -2 + \lambda = -3 + 3\mu \qquad [2]$$

Solving [1] and [2] simultaneously gives $\lambda = 2$ and $\mu = 1$.

Check this for yourself.

Now check these values in the z components:

$$0 - 2\lambda = -2(2) = -4$$
$$4 - 2\mu = 4 - 2(1) = 2$$

Hence the lines do not meet; they are skew.

b) To find the angle between the skew lines calculate the angle between the direction vectors $\begin{pmatrix}-4\\1\\-2\end{pmatrix}$ and $\begin{pmatrix}1\\3\\-2\end{pmatrix}$.

Working out the scalar product gives

$$\begin{pmatrix}-4\\1\\-2\end{pmatrix}.\begin{pmatrix}1\\3\\-2\end{pmatrix} = (-4 \times 1) + (1 \times 3) + (-2 \times -2) = 3$$

where $\left|\begin{pmatrix}-4\\1\\-2\end{pmatrix}\right| = \sqrt{(-4)^2 + 1^2 + (-2)^2} = \sqrt{21}$

and $\left|\begin{pmatrix}1\\3\\-2\end{pmatrix}\right| = \sqrt{1^2 + 3^2 + (-2)^2} = \sqrt{14}$

So $3 = \sqrt{21}\sqrt{14}\cos\theta$

$$\therefore \quad \cos\theta = \frac{3}{7\sqrt{6}} = \frac{\sqrt{6}}{14}$$

So the angle between the lines is $\cos^{-1}\left(\frac{\sqrt{6}}{14}\right)$.

Exercise 12F

1 Three lines, l_1, l_2 and l_3, have equations

$$l_1: \ \mathbf{r}_1 = \begin{pmatrix} 1 \\ 1 \end{pmatrix} + \lambda \begin{pmatrix} 3 \\ 1 \end{pmatrix}$$

$$l_2: \ \mathbf{r}_2 = \begin{pmatrix} 6 \\ -4 \end{pmatrix} + \mu \begin{pmatrix} 1 \\ 2 \end{pmatrix}$$

and

$$l_3: \ \mathbf{r}_3 = \begin{pmatrix} 12 \\ -8 \end{pmatrix} + \nu \begin{pmatrix} -1 \\ 6 \end{pmatrix}$$

Show that l_1, l_2 and l_3 intersect at a single point, and find the position vector of their point of intersection.

2 Show that the lines $\mathbf{r}_1 = \begin{pmatrix} 6 \\ -5 \end{pmatrix} + \lambda \begin{pmatrix} -2 \\ 1 \end{pmatrix}$, $\mathbf{r}_2 = \begin{pmatrix} 0 \\ 3 \end{pmatrix} + \mu \begin{pmatrix} 1 \\ -3 \end{pmatrix}$ and $\mathbf{r}_3 = \begin{pmatrix} 5 \\ -9 \end{pmatrix} + \nu \begin{pmatrix} -1 \\ 2 \end{pmatrix}$ intersect at a single point, and find the position vector of their point of intersection.

3 Two lines, l_1 and l_2, have equations

$$l_1: \ \mathbf{r}_1 = \begin{pmatrix} 0 \\ -1 \\ -3 \end{pmatrix} + \lambda \begin{pmatrix} 1 \\ 3 \\ 6 \end{pmatrix}$$

and

$$l_2: \ \mathbf{r}_2 = \begin{pmatrix} -2 \\ 1 \\ 1 \end{pmatrix} + \mu \begin{pmatrix} 1 \\ 1 \\ 2 \end{pmatrix}$$

a) Show that l_1 and l_2 intersect at a single point, and find the position vector of their point of intersection.

b) Find also the angle between l_1 and l_2.

4 Two lines, l and m, have vector equations

$$\mathbf{r}_l = \begin{pmatrix} 3 \\ -5 \\ -1 \end{pmatrix} + \lambda \begin{pmatrix} -1 \\ 4 \\ 3 \end{pmatrix}$$

and

$$\mathbf{r}_m = \begin{pmatrix} -5 \\ 0 \\ -1 \end{pmatrix} + \mu \begin{pmatrix} 2 \\ 1 \\ 2 \end{pmatrix}$$

a) Show that l and m intersect at a single point, and find the position vector of their point of intersection.

b) Find also the angle between l and m.

5 a) Show that these two lines are skew:

$$\mathbf{r}_1 = \begin{pmatrix} 2 \\ 1 \\ 4 \end{pmatrix} + \lambda \begin{pmatrix} 5 \\ -1 \\ 6 \end{pmatrix} \quad \text{and} \quad \mathbf{r}_2 = \begin{pmatrix} 5 \\ 4 \\ 2 \end{pmatrix} + \mu \begin{pmatrix} 12 \\ -6 \\ 3 \end{pmatrix}$$

b) Calculate the angle between the two lines.

6 a) Show that these two lines are skew:

$$\mathbf{r}_1 = \begin{pmatrix} -1 \\ 2 \\ 3 \end{pmatrix} + \lambda \begin{pmatrix} -4 \\ 3 \\ 5 \end{pmatrix} \quad \text{and} \quad \mathbf{r}_2 = \begin{pmatrix} 2 \\ 1 \\ 5 \end{pmatrix} + \mu \begin{pmatrix} 1 \\ -1 \\ 4 \end{pmatrix}$$

b) Calculate the angle between the two lines.

7 a) Show that these two lines are skew:

$$\mathbf{r}_1 = \begin{pmatrix} 0 \\ 5 \\ -6 \end{pmatrix} + \lambda \begin{pmatrix} 1 \\ 2 \\ -1 \end{pmatrix} \quad \text{and} \quad \mathbf{r}_2 = \begin{pmatrix} 1 \\ -2 \\ 1 \end{pmatrix} + \mu \begin{pmatrix} -1 \\ -5 \\ 6 \end{pmatrix}$$

b) Calculate the angle between the two lines.

8 The line l_1 has equation $\mathbf{r} = \begin{pmatrix} 2 \\ 1 \\ -1 \end{pmatrix} + \lambda \begin{pmatrix} 3 \\ 1 \\ -1 \end{pmatrix}$.

C4

The line l_2 passes through the points $A(-2, 0, 4)$ and $B(4, 2, 2)$.

a) Find the equation of l_2.

b) Show that l_1 and l_2 are parallel lines.

9 The line l_1 has equation $\mathbf{r} = \begin{pmatrix} -1 \\ 0 \\ 3 \end{pmatrix} + \lambda \begin{pmatrix} 2 \\ -5 \\ 1 \end{pmatrix}$.

The line l_2 passes through the points $A(3, 5, -2)$ and $B(-3, 20, -5)$.

a) Find the equation of l_2.

b) Show that l_1 and l_2 are parallel lines.

12.6 The perpendicular distance from a point to a line

You can use vector geometry to find the coordinates of the point where a perpendicular from another point meets a straight line, and the length of the perpendicular.

Example 21

A line l has equation $\mathbf{r} = \begin{pmatrix} x \\ y \\ z \end{pmatrix} = \begin{pmatrix} 3 \\ -2 \\ 1 \end{pmatrix} + t \begin{pmatrix} 4 \\ 4 \\ 3 \end{pmatrix}$. Q is the point (5, 9, 11).

a) Find the coordinates of the foot of the perpendicular from the point Q to l.

b) Find the distance from Q to l.

a) The diagram shows the line l and the point Q.
Let $F(a, b, c)$ be the foot of the perpendicular from Q.

Since F is on l, $\overrightarrow{OF} = \begin{pmatrix} a \\ b \\ c \end{pmatrix} = \begin{pmatrix} 3 \\ -2 \\ 1 \end{pmatrix} + t\begin{pmatrix} 4 \\ 4 \\ 3 \end{pmatrix}$

$$= \begin{pmatrix} 3+4t \\ -2+4t \\ 1+3t \end{pmatrix} \text{ for some value of } t$$

$$\therefore \quad \overrightarrow{QF} = \overrightarrow{OF} - \overrightarrow{OQ}$$

$$= \begin{pmatrix} 3+4t \\ -2+4t \\ 1+3t \end{pmatrix} - \begin{pmatrix} 5 \\ 9 \\ 11 \end{pmatrix}$$

$$= \begin{pmatrix} -2+4t \\ -11+4t \\ -10+3t \end{pmatrix} \qquad [1]$$

And since $\overrightarrow{QF}$ is perpendicular to l, $\begin{pmatrix} -2+4t \\ -11+4t \\ -10+3t \end{pmatrix} . \begin{pmatrix} 4 \\ 4 \\ 3 \end{pmatrix} = 0$

Remember:
The scalar product of two perpendicular vectors is 0.

Working out the scalar product gives

$$(-2+4t)(4) + (-11+4t)(4) + (-10+3t)(3) = 0$$

$$\therefore \quad -8 + 16t - 44 + 16t - 30 + 9t = 0$$

$$41t - 82 = 0$$

$$\therefore \quad t = 2$$

So $\overrightarrow{OF} = \begin{pmatrix} 3+4t \\ -2+4t \\ 1+3t \end{pmatrix}$ where $t = 2$

that is, $\overrightarrow{OF} = \begin{pmatrix} 3+4(2) \\ -2+4(2) \\ 1+3(2) \end{pmatrix} = \begin{pmatrix} 11 \\ 6 \\ 7 \end{pmatrix}$

The coordinates of the foot of the perpendicular are (11, 6, 7).

b) The distance from Q to l is just the distance QF.
From [1]

$$\overrightarrow{QF} = \begin{pmatrix} -2+4t \\ -11+4t \\ -10+3t \end{pmatrix}, \text{ where } t = 2$$

So

$$\overrightarrow{QF} = \begin{pmatrix} -2+4(2) \\ -11+4(2) \\ -10+3(2) \end{pmatrix} = \begin{pmatrix} 6 \\ -3 \\ -4 \end{pmatrix}$$

And $|\overrightarrow{QF}| = \sqrt{6^2 + (-3)^2 + (-4)^2} = \sqrt{36 + 9 + 16} = \sqrt{61}$
So the distance from Q to l is $\sqrt{61}$.

Exercise 12G

1 A line l has equation $\begin{pmatrix} x \\ y \\ z \end{pmatrix} = \begin{pmatrix} 2 \\ 1 \\ 5 \end{pmatrix} + t\begin{pmatrix} 3 \\ 2 \\ -2 \end{pmatrix}$. Q is the point (8, 8, 4).

a) Find the coordinates of the foot of the perpendicular from the point Q to l.
b) Find the distance from Q to l.

2 A line l has equation $\begin{pmatrix} x \\ y \\ z \end{pmatrix} = \begin{pmatrix} -1 \\ 3 \\ 0 \end{pmatrix} + t\begin{pmatrix} 3 \\ -1 \\ 4 \end{pmatrix}$. P is the point (8, 2, 19).

a) Find the coordinates of the foot of the perpendicular from the point P to l.
b) Find the distance from P to l.

3 A line l has equation $\begin{pmatrix} x \\ y \\ z \end{pmatrix} = \begin{pmatrix} 2 \\ 5 \\ -3 \end{pmatrix} + t\begin{pmatrix} 4 \\ 2 \\ -1 \end{pmatrix}$. S is the point (−15, 14, 10).

a) Find the coordinates of the foot of the perpendicular from the point S to l.
b) Find the distance from S to l.

4 Points A, B and C have coordinates (0, 5), (9, 8) and (4, 3), respectively.

a) Find a vector equation for the line, l, joining A and B.
b) Show that the perpendicular distance from the point C to the line AB is $\sqrt{10}$.

5 Points A, B and C have coordinates (−1, −2), (5, 10) and (0, 5) respectively.

a) Find a vector equation for the line, l, joining A and B.
b) Show that the perpendicular distance from the point C to the line AB is $\sqrt{5}$.

6 Points P, Q and R have coordinates (−1, 1), (4, 6) and (7, 3) respectively.

a) Find a vector equation for the line joining P and Q.
b) Show that the perpendicular distance from the point R to the line PQ is $3\sqrt{2}$.

7 The points A, B and C have position vectors $\begin{pmatrix} 4 \\ 1 \\ -4 \end{pmatrix}$, $\begin{pmatrix} 3 \\ 2 \\ -3 \end{pmatrix}$ and $\begin{pmatrix} 2 \\ 3 \\ -5 \end{pmatrix}$, respectively.

a) Given that the angle between $\overrightarrow{AB}$ and $\overrightarrow{AC}$ is θ,
   i) find the value of $\cos\theta$
   ii) deduce that $\sin\theta = \dfrac{\sqrt{6}}{3}$.
b) Hence show that the perpendicular distance from the point C to the line AB is $\sqrt{6}$.

8 The points P, Q and R have position vectors $\begin{pmatrix} 2 \\ 5 \\ -3 \end{pmatrix}$, $\begin{pmatrix} 1 \\ 4 \\ -2 \end{pmatrix}$ and $\begin{pmatrix} 3 \\ 3 \\ -2 \end{pmatrix}$, respectively.

a) Given that the angle between $\overrightarrow{PQ}$ and $\overrightarrow{PR}$ is θ,
   i) find the value of $\cos \theta$
   ii) deduce that $\sin \theta = \frac{\sqrt{7}}{3}$.

b) Hence show that the perpendicular distance from the point R to the line PQ is $\frac{\sqrt{42}}{3}$.

9 Points A, B and C have position vectors $\begin{pmatrix} -1 \\ 3 \\ 5 \end{pmatrix}$, $\begin{pmatrix} 5 \\ 6 \\ -4 \end{pmatrix}$ and $\begin{pmatrix} 4 \\ 7 \\ 5 \end{pmatrix}$ respectively. P is the point on AB such that $\overrightarrow{AP} = \lambda\overrightarrow{AB}$.

a) Find $\overrightarrow{AB}$.

b) Find $\overrightarrow{CP}$.

c) By considering the scalar product $\overrightarrow{AB}.\overrightarrow{CP}$ find the position vector of the point on the line AB which is closest to C.

d) Deduce that the perpendicular distance from the point C to the line AB is $3\sqrt{3}$.

10 Points A, B and C have position vectors $\begin{pmatrix} 3 \\ -2 \\ 5 \end{pmatrix}$, $\begin{pmatrix} 7 \\ 6 \\ 1 \end{pmatrix}$ and $\begin{pmatrix} 8 \\ 6 \\ 8 \end{pmatrix}$ respectively. P is the point on AB such that $\overrightarrow{AP} = \lambda\overrightarrow{AB}$.

a) Find $\overrightarrow{AB}$.

b) Find $\overrightarrow{CP}$.

c) By considering the scalar product $\overrightarrow{AB}.\overrightarrow{CP}$ find the position vector of the point on the line AB which is closest to C.

d) Deduce that the perpendicular distance from the point C to the line AB is $2\sqrt{11}$.

Summary

You should know how to ...	Check out
1 Calculate with vectors.	**1** A is the point $(2, -3, 1)$ and B is the point $(5, 2, -3)$. O is the origin. a) Find the vector $\overrightarrow{OA} + \overrightarrow{OB}$. b) i) Find the direction vector $\overrightarrow{AB}$. ii) Which of these vectors is parallel to the line AB? $\mathbf{p}\begin{pmatrix} 9 \\ 15 \\ -12 \end{pmatrix}$ $\mathbf{q}\begin{pmatrix} -4 \\ 5 \\ 3 \end{pmatrix}$ $\mathbf{r}\begin{pmatrix} -3 \\ -5 \\ 4 \end{pmatrix}$ $\mathbf{s}\begin{pmatrix} 6 \\ 10 \\ 8 \end{pmatrix}$ $\mathbf{t}\begin{pmatrix} 4 \\ 6 \\ -3 \end{pmatrix}$
2 Calculate the vector equation of a line.	**2** a) Using the points A and B from Q1, find the vector equation of the line AB. b) The line l has equation $\mathbf{r} = \begin{pmatrix} 2 \\ 7 \\ -4 \end{pmatrix} + \mu\begin{pmatrix} -2 \\ 0 \\ 1 \end{pmatrix}$. i) Show that AB intersects l. ii) Find the coordinates of the point of intersection.
3 Solve problems involving parallel and skew lines.	**3** Find the angle between the vectors $\begin{pmatrix} 3 \\ 5 \\ -4 \end{pmatrix}$ and $\begin{pmatrix} -2 \\ 0 \\ 1 \end{pmatrix}$.
4 Find the perpendicular distance from a point to a line.	**4** Find the distance from the point $(1, -3, 2)$ to the line $\mathbf{r} = \begin{pmatrix} 0 \\ 1 \\ 4 \end{pmatrix} + \lambda\begin{pmatrix} 1 \\ -1 \\ 3 \end{pmatrix}$.

C4

Revision exercise 12

1 a) Find the vector equation of the line l_1, which passes through the points $A(3, -1, 2)$ and $B(2, 0, 2)$.

b) The line l_2 has vector equation $\mathbf{r} = \begin{pmatrix} 4 \\ 1 \\ -1 \end{pmatrix} + \mu \begin{pmatrix} 1 \\ 0 \\ -1 \end{pmatrix}$.

Show that the lines l_1 and l_2 intersect and find the coordinates of their point of intersection.

c) Show that the point $C(9, 1, -6)$ lies on the line l_2.

d) Find the coordinates of the point D on l_1, such that CD is perpendicular to l_1.

(*AQA, 2004*)

2 The line l_1 has equation $\mathbf{r} = \begin{pmatrix} 1 \\ 0 \\ -2 \end{pmatrix} + \lambda \begin{pmatrix} 1 \\ 4 \\ 3 \end{pmatrix}$. The line l_2 has equation $\mathbf{r} = \begin{pmatrix} 5 \\ 5 \\ 10 \end{pmatrix} + \lambda \begin{pmatrix} 2 \\ -3 \\ 6 \end{pmatrix}$.

a) Show that the lines l_1 and l_2 intersect at a point P and find the position vector of P.

b) Find the acute angle between the lines l_1 and l_2, giving your answer to the nearest degree.

c) The line l_3 passes through the points $(0, 0, 0)$ and $(2, 8, 6)$. Show that l_1 and l_3 are parallel lines.

(*AQA, 2003*)

C4

3 The line l_1 has equation $\begin{pmatrix} x \\ y \\ z \end{pmatrix} = \begin{pmatrix} 3 \\ -2 \\ 1 \end{pmatrix} + t \begin{pmatrix} 4 \\ 4 \\ 3 \end{pmatrix}$. The line l_2 has equation $\begin{pmatrix} x \\ y \\ z \end{pmatrix} = \begin{pmatrix} 8 \\ -1 \\ 2 \end{pmatrix} + s \begin{pmatrix} -1 \\ 3 \\ 2 \end{pmatrix}$.

a) Show that the lines l_1 and l_2 intersect and find the coordinates of their point of intersection.

b) Show that the vector $\begin{pmatrix} 1 \\ 11 \\ -16 \end{pmatrix}$ is perpendicular to both l_1 and l_2.

(*AQA, 2002*)

4 The line l_1 passes through the point $A(2, 0, 2)$ and has equation $\mathbf{r} = \begin{pmatrix} 2 \\ 0 \\ 2 \end{pmatrix} + t \begin{pmatrix} 2 \\ 6 \\ -3 \end{pmatrix}$.

a) Show that the line l_2 which passes through the points $B(4, 4, -5)$ and $C(0, -8, 1)$ is parallel to the line l_1.

b) P is the point on line l_1 such the angle CPA is a right angle.

i) Find the value of the parameter t at P.

ii) Hence show that the shortest distance between the lines l_1 and l_2 is $2\sqrt{5}$.

(*AQA, 2002*)

5 A pyramid has a square base $OABC$ with sides of length 6 units. The coordinates of the corners of the square are $O(0, 0, 0)$, $A(6, 0, 0)$, $B(6, 6, 0)$ and $C(0, 6, 0)$ and the vertex of the pyramid is $V(3, 3, 2)$.

a) Calculate the size of angle OVB giving your answer to the nearest degree. *(AQA, 2002)*

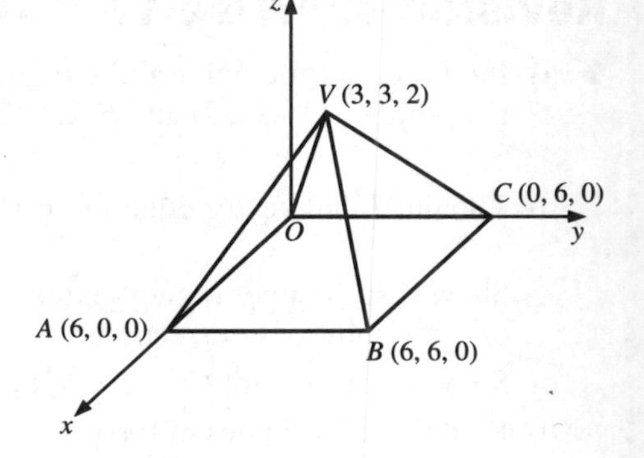

6 The line l_1 has equation $\mathbf{r} = \begin{pmatrix} 2 \\ 0 \\ -1 \end{pmatrix} + \lambda \begin{pmatrix} 3 \\ 4 \\ 5 \end{pmatrix}$.

a) Find the acute angle between the line l_1 and the vector $\begin{pmatrix} 3 \\ -1 \\ -2 \end{pmatrix}$.

b) The line l_2 passes through the points $A(3, -2, 4)$ and $B(5, 1, 7)$. Show that the lines l_1 and l_2 intersect. *(AQA, 2001)*

7 The diagram shows the points O, A and B in a plane. The point O is the origin and $\overrightarrow{OA} = \mathbf{a}$, $\overrightarrow{OB} = \mathbf{b}$.

a) On separate copies of the diagram draw vectors representing
i) $\mathbf{a} + \mathbf{b}$ ii) $\mathbf{a} - \mathbf{b}$.

b) Write down a vector equation for the line which passes through A and is parallel to the line OB. *(AQA, 2001)*

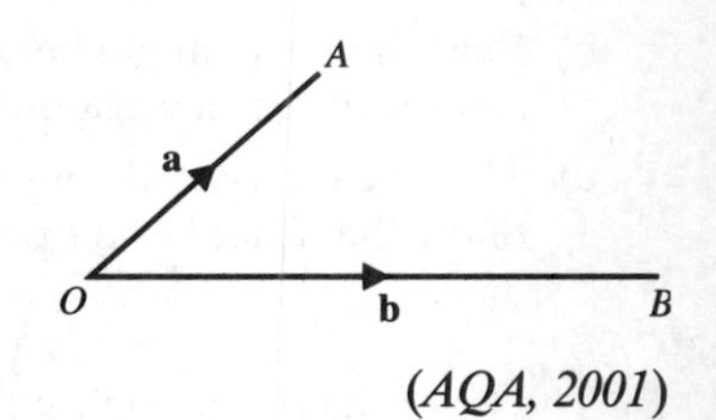

8 a) Given that p is a constant, and that the lines with equations

$$\mathbf{r} = \begin{pmatrix} 24 \\ 8 \\ p \end{pmatrix} + \lambda \begin{pmatrix} 3 \\ 2 \\ -2 \end{pmatrix} \text{ and } \mathbf{r} = \begin{pmatrix} 13 \\ -6 \\ 11 \end{pmatrix} + \mu \begin{pmatrix} 1 \\ 4 \\ -7 \end{pmatrix} \text{ intersect at the}$$

point X, determine the value of p and the position vector of X.

b) Calculate the acute angle between the two lines, giving your answer to the nearest degree. *(AQA, 2001)*

C4 Practice Paper

90 minutes *75 marks* *You may use a calculator*

1 a) Write $\sin 2x$ in the form $a \sin x \cos x$. *(1 mark)*

b) Hence, or otherwise, evaluate $\int_0^{\frac{\pi}{6}} 6 \sin x \cos x \, dx$. *(5 marks)*

2 The function f is defined by $f(x) = 4x^3 - 13x + 6$.

a) Show that

i) $f(-2) = 0$ *(1 mark)*

ii) $(2x - 3)$ is a factor of $f(x)$. *(2 marks)*

iii) Hence, or otherwise, write $f(x)$ as the product of three linear factors. *(2 marks)*

b) It is given that $\dfrac{4x^3 - 13x + 6}{4x(x + 2)} = Ax + B + \dfrac{C}{x} + \dfrac{D}{x + 2}$.

Find the values of A, B, C and D. *(3 marks)*

3 a) i) Obtain the binomial expansion of $(1 + x)^{\frac{1}{3}}$ up to and including the term in x^2. *(2 marks)*

ii) Hence, or otherwise, show that $(8 + 5x)^{\frac{1}{3}} \approx 2 + \dfrac{5}{12}x - \dfrac{25}{288}x^2$. *(3 marks)*

b) By choosing a suitable value of x, use the result from part a) ii) to find $\sqrt[3]{7.95}$, giving your answer to 6 decimal places. *(3 marks)*

C4

4 a) Express $2 \cos \theta + 5 \sin \theta$ in the form $R \cos(\theta - \alpha)$ where R and α are constants. $R > 0$ and $0° < \alpha < 90°$.

i) Find the values of R and α, giving your answer for α to the nearest 0.1°. *(3 marks)*

ii) Hence, or otherwise, solve the equation

$$2 \cos \theta + 5 \sin \theta = 1 \text{ for } 0° \leqslant \theta \leqslant 360°$$ *(4 marks)*

b) The parametric equations of a curve are $x = 2 \cos \theta$ and $y = 5 \sin \theta$. Show that the cartesian equation of the curve is given by

$$25x^2 + 4y^2 = 100$$ *(3 marks)*

5 a) Use the identity $\tan(A + B) = \dfrac{\tan A + \tan B}{1 - \tan A \tan B}$ to

show that $\tan(2x) = \dfrac{2 \tan x}{1 - \tan^2 x}$ *(2 marks)*

b) Given that $\tan 45° = 1$, use the identity for $\tan 2x$ to show that the exact value of $\tan 22\frac{1}{2}°$ is $\sqrt{2} - 1$. *(5 marks)*

6 The equation of a curve is given implicitly by $2x^2y + (y - 1)^2 = 5$.

a) Show that the coordinates of the two points on the curve where $x = 1$ are $(1, 2)$ and $(1, -2)$. *(2 marks)*

b) i) Differentiate the equation $2x^2y + (y - 1)^2 = 5$. *(5 marks)*

ii) Hence show that the tangents to the curve at $(1, 2)$ and $(1, -2)$ are parallel. *(5 marks)*

7 The points A and B have coordinates (1, 4, 3) and (3, 3, 5) respectively.

a) i) Find the vector $\overrightarrow{AB}$. *(2 marks)*

ii) Calculate the distance between A and B. *(2 marks)*

iii) Find a vector equation of the line AB. *(2 marks)*

b) The line l_1 has equation $\mathbf{r} = \begin{pmatrix} 3 \\ 0 \\ -2 \end{pmatrix} + \lambda \begin{pmatrix} 1 \\ 3 \\ -2 \end{pmatrix}$.

Show that the line l_1 and the line AB are skew lines (they do not intersect). *(5 marks)*

8 a) i) Express $\dfrac{1}{x(1-x)}$ in the form $\dfrac{A}{x} + \dfrac{B}{1-x}$. *(2 marks)*

ii) Integrate the differential equation $\dfrac{dx}{dt} = ax(1-x)$ where a is a constant, to show that $\dfrac{x}{1-x} = Ke^{at}$ where K is a constant. *(5 marks)*

b) Fungus is spreading on an old wooden log at a rate given by the differential equation

$$\frac{dx}{dt} = ax(1-x)$$

where a is a constant and x is the proportion of the log covered in the fungus after t hours. Initially one tenth of the log is covered in the fungus. 24 hours later one quarter of the log is covered.

i) Find the value of K. *(1 mark)*

ii) Show that $a = \dfrac{\ln 3}{24}$ *(3 marks)*

iii) Find how long it will be before half the log is covered in fungus. *(2 marks)*

Answers

Answers are given only for questions with a numerical or algebraic solution.

Chapter 1

Check in

1 a) i) 10 ii) 0.985 (3 s.f.) iii) 4096 b) i) 9 ii) 0.005 (3 s.f.) iii) 8 **2** a) C b) B c) A d) D **3** a) 3 b) $\frac{1}{2}$

4 a) $x = \frac{1}{2}$ or $x = -3$ b) $x = 4.54$ or $x = -1.54$ (3 s.f.) **5** a) $x = \frac{1}{2}\left(\frac{3}{y} + 1\right)$ b) $x = \pm\frac{\sqrt{9 - y^2}}{2}$

Exercise 1A

1 a) 2 b) 1.5 **2** a) 9.2 b) 4 **3** 6 **4** a) $\frac{1}{2}$ b) $\pm\frac{\pi}{2}$

5 a) $f(x) \in \mathbb{R}, 4 < f(x) < 9$ b) $f(x) \in \mathbb{R}, f(x) \geqslant 7$ c) $f(x) \in \mathbb{R}, 1 < f(x) \leqslant 9$ d) $f(x) \in \mathbb{R}, \frac{1}{18} \leqslant f(x) \leqslant \frac{1}{3}$ e) $f(x) \in \mathbb{R}, f(x) \geqslant 9$

5 f) $f(x) \in \mathbb{R}, 4 < f(x) < 134$ g) $f(x) \in \mathbb{R}, -9 \leqslant f(x) \leqslant 0$ h) $f(x) \in \mathbb{R}, \frac{1}{10} < f(x) \leqslant \frac{1}{2}$ i) $f(x) \in \mathbb{R}, -4 < f(x) < \infty$

5 j) $f(x) \in \mathbb{R}, 2 \leqslant f(x) \leqslant 5$ k) $f(x) \in \mathbb{R}, 0 < f(x) \leqslant 20$ l) $f(x) \in \mathbb{R}, 0 < f(x) \leqslant \frac{1}{3}$

6 a) $f(x) \in \mathbb{R}, \frac{1}{2} < f(x) < 1$ b) $f(x) \in \mathbb{R}, -1 \leqslant f(x) < 1$ c) $f(x) \in \mathbb{R}, \frac{1}{2} \leqslant f(x) \leqslant 1$ d) $f(x) \in \mathbb{R}, 2 < f(x) \leqslant 2.5$

6 e) $f(x) \in \mathbb{R}, 3 \leqslant f(x) \leqslant 4$ f) $f(x) \in \mathbb{R}, 4 < f(x) < \infty$

Exercise 1B

1 a) one-to-one b) two-to-one c) one-to-one d) two-to-one e) two-to-one f) one-to-one g) one-to-one

1 h) one-to-one i) two-to-one j) one-to-one k) two-to-one l) two-to-one

2 a) two-to-one b) one-to-one c) many-to-one d) one-to-one e) one-to-one f) one-to-one g) many-to-one

2 h) one-to-one

Exercise 1C

1 a) $-2, 6$ b) $-9, 1$ c) $-3, 9$ d) $2, 6$ e) $-\frac{5}{3}, 1$ f) $-\frac{4}{5}, 2$ **2** $-1, \frac{7}{3}$ **3** $-1, 2$ **4** $2, 10$

5 a) ± 2 b) ± 1 or ± 3 c) ± 2 or $\pm\sqrt{3}$ d) ± 3 or $\pm\sqrt{15}$ e) ± 1 or $\pm\sqrt{\frac{7}{2}}$ f) ± 2 or $\pm\frac{4}{\sqrt{3}}$ **6** 60°, 120°, 240°, 300°

7 30°, 150°, 210°, 330° **8** ±22.5°, ±67.5° **9** 20°, 40°, 80°, 100°, 140°, 160° **10** a) 1 b) 5 c) $\frac{1}{3}, 1$ d) $-\frac{5}{4}, \frac{3}{2}$

10 e) $-3, \frac{7}{5}$ f) $-2, -\frac{1}{3}$ **11** 4 **12** $-6 < x < 1$ **13** $x < -\frac{3}{4}$ or $x > \frac{3}{2}$ **14** $x \leqslant -24$ or $x \geqslant 0$ **15** $-\frac{8}{3} \leqslant x \leqslant 8$

16 a) $x > 1$ b) $x \leqslant -\frac{1}{2}$ c) $-\frac{1}{3} \leqslant x \leqslant 7$ d) $x \leqslant -\frac{7}{2}$ or $x \geqslant \frac{3}{4}$ e) $-6 < x < -\frac{4}{3}$ f) $x < \frac{7}{2}$ **17** $-2\sqrt{5} < x < -2$ or $2 < x < 2\sqrt{5}$

18 $x \leqslant -\frac{\sqrt{13}}{2}$ or $x \geqslant \frac{\sqrt{13}}{2}$ **19** $0 \leqslant x \leqslant 60°$ or $120° \leqslant x \leqslant 240°$ or $300° \leqslant x \leqslant 360°$

20 $-180° \leqslant x < -157.5°$ or $-112.5° < x < 67.5°$ or $-22.5° < x < 22.5°$ or $67.5° < x < 112.5°$ or $157.5° < x \leqslant 180°$

Exercise 1D

1 a) 7 b) 4 c) $\frac{1}{4}$ d) 19 e) 9 f) $\frac{1}{4}$ g) $\frac{1}{9}$ h) 23 i) 81 j) 12 k) $\frac{3}{2}$ l) $\frac{1}{33}$

2 a) $3x^2 - 1$ b) $(3x - 1)^2$ c) $\frac{6}{x} - 1$ d) $\frac{2}{x^2}$ e) x^4 f) $9x - 4$ **3** a) i) $x^2 + 10x + 28$ ii) $x^2 + 8$ b) -2

4 a) $\frac{3}{x + 5}, -2$ b) $\frac{3}{x} + 5, 3$ **5** a) $-4, -1$ b) $\frac{3}{2}$ **6** $2x^2 + 5, 1 \leqslant x \leqslant 5, 7 \leqslant gf(x) \leqslant 55$ **7** $9x^4 + 6x^2 - 1, -1 \leqslant gf(x) \leqslant 167$

8 $\frac{1}{x^2 + 1}, 0 < gf(x) \leqslant \frac{1}{17}$ **9** a) $x^2 - 10|x| + 28, fg(x) \in \mathbb{R}, fg(x) \geqslant 3$ b) $x^4 + 6x^2 + 12, ff(x) \in \mathbb{R}, ff(x) \geqslant 12$

10 a) $\sqrt{x^2 + 1}, fg(x) \in \mathbb{R}, fg(x) > 1$ b) $x + 1, gf(x) \in \mathbb{R}, gf(x) > 1$

Exercise 1E

1 a) $\frac{x-2}{3}$ b) $\frac{1+x}{5}$ c) $\frac{4-x}{3}$ d) $\frac{2}{x}, x \neq 0$ e) $\frac{3+x}{x}, x \neq 0$ f) $\frac{2x-5}{3x}, x \neq 0$ g) $\frac{2x}{1-x}, x \neq 1$ h) $\frac{5x}{2+x}, x \neq -2$

2 a) $\sqrt{x}, x \in \mathbb{R}, x > 4$ b) $\frac{1-2x}{x}, x \in \mathbb{R}, 0 < x < \frac{1}{2}$ c) $2 + x^2, x \in \mathbb{R}, x > 1$ d) $\sqrt{\frac{x+1}{3}}, x \in \mathbb{R}, 2 < x < 47$ e) $\frac{x^2-3}{2}, x \in \mathbb{R}, x \geqslant 5$

2 f) $\frac{1}{x+3}, x \in \mathbb{R}, -\frac{14}{5} < x < -\frac{5}{2}$ g) $-2 + \sqrt{x-3}, x \in \mathbb{R}, x \geqslant 3$ h) $\sqrt[3]{x-1}, x \in \mathbb{R}$

3 a) $\frac{x+4}{3}, x \in \mathbb{R}$ c) 2 **4** b) $\frac{10-x}{2}, x \in \mathbb{R}, x \leqslant 10$ c) $\frac{10}{3}$

5 a) $\sqrt{x+6}, x \in \mathbb{R}, x > -6$ c) 3 **6** b) $2 + \sqrt{x}, x \in \mathbb{R}, x > 0$ c) 4 **7** a) 2 b) $\frac{x+5}{2}, x \in \mathbb{R}; \frac{7-x}{4}, x \in \mathbb{R}$

7 c) -1, from a) $f(2) = g(2) = -1$ **8** a) $0 < f(x) \leqslant \frac{4}{3}$ b) $f^{-1}(x) = \frac{4}{x} - 3, x \in \mathbb{R}, x \neq 0$ c) 1

9 a) $3x - 5$ b) i) $\frac{x-1}{3}, x \in \mathbb{R}$ ii) $x + 2, x \in \mathbb{R}$ iii) $\frac{x+5}{3}, x \in \mathbb{R}$ **10** a) $g(x) \geqslant -3$ b) g^{-1} exists because g is one-to-one

10 c) $x \geqslant -\frac{1}{9}$ **11** a) $\frac{x-3}{2}, x \in \mathbb{R}$ b) $\frac{1}{2(x+1)}, x \in \mathbb{R}, x \neq -1$ c) $\pm\sqrt{2}$

Exercise 1F

1 b) 1 **2** b) 1 **3** b) $-\frac{1}{2}$ **4** b) 1

Check out

1 a) $fg(x) = \frac{1}{8-x^2}$ b) $f^{-1}(x) = \sqrt{\frac{1}{x} - 4}, \quad x \neq 0$ c) $g^{-1}(x) = \sqrt{4-x^2}$

2 a) $0 \leqslant x \leqslant 2, \frac{1}{8} \leqslant fg(x) \leqslant \frac{1}{4}$ b) $0 < x \leqslant \frac{1}{4}, 0 \leqslant f^{-1}(x)$ c) $0 \leqslant x \leqslant 2, 0 \leqslant g^{-1}(x) \leqslant 2$ **3** a) i) 2 ii) 3 iii) 5 c) $1 < x < 5$

Revision exercise 1

1 a) i) $fg(x) = \sqrt{x-1}$, $gf(x) = \sqrt{x} - 1$ b) i) Translation by $\begin{pmatrix}1\\0\end{pmatrix}$ ii) $0 \leqslant h(x) \leqslant 2$ iii) domain $0 \leqslant x \leqslant 2$, range $1 \leqslant x \leqslant 5$

1 iv) $h^{-1}(x) = x^2 + 1$ **2** a) Graph meets axes at (2, 0) and (0, 4). b) i) $P(\frac{4}{3}, \frac{4}{3})$, $Q(4, 4)$ ii) $x < \frac{4}{3}, x > 4$ c) 2

3 a) i) Graph crosses x-axis at (2, 0) and $(-1, 0)$ and y-axis at (0, 2). b) $f(x) \leqslant 3$ c) $x > \frac{4}{3}, x < -2$

4 i) Reflection in the line $y = x$ **5** a) i) Stretch parallel to Ox, scale factor $\frac{1}{2}$ ii) $x = \frac{\pi}{2}$

5 b) i) $-1 \leqslant f(x) \leqslant 1$ ii) domain $-1 \leqslant x \leqslant 1$, range $0 \leqslant f^{-1}(x) \leqslant \frac{\pi}{2}$ c) i) $gf(x) = |\cos 2x|$

6 a) Graph crosses x-axis at (1, 0) and (2, 0) and y-axis at (0, 2). b) $-1 \leqslant f(x)$ c) $x = 4, x = \frac{2}{3}$

7 a) $y = f(x)$: translation by $\begin{pmatrix}2\\0\end{pmatrix}$ $y = g(x)$: translation by $\begin{pmatrix}0\\-2\end{pmatrix}$ c) i) f does not have an inverse function ii) $0 \leqslant h(x) \leqslant 4$

7 d) i) $0 < x < 4$ ii) $-4 < x < 4$ iii) $x < 2$

Chapter 2

Check in

1 $3x^{\frac{1}{2}} - 2x$ **2** 0.243 (3 s.f.) **3** $(\frac{9}{4}, 1\frac{11}{16})$ **4** maximum

5 $0 < x < \frac{9}{4}$ (function is not defined at $x < 0$ so there is not a turning point at $x = 0$)

Exercise 2A

1 a) $6(2x-1)^2$ b) $6(3x+4)$ c) $20(5x-3)^3$ d) $-5(3-x)^4$ e) $-18(4-3x)^5$ f) $8x(x^2+1)^3$ g) $6x^2(x^3-6)$

1 h) $-12x(1-2x^2)^2$ i) $-8x^3(4-x^4)$ j) $-90x^2(7-5x^3)^5$ k) $48x(6x^2-5)^3$ l) $-42x(9-7x^2)^2$ **2** a) $-6(2x-5)^{-4}$

2 b) $-3(3x+2)^{-2}$ c) $-4x(x^2+3)^{-3}$ d) $6x^2(5-2x^3)^{-2}$ e) $-\dfrac{4}{(3+4x)^2}$ f) $\dfrac{2x}{(4-x^2)^2}$ g) $\dfrac{10}{(3-2x)^2}$ h) $-\dfrac{6}{(x+1)^3}$

2 i) $\dfrac{70x}{(2-x^2)^6}$ j) $\dfrac{6x}{(3x^2+8)^2}$ k) $-60x^2(5x^3-4)^{-5}$ l) $\dfrac{12x^3}{(5-3x^4)^3}$ **3** a) $(2x-1)^{-\frac{1}{2}}$ b) $-\frac{1}{3}(6-x)^{-\frac{2}{3}}$ c) $2x^2(x^3-2)^{-\frac{1}{3}}$

3 d) $x^4(4-x^5)^{-\frac{6}{5}}$ e) $\dfrac{2}{\sqrt{4x-5}}$ f) $\dfrac{2x}{3\sqrt[3]{(x^2+3)^2}}$ g) $\dfrac{1}{\sqrt{(5-2x)^3}}$ h) $-\dfrac{4x}{\sqrt[3]{(x^2+5)^4}}$ i) $\dfrac{2}{\sqrt[6]{(4x-7)^7}}$ j) $-\dfrac{10(5-4\sqrt{x})^4}{\sqrt{x}}$

3 k) $\dfrac{1}{4\sqrt{x}\sqrt{3+\sqrt{x}}}$ l) $-\dfrac{1}{3\sqrt[3]{x^2}(4-\sqrt[3]{x})^2}$

Exercise 2B

The constant of integration is omitted from the answers.

1 a) $\frac{1}{10}(2x-3)^5$ b) $\frac{1}{15}(5x+8)^3$ c) $\frac{1}{18}(3x-4)^6$ d) $(x-7)^3$ e) $-\frac{1}{6}(4-x)^6$ f) $\frac{1}{28}(6-7x)^4$ g) $-\frac{1}{6}(3x-4)^{-2}$ h) $\frac{2}{3}(5-9x)^{-1}$

1 i) $-\dfrac{1}{12(2x-1)^6}$ j) $\dfrac{3}{1-x}$ k) $\frac{1}{3}\sqrt{(2x-3)^3}$ l) $18\sqrt[3]{(x-4)^2}$ **2** a) $\frac{1}{12}(2x-7)^6$ b) $\sqrt{2x-1}$ c) $\frac{1}{8}(x^2+2)^4$ d) $-\frac{1}{18}(4-3x^2)^6$

2 e) $\frac{1}{9}(x^3-4)^3$ f) $\dfrac{2}{3-x^2}$ g) $\frac{1}{6}\sqrt{(x^4-1)^3}$ h) $-\frac{1}{2}\sqrt[3]{(2-3x^2)^4}$ i) $\frac{1}{4}(x^{\frac{4}{3}}-2)^3$ j) $\dfrac{5}{3-x^5}$ k) $\frac{1}{6}(x^2+1)^3$ l) $\frac{1}{5}x^5+\frac{2}{3}x^3+x$

Exercise 2C

1 a) $2x-1$ b) $4x-9$ c) $12x+7$ d) $-(1+2x)$ e) $-(5+12x)$ f) $3x^2+8x-2$ g) $12x^2-2x-20$ h) $3x^2+6x-15$

1 i) $9x^2-22x+31$ j) $x(5x^3+3x-10)$ k) $25x^4+80x^3-15x^2-6x-12$ l) $27x^8-8x^7+24x^5-18x^2+4x$

2 a) $6x(x+1)(x+3)^3$ b) $x^2(2+x)(6+5x)$ c) $x^3(21x-4)(3x-1)^2$ d) $6x(2x+5)(4x+5)$ e) $3x^2(12x^2-1)(4x^2-1)^2$

2 f) $20x(2-x^3)(1-2x^3)$ g) $6x(25x^2+1)(5x^2+1)^3$ h) $x^6(14-95x^3)(2-5x^3)^3$ i) $x(8x^2+5x-2)(x^2+x-1)^2$

2 j) $x^2(12-7x+22x^2)(4-x+2x^2)^3$ k) $4x^3(15x^2-21x+4)(3x^2-6x+2)^2$ l) $5x(11x^3-5x+2)(x^3-x+1)^2$

3 a) $(5x-4)(x+2)(x-5)^2$ b) $2(5x+11)(2x-1)^2(x+4)$ c) $4(35x-9)(5x+2)^3(4x-3)^2$ d) $-2(7+10x)(2-x)^5(5+2x)^3$

3 e) $-(107+315x)(3+5x)(4-7x)^6$ f) $4(4x^2-3x+2)(x^2+1)(2x-3)^3$ g) $3(15x^2+18x-10)(5x+9)^2(x^2-2)^2$

3 h) $4(32x^2-35x-18)(2x^2-3)^4(4x-7)^5$ i) $x(52x^3+45x-16)(x^3-1)^2(4x^2+5)$ j) $4x(25x^2-87)(5-x^2)^3(6-5x^2)^5$

3 k) $(26x^2-78x+51)(x^2-3x+1)^4(2x-3)^2$ l) $15(7x^4-12x^3+12x^2-12x+12)(5x^2-10x+12)^2(x^3-6)^4$

4 a) $\dfrac{3x+2}{2\sqrt{x+1}}$ b) $\dfrac{3(2-x)}{\sqrt{3-x}}$ c) $\dfrac{3(3x+5)}{\sqrt{5+2x}}$ d) $\dfrac{x(5x+12)}{2\sqrt{x+3}}$ e) $\dfrac{2x(3-5x)}{\sqrt{3-4x}}$ f) $\dfrac{6x+11}{2\sqrt{x+3}}$ g) $-\dfrac{9x+14}{\sqrt{2x+5}}$

4 h) $\dfrac{(35x-4)(5x-4)^2}{2\sqrt{x}}$ i) $\dfrac{(15x-19)(3x+5)}{2\sqrt{x-2}}$ j) $\dfrac{8x-5}{\sqrt{(2x-3)(4x+1)}}$ k) $-\dfrac{9+4x}{2\sqrt{(6+x)(3-2x)}}$ l) $\dfrac{9x^2-x-6}{\sqrt{(x^2-2)(6x-1)}}$

Exercise 2D

1 a) $-\dfrac{2}{(x-2)^2}$ b) $-\dfrac{4}{(x-1)^2}$ c) $-\dfrac{7}{(4+x)^2}$ d) $\dfrac{11}{(x+2)^2}$ e) $\dfrac{13}{(x+4)^2}$ f) $\dfrac{10}{(x+2)^2}$ g) $\dfrac{11}{(2-5x)^2}$ h) $-\dfrac{10}{(2x-1)^2}$

1 i) $\dfrac{x(x+6)}{(x+3)^2}$ j) $\dfrac{x(8-x)}{(4-x)^2}$ k) $\dfrac{x^2(4x-9)}{(2x-3)^2}$ l) $\dfrac{x^4(15-4x)}{(3-x)^2}$ **2** a) $\dfrac{(9x+2)(3x-2)}{2\sqrt{x^3}}$ b) $\dfrac{(25x-1)(5x+1)^2}{2\sqrt{x^3}}$

2 c) $\dfrac{(19x^2+4)(x^2-4)^4}{2\sqrt{x^3}}$ d) $-\dfrac{2x+1}{2\sqrt{x}(2x-1)^2}$ e) $\dfrac{3x-12\sqrt{x}-2}{2\sqrt{x}(2+x)^3}$ f) $\dfrac{5(4x+12\sqrt{x}+1)}{\sqrt{x}(5-4x)^4}$ g) $\dfrac{(45x^2-24x-2)(3x^2+2)^3}{\sqrt{(2x-1)^3}}$

2 h) $\dfrac{(3x^2+2x-6)(2-3x)}{\sqrt{(1-x^2)^3}}$ i) $\dfrac{3}{2\sqrt{x-2}\sqrt{(x+1)^3}}$ j) $\dfrac{11}{2\sqrt{x-3}\sqrt{(2x+5)^3}}$ k) $\dfrac{11}{2\sqrt{3+x}\sqrt{(2-3x)^3}}$ l) $-\dfrac{x(x^3+3x+6)}{2\sqrt{x^2+1}\sqrt{(x^3-3)^3}}$

3 a) $x^2(9-5x)(3-x)$ b) $-\dfrac{1}{(2x-1)^2}$ c) $\dfrac{(25x-1)(5x-1)}{2\sqrt{x}}$ d) $\dfrac{\sqrt{x}+2}{(\sqrt{x}+1)^2}$ e) $(25x-24)(5x+3)^2(x-2)$

3 f) $-\dfrac{7}{2\sqrt{3x-2}\sqrt{(x-3)^3}}$ g) $\dfrac{7(3-x)x^2}{\sqrt{7-2x}}$ h) $\dfrac{x(4-x)}{(2-x)^2}$ i) $-\dfrac{1}{\sqrt{x}(\sqrt{x}-1)^2}$ j) $(7-9x)(3-x)^3(2+x)^4$ k) $\dfrac{3x^2-2x-3}{(3x-1)^2}$

3 l) $\dfrac{4x-1}{2\sqrt{(x-1)(2x+1)}}$

Exercise 2E

1 a) $\dfrac{1}{2y-2}$ b) $-\dfrac{1}{15y^2}$ c) $\dfrac{1}{4-18y^2}$ d) $2\sqrt{y}$ e) $\dfrac{1}{6(2y-1)^2}$ f) $\dfrac{y^2}{y^2-1}$ g) $\dfrac{1}{y(5y+2)(1+y)^2}$ h) $(1+y)^2$ i) $-\sqrt{y}(1+\sqrt{y})^2$

2 a) $\frac{1}{25}$ b) $\frac{1}{11}$ c) $\frac{4}{3}$ d) -1 e) $\frac{25}{26}$ f) $\frac{2}{3}$ g) $-\frac{1}{4}$ h) -5

Exercise 2F

1 $y+4x=16$, $4y-x=30$ **2** $2y+x=9$, $y=2x-3$ **3** $(-4, \frac{4}{3})$, $(2, \frac{2}{3})$ **4** $(2, 2)$ **5** $(0, 0)$, $(4, -8)$ **6** $(-1, -1)$

7 $(-3, 0)$, min; $(\frac{1}{3}, \frac{500}{27})$, max **8** $(-2, \frac{1}{4})$, max **9** $(10, 2\sqrt{5})$, min **10** 100, £15 000 **11** 25 cm^2

Check out

1 a) $6x(x^2+5)^2$ b) $\dfrac{-1}{(3+2x)^{\frac{3}{2}}}$ **2** a) $\sqrt{3+2x}+\dfrac{x}{\sqrt{3+2x}}$ b) $2x^2(3-4x)(2-x)^4$ **3** a) $\dfrac{1}{(x+2)^2}$ b) $\dfrac{x^4-3x^2}{(x^2-1)^2}$

Revision exercise 2

1 a) $16x(1+x^2)^7$ b) $\dfrac{10-10x^3}{(x^3+2)^2}$ **2** a) -2 b) i) -1.6 ii) Plane is descending **3** 1.08 **4** $\frac{5}{2}$ **5** a) $x=5$, $x=-1$

5 b) $9y=x-2$ **6** a) $30x^2(x^3+1)^9$ b) $\frac{1023}{30}$ **7** a) $-\frac{3}{5}$ b) -1, max **8** b) $x=\frac{1}{4}, \frac{3}{2}$ **9** a) $\dfrac{1-2x^2}{(1-x^2)^{\frac{1}{2}}}$ **10** $y=\frac{1}{2}x-1$

Chapter 3

Check in

1 a) i) $y=\cos\left(\dfrac{x}{2}\right)$ ii) Stretch, parallel to x-axis, scale factor 2 **2** a) 9.74°, 80.26° (2 d.p.) b) -0.983, 2.159 (3 d.p.)

2 c) 66.42°, 293.58° (2 d.p.), 180° d) 2.558, 5.700 (3 d.p.) **3** a) $15x^2(x^3-2)^4$ b) $2(1-x^2)^5(1-13x^2)$ c) $\dfrac{-x(x^3+3x+2)}{(1+x^2)^2}$

Exercise 3B

1 30°, 150° b) ±70.5° c) $-116.6°$, 63.4° d) $-18.4°$, 161.6° e) ±80.4° f) 11.5°, 168.5° g) ±180° h) $-5.7°$, $-174.3°$

2 a) $\dfrac{\pi}{3}, \dfrac{2\pi}{3}, \dfrac{4\pi}{3}, \dfrac{5\pi}{3}$ b) $\dfrac{\pi}{2}, \dfrac{3\pi}{2}$, 2.82, 5.96 c) $\dfrac{\pi}{4}, \dfrac{5\pi}{4}$, 0.46, 3.61 d) 0, 2π, 1.77, 4.51 e) 0.59, 3.73, 0.46, 3.61 f) 0.84, 5.44

2 g) $\dfrac{3\pi}{2}$, 0.85, 2.29 h) 1.82, 4.46 **3** a) 35.3°, 144.7° b) 4.7°, 64.7°, 124.7° c) 7.5°, 37.5°, 97.5°, 127.5° d) 55.9°, 145.9°

3 e) 105°, 165° f) 22.7°, 82.7°, 142.7° g) 24.3°, 65.7°, 114.3°, 155.7° h) 63.8°, 116.2° **4** a) 145.5° b) 159.5° c) 21.3°

4 d) 171.6° e) 132.5° f) 94.8° g) 29.0° h) 164.5°

Exercise 3C

1 a) 51.3°, 231.3° b) 20.6°, 200.6° c) 33.7°, 213.7° d) 21.8°, 201.8° e) 45°, 225° f) 146.3°, 326.3°

1 g) 63.4°, 243.4°, 116.6°, 296.6° h) 52.2°, 232.2°, 127.8°, 307.8° **2** a) 1.23, 5.05 b) $\dfrac{3\pi}{2}$, 0.34, 2.80 c) 1.25, 4.39, 2.03, 5.18

2 d) $\dfrac{\pi}{3}, \dfrac{5\pi}{3}$, 1.37, 4.91 e) 1.11, 4.25, 2.90, 6.04 f) 0.34, 2.80, 3.55, 5.87 g) 0.98, 4.14, 1.37, 4.51 h) 1.11, 4.25, 2.50, 5.64

3 a) $(x-1)(x-2)(3x-1)$ b) 90°, 19.5°, 160.5°

Exercise 3E

The constant of integration is omitted in the answers to Question **3**.

1 a) $3\cos 3x$ b) $-2\sin 2x$ c) $5\sec^2 5x$ d) $-6\cos 6x$ e) $-14\sin 7x$ f) $30\sin 5x$ g) $4\cos\frac{1}{2}x$ h) $\sec^2(x+3)$ i) $\cos(x-4)$

1 j) $3\cos\left(x+\dfrac{\pi}{4}\right)$ k) $-8\sec^2(4x-7)$ l) $12\cos\left(\dfrac{3x-\pi}{2}\right)$ **2** a) $2x\cos(x^2)$ b) $3x^2\sec^2(x^3)$ c) $-4x\sin(x^2-1)$

2 d) $18x^2\cos(2x^3+3)$ e) $8x\cos(1-x^2)$ f) $72x^3\sin(4-3x^4)$ g) $2x\sin(x^2-2)$ h) $3x^2\sec^2(x^3-3)$

2 i) $-6x\sin(6x^2+1)$ j) $28x^3\sec^2(2-x^4)$ k) $\dfrac{3\cos\sqrt{x}}{\sqrt{x}}$ l) $\dfrac{1}{x^2}\sin\left(\dfrac{1}{x}\right)$ **3** a) $\sin 2x$ b) $-\frac{1}{4}\cos 4x$

3 c) $\frac{1}{4}\tan 4x$ d) $\frac{1}{2}\tan(2x-1)$ e) $2\cos(3x+2)$ f) $-\frac{4}{5}\cos\left(\frac{5x-\pi}{4}\right)$ g) $\frac{1}{2}\sin(x^2)$ h) $2\tan(x^4)$ i) $\frac{3}{2}\sin(x^2-7)$

3 j) $6\sin(x^2-4)$ k) $\frac{1}{3}\cos(3-x^3)$ l) $-2\cos(\sqrt{x})$ **4** a) $2\sin x\cos x$ b) $-3\sin x\cos^2 x$ c) $-\frac{\sin x}{2\sqrt{\cos x}}$ d) $6\sec^2 x\tan^5 x$

4 e) $14\cos x\sin^6 x$ f) $18\sin x\cos^5 x$ g) $20\cos 5x\sin^3 5x$ h) $12\sec^2 3x\tan^3 3x$ i) $-\frac{4\sin 4x}{\sqrt{\cos 4x}}$

Exercise 3F

1 a) $\tan x + x\sec^2 x$ b) $x^2(3\cos x - x\sin x)$ c) $2x(\sin 2x + x\cos 2x)$ d) $4x^3\tan 3x + 3x^4\sec^2 3x$ e) $2\cos 2x\sin 3x + 3\sin 2x\cos 3x$

1 e) $2\cos 2x\sin 3x + 3\sin 2x\cos 3x$ f) $\cos x\cos 5x - 5\sin x\sin 5x$ g) $2(2\sec^2 4x\tan 6x + 3\sec^2 6x\tan 4x)$ h) $3\sin x + 3x\cos x$

1 i) $5(\tan 2x + 2x\sec^2 2x)$ j) $2x(2\cos 3x - 3x\sin 3x)$ k) $\cos x - \sin x + \cos^2 x - \sin^2 x (\equiv (\cos x - \sin x)(\cos x + \sin x + 1))$

1 l) $2x^3(2 + 2\sin 2x + x\cos 2x)$ **2** a) $\frac{\sin x - x\cos x}{\sin^2 x}$ b) $\frac{2x\tan x - x^2\sec^2 x}{\tan^2 x}$ c) $\frac{x\sec^2 x - \tan x}{x^2}$ d) $\frac{\cos 2x + 2x\sin 2x}{\cos^2 2x}$

2 e) $\frac{2x\cos 2x - 3\sin 2x}{x^4}$ f) $-\frac{1}{\sin^2 x}$ g) $\frac{3\sin x\cos 3x - \cos x\sin 3x}{\sin^2 x}$ h) $-\frac{2\sin 2x\tan 3x + 3\cos 2x\sec^2 3x}{\tan^2 3x}$

2 i) $\frac{1 + \sin x - x\cos x}{(1+\sin x)^2}$ j) $\frac{x\sec^2 x - 2(1+\tan x)}{x^3}$ k) $-\frac{2\sin x}{(1-\cos x)^2}$ l) $-\frac{2}{(\sin x - \cos x)^2}$

3 a) $\sin x + x\cos x$ b) $x(2\cos x - x\sin x)$ c) $\cos 3x - 3x\sin 3x$ d) $3x^2(\tan 6x + 2x\sec^2 6x)$ e) $\sin^4 x(\sin x + 5x\cos x)$

3 f) $6x\cos^3 x(\cos x - 2x\sin x)$ g) $\frac{\tan x - x\sec^2 x}{\tan^2 x}$ h) $-\frac{2(x+1)\sin 2x + \cos 2x}{(x+1)^2}$ i) $-\frac{\sec^2 x}{(1+\tan x)^2}$ j) $\frac{2 + 2\sin 2x}{\cos^2 2x}$

3 k) $\frac{1 + \cos x + x\sin x}{(1+\cos x)^2}$ l) $\frac{1 + \sin x + \cos x}{(1+\cos x)^2}$

Exercise 3G

1 $y = x + 1$ **2** $y = -x; y = x - 2\pi$ **3** a) $2y = x - \frac{\pi}{4}$; $2y = -x + \frac{3\pi}{4}$ b) $\left(\frac{\pi}{2}, \frac{\pi}{8}\right)$ c) $\frac{\pi^2}{32}$

4 $y + 2\pi = 2x + 1, 2y + x = \pi + 2$ **5** $-\frac{\pi}{3}, \frac{\pi}{3}$ **6** b) $\frac{\pi}{3}, \frac{5\pi}{3}$ **7** b) $0, \frac{2\pi}{3}$ **8** b) $(0, 0), (2\pi, \pi)$

9 a) $\left(\frac{\pi}{2}, \frac{\pi}{2}\right)$ b) $\left(\frac{\pi}{2}, 1\right), \left(\frac{3\pi}{2}, -\frac{1}{3}\right)$ **10** a) $\frac{\pi}{2} - 1$ b) $\frac{1}{3}$ c) $\frac{2+\pi}{4}$ d) 2 **11** 2 **12** 5

13 a) $A\left(\frac{\pi}{6}, \frac{1}{2}\right), B\left(\frac{5\pi}{6}, \frac{1}{2}\right)$ b) $\sqrt{3} - \frac{\pi}{3}$

Check out

2 a) 60°, 300° b) −0.34, −2.80 (2 d.p.) c) −45°, 135° **4** a) $3x^2\cos 3x + 2x\sin 3x$ b) $\frac{1}{2}\sin 2x$

Revision exercise 3

1 a) $\frac{2\sin x - 2x\cos x}{\sin^2 x}$ b) ii) $y = -\frac{1}{2}x + \frac{5\pi}{4}$ **2** b) $\theta = 104.5°, 255.5°, 70.5°, 289.5°$ **3** a) $2 + 2\cos 2x$

4 a) i) $3x\sec^2 3x + \tan 3x$ ii) $\frac{x\cos x - \sin x}{x^2}$ **5** a) $(2x+5)\cos 3x - 3(x^2+5x+4)\sin 3x$ b) $y = 5x + 4$ **6** $y = x + 2$

7 a) $6\ln(\sec\theta) - \tan\theta$ b) $\theta = 1.1, 1.3, 4.2, 4.5$ **8** 67.5°, 292.5°

Chapter 4

Check in

2 a) $\log_a 6$ b) $\log_2(\frac{2}{9})$ **3** a) $x = 2.5$ b) $x = -0.683$ (3 d.p.)

Exercise 4A

The constant of integration is omitted in the answers to question **3**.

1 $2e^{2x}$ b) $6e^{6x}$ c) $-e^{-x}$ d) $6e^{3x}$ e) $-21e^{-3x}$ f) $2e^{2x+5}$ g) $3e^{3x-1}$ h) $8e^{4x+3}$ i) $15e^{3x-8}$ j) $-4e^{2-x}$ k) $-12e^{1-2x}$

1 l) $-35e^{2-7x}$ **2** a) $3x^2e^{x^3}$ b) $6x^5e^{x^6}$ c) $10x\,e^{x^2}$ d) $-6x\,e^{-x^2}$ e) $2x\,e^{x^2+1}$ f) $6x^2\,e^{2x^3-3}$ g) $-8x^3\,e^{-2x^4}$ h) $-12x^2\,e^{6-x^3}$

2 i) $-8x^3\,e^{2x^4}$ j) $-\dfrac{1}{x^2}e^{\frac{1}{x}}$ k) $\dfrac{e^{\sqrt{x}}}{2\sqrt{x}}$ l) $\dfrac{10}{x^3}e^{\frac{1}{x^2}}$ **3** a) $\frac{1}{2}e^{2x}$ b) $\frac{1}{5}e^{5x}$ c) $-e^{-x}$ d) $2e^{4x}$ e) $\frac{2}{3}e^{3x}$ f) $-e^{5-x}$ g) $\frac{1}{4}e^{4x+3}$

3 h) $2e^{3x+1}$ i) $\frac{1}{2}e^{x^2}$ j) $\frac{1}{5}e^{x^5}$ k) $\frac{2}{3}e^{x^3}$ l) $-\frac{1}{2}e^{-x^2}$

Exercise 4B

The constant of integration is omitted in the answers to questions **2** and **3**.

1 a) $\dfrac{2}{1+2x}$ b) $-\dfrac{4}{1-4x}$ c) $\dfrac{2x}{1+x^2}$ d) $\dfrac{3x^2}{x^3-2}$ e) $\dfrac{3(x^2-1)}{x^3-3x}$ f) $\dfrac{e^x}{e^x+4}$ g) $\dfrac{6e^{6x}}{1+e^{6x}}$ h) $\dfrac{1}{2x}$ **2** a) $\ln(1+x)$

2 a) $\ln(1+x)$ b) $2\ln x$ c) $2\ln(2x-1)$ d) $\frac{1}{2}\ln(5+6x)$ e) $\ln(x^2+1)$ f) $-2\ln(2-x^2)$ g) $\ln(x^2-x)$ h) $\ln(1+e^x)$

3 a) $\ln 2$ b) $\ln(\frac{3}{2})$ c) $\frac{1}{2}\ln(\frac{5}{3})$ d) $\frac{1}{2}\ln 3$

Exercise 4C

The constant of integration is omitted in the answers to question **4**.

1 a) $1+\ln x$ b) $x\,e^x(2+x)$ c) $3x^2\,e^{3x}(1+x)$ d) $\dfrac{x}{1+x}+\ln(1+x)$ e) $e^x\left(\dfrac{1}{x}+\ln x\right)$ f) $e^{2x}(2-2x^3-3x^2)$

1 g) $\dfrac{2\ln(1+3x)}{1+2x}+\dfrac{3\ln(1+2x)}{1+3x}$ h) $e^{5x}\left(5\ln(2+x^2)+\dfrac{2x}{2+x^2}\right)$ **2** a) $\dfrac{1-x}{e^x}$ b) $\dfrac{x(2\ln x-1)}{(\ln x)^2}$ c) $\dfrac{1-\ln x}{x^2}$ d) $\dfrac{1-3x}{e^{3x}}$

2 e) $-\dfrac{e^{-4x}(4x+3)}{x^4}$ f) $\dfrac{1-2x\ln x}{xe^{2x}}$ g) $\dfrac{2e^x}{(1-e^x)^2}$ h) $\dfrac{5}{x(3-\ln x)^2}$ **3** a) $-\sin x\,e^{\cos x}$ b) $e^x(\sin x+\cos x)$ c) $-6e^{2x}(1-e^{2x})^2$

3 d) $\dfrac{\sec^2 x}{(1+\tan x)}$ e) $\dfrac{\cos x-3\sin x-3}{e^{3x}}$ f) $\cos x\ln(1-2x)-\dfrac{2\sin x}{1-2x}$ g) $-2\sin 2x\,e^{\cos 2x}$ h) $\sec^2 x\ln x+\dfrac{\tan x}{x}$

3 i) $e^{-3x}(2\cos 2x-3\sin 2x)$ j) $\dfrac{1+\tan x-x\sec^2 x\ln x}{x(1+\tan x)^2}$ k) $\dfrac{-\sin(1+\ln x)}{x}$ l) $-e^{-5x}(5\cos 3x+3\sin 3x)$

4 a) $-e^{\cos x}$ b) $-\cos(e^x)$ c) $-\ln(1+\cos x)$ d) $\frac{1}{2}e^{\tan 2x}$ e) $\ln(x+\tan x)$ f) $-\frac{1}{2}\ln(\sin 2x+\cos 2x)$

Exercise 4D

1 $y=3x+1$ **2** $3y-x=3\ln 3-2$; $y+3x=6+\ln 3$ **3** $y=e(2x-1)$; $2ey+x=2e^2+1$ **4** a) $y=2x-e$ b) $\dfrac{\sqrt{5}e}{2}$

5 $y=e(x+1)$; $ey+x=1$ **6** $(\frac{1}{2},\frac{1}{4}-\ln 2)$, $(1,1)$ **7** $\left(\frac{1}{5},\ln\left(\frac{26}{25}\right)\right)$, $(5,\ln 26)$ **8** $\left(\ln 2,\ln\left(\frac{5}{2}\right)\right)$ **9** $(0,0)$, $\left(2,\dfrac{4}{e^2}\right)$ **10** $\left(e,\dfrac{1}{e}\right)$

11 $\left(-1,-\dfrac{1}{2e}\right)$, $\left(3,\dfrac{e^3}{6}\right)$ **12** a) $(1,\ln 4)$, min b) $\left(3,\dfrac{e^3}{27}\right)$, min c) (e^2,e^2), max d) $\left(-1,\dfrac{4}{e}\right)$, max; $(1,0)$, min **13** $\dfrac{e^6-1}{2}$

14 $2\ln 2$ **15** $2e^4-5e^2+3e$ **16** a) $(2,\frac{1}{3})$ b) $\ln 3-\frac{2}{3}$ **17** a) $P(1,2)$, $Q(3,4)$

Check out

1 a) $6e^{3x}$ b) $-e^{-x}$ c) $-\dfrac{12}{e^{4x}}$ **2** a) $\frac{1}{3}e^{3x}+c$ b) $-e^{-x}+c$ c) $-\dfrac{1}{4e^{4x}}+c$ **3** a) i) $\dfrac{3}{x}$ ii) $\dfrac{4}{x}$

3 b) i) $\frac{1}{3}\ln 3x$ ii) $\ln(x^2-4)$ **4** c) i) $\frac{1}{2}\ln x$

Revision exercise 4

1 a) i) $\frac{1}{2}e^{2x}+x+c$ b) i) 2 ii) 5 c) i) $2\leqslant f(x)\leqslant 5$ iii) $f^{-1}(x)=\frac{1}{2}\ln(x-1)$ **2** a) i) $e^x(2\cos 2x+\sin 2x)$ ii) $y=2x$

3 a) iv) 2 b) i) $\dfrac{1}{x}-\dfrac{1}{4}$ ii) $-\dfrac{1}{x^2}$ iii) (4, 0.386) iv) max. **4** a) $y=7$ b) $x=\frac{1}{3}\ln 8$ d) i) $8x-\frac{1}{3}e^{3x}+c$ e) ii) $x=\ln 3$

5 $2e^{2x}-2x^{-2}$ e) $\frac{1}{2}e^{2x}+2\ln x+c$ **6** a) 14 m^3 b) $-0.184\ m^3\,h^{-1}$; tank emptying c) 8 h 19 min

7 a) i) $\frac{3}{4}x^2-\dfrac{6}{x}$ b) i) $x=2$ ii) $\frac{3}{2}x+\dfrac{6}{x^2}$ iii) min. c) $28-\frac{3}{2}\ln 2$ **8** a) i) $\dfrac{1}{x}$ ii) $\dfrac{1}{e}$ b) Translation by $\begin{pmatrix}0\\2\end{pmatrix}$

8 c) i) all real numbers ii) domain: all real numbers, range: $f^{-1}(x) > 0$ iii) $f^{-1}(x) = e^{x-2}$

9 a) i) $-2e^{-2x} - \frac{3}{x^2}$ b) $f(x) > 3$ **10** a) $6x(x+1)$ b) $k = \frac{1}{6}$ **11** a) i) $(0, 5)$ ii) $f(x) > 0$ iii) $\frac{5}{6}$ b) ii) $gf(x) > 10$

11 c) i) 15 °C ii) 1.6 min **12** a) $(0, 3)$ b) i) $-2e^{2x}$ ii) -8 c) i) $\ln 2$ ii) $4x - \frac{1}{2}e^{2x} + c$ iii) $2\ln 2 - 1$

13 a) i) $2x - 3 + \frac{1}{x}$ b) ii) $x = \frac{1}{2}, x = 1$ iii) $2 - \frac{1}{x^2}$ iv) $-2, 1$ **14** b) i) $\frac{1}{2}e^{2x} - 2x + c$ **15** a) $y = -2$ b) $x = \frac{1}{2}\ln 3$ c) i) $2e^{2x}$

15 d) $1 - \frac{3}{2}\ln 3$ **16** a) $2e^{2x}\sin 3x + 3e^{2x}\cos 3x$ b) $20x(2x^2+1)^4$ **17** a) 32 m^3 b) $-1.25\ m^3\ h^{-1}$; tank emptying

17 c) 24.5 hours (3 s.f.) **18** a) Stretch, parallel to Ox, scale factor $\frac{1}{3}$ b) Translation $\begin{pmatrix}3\\0\end{pmatrix}$ c) Reflection in the line $y = x$

Chapter 5

Check in

1 a) $\frac{1}{2}x^4 - \frac{3}{2}x^2 + c$ b) $-\frac{2}{3x\sqrt{x}} + c$ c) $-\frac{2}{3x} - \ln x + c$ d) $-\frac{1}{3}\cos 3x + c$ e) $\frac{1}{4}e^{4x} + c$ **2** 3.627 (3 d.p.)

Exercise 5A

The constant of integration is omitted in the answers to all of these questions.

1 a) $\frac{(2x-1)^4}{8}$ b) $\frac{(3x+5)^5}{15}$ c) $\frac{(4x-3)^6}{24}$ d) $-\frac{(2-5x)^3}{15}$ e) $\frac{(x^2-1)^4}{8}$ f) $-\frac{(1-2x^2)^4}{16}$ g) $\frac{(3x^2+5)^3}{9}$ h) $\frac{(x^3+1)^3}{9}$

2 a) $\frac{1}{4}(x+1)(x-3)^3$ b) $\frac{1}{5}(x-1)(x+4)^4$ c) $\frac{1}{20}(4x-19)(x-1)^4$ d) $\frac{1}{16}(2x+1)(2x-3)^3$ e) $\frac{1}{48}(18x+23)(2x-5)^3$

2 f) $x - 3\ln(x+3)$ g) $\frac{1}{x+1} + \ln(x+1)$ h) $-\frac{4x+1}{8(2x-3)^2}$ **3** a) $\frac{2}{15}(3x-2)\sqrt{(x+1)^3}$ b) $\frac{2}{15}(3x+2)\sqrt{(x-1)^3}$

3 c) $\frac{2}{5}(x-10)\sqrt{(x+5)^3}$ d) $\frac{1}{15}(7-9x)\sqrt{(1-2x)^3}$ e) $\frac{2}{3}(x-2)\sqrt{x+1}$ f) $\frac{2}{3}(x+6)\sqrt{x-3}$ g) $\frac{2}{3}(x+2)\sqrt{x-4}$ h) $-\frac{2}{3}(x+19)\sqrt{5-x}$

4 a) $\frac{1}{20}(4x-3)(x+3)^4$ b) $x + 3\ln(x-1)$ c) $-\frac{2}{15}(3x+10)\sqrt{(5-x)^3}$ d) $\frac{5}{x+2} + \ln(x+2)$ e) $\frac{1}{3}(x-1)\sqrt{2x+1}$

4 f) $\frac{1}{120}(31-10x)(5-2x)^5$ g) $2x - 15\ln(x+7)$ h) $\frac{2(x+2)}{\sqrt{x+1}}$ i) $-\frac{2}{3}(x+14)\sqrt{4-x}$ j) $\frac{6}{3-x} + \ln(3-x)$

4 k) $\frac{1}{105}(15x^2+5x+1)(x-1)^5$ l) $-\frac{2}{35}(5x+2)\sqrt{(1-x)^5}$

Exercise 5B

1 a) $\frac{182}{3}$ b) $\frac{13}{3}$ c) 0 d) $4\ln 2$ e) $4\ln 3$ f) $\frac{16}{3}\ln 2$ **2** a) $1 + 2\ln 2$ b) 12 c) 46.4 d) $2.5 + \frac{1}{8}\ln 5$ e) $-\frac{2}{3}$ f) 2.8

Exercise 5C

The constant of integration is omitted in the answers to questions 1–3 and 5.

1 a) $\frac{1}{12}(3x+1)(x-1)^3$ b) $\frac{1}{20}(4x-1)(x+1)^4$ c) $-\frac{1}{5}(1+x)(4-x)^4$ d) $\frac{1}{56}(4x-1)(2x+3)^6$ e) $\frac{1}{4}(x-2)(x+2)^3$

1 f) $\frac{1}{42}(6x+25)(x-4)^6$ g) $\frac{1}{48}(18x-17)(2x+3)^3$ h) $\frac{1}{30}(25x+8)(4-x)^5$ **2** a) $\ln(x-1) - \frac{x}{x-1}$ b) $\frac{1}{x+1} + \ln(x+1)$

2 c) $\frac{1}{4}\ln(2x-3) + \frac{1}{4(2x-3)}$ d) $\frac{2-9x}{6(x+2)^3}$ e) $\frac{1}{3}(x+3)\sqrt{2x-3}$ f) $\frac{2}{27}(3x+40)\sqrt{3x-2}$ g) $-(2+x)\sqrt{1-2x}$

2 h) $-\frac{2}{15}(3x+8)\sqrt{(4-x)^3}$ i) $\frac{1}{3}\sqrt{(3-2x)^3}\,(1+3x)$ **3** a) $x\sin x + \cos x$ b) $-\frac{x}{2}\cos 2x + \frac{1}{4}\sin 2x$ c) $\frac{e^{3x}}{9}(3x-1)$

3 d) $-e^{-x}(x+1)$ e) $\frac{(6x-1)}{3}\sin 3x + \frac{2}{3}\cos 3x$ f) $\frac{x^2}{4}(2\ln x - 1)$ g) $\frac{x^3}{9}(3\ln x - 1)$ h) $\frac{2}{9}\sqrt{x^3}(3\ln x - 2)$ **4** a) $\frac{26}{3}$ b) $\frac{\pi}{4}$

4 c) $\frac{e^2+1}{e}$ d) $8\frac{2}{3}$ e) $124\ln 2 - 15$ f) $-\frac{2+\pi}{18}$ **5** a) $-x^2\cos x + 2x\sin x + 2\cos x$ b) $\frac{1}{60}(10x^2 - 12x + 9)(x+3)^4$

5 c) $e^x(x^2 - 2x + 2)$ d) $\frac{1}{2}x^2\sin 2x + \frac{1}{2}x\cos 2x - \frac{1}{4}\sin 2x$ e) $-\frac{e^{-2x}}{4}(2x^2 + 2x + 1)$ f) $(1 - 2x - x^2)\cos x + 2(x+1)\sin x$

6 a) $\frac{e^2-1}{4}$ b) $\pi^2 - 4$ c) $\frac{108}{5}$ d) $\frac{\pi^2-8}{32}$ e) $\frac{586}{15}$ f) $\pi^2 - 4$

Exercise 5D

1 a) $\sin^{-1}\left(\frac{x}{2}\right)$ b) $\sin^{-1}\left(\frac{x}{5}\right)$ c) $\frac{1}{3}\sin^{-1}(3x)$ d) $\sin^{-1}(6x)$ e) $\frac{1}{3}\sin^{-1}\left(\frac{3x}{2}\right)$ f) $\frac{1}{4}\sin^{-1}\left(\frac{4x}{5}\right)$ g) $2\tan^{-1}\left(\frac{x}{2}\right)$ h) $\frac{1}{10}\tan^{-1}\left(\frac{x}{10}\right)$

1 i) $\frac{1}{3}\tan^{-1}(3x)$ j) $\frac{1}{5}\tan^{-1}(5x)$ k) $\frac{1}{6}\tan^{-1}\left(\frac{2x}{3}\right)$ l) $\frac{1}{2}\tan^{-1}\left(\frac{4x}{7}\right)$ **2** a) $\frac{\pi}{2}$ b) $\frac{\pi}{4}$ c) $\frac{\pi}{12}$ d) 3π e) $\frac{\pi}{2}$ f) 2π

Exercise 5E

1 a) 72π b) 625π c) 8π d) $\frac{7\pi}{24}$ e) 54π f) $\frac{158\pi}{3}$ g) 9π h) $\frac{348\pi}{5}$ i) $\frac{352\pi}{3}$ j) 42π k) $\frac{803\pi}{12}$ l) $\frac{3\pi}{2}$

2 a) 288π b) $\frac{81\pi}{2}$ c) $\frac{96\pi}{5}$ d) $\frac{243\pi}{5}$ e) 7π f) $\frac{14\pi}{3}$ g) $\frac{\pi}{4}$ h) 63π i) $\frac{104\pi}{3}$ j) 16π k) $\frac{20\pi}{3}$ l) $\frac{115\pi}{3}$

3 a) $P(-2, 4), Q(2, 4)$ b) $\frac{64\pi}{5}$ **4** a) $A(-1, 2), B(1, 2)$ b) $\frac{56\pi}{15}$ **5** $\frac{28\pi}{3}$ **6** 9π **7** $\frac{128\pi}{5}$ **8** $\frac{5\pi}{2}$ **9** $\frac{25\pi}{2}$

10 a) $P(3, 9)$ b) i) $\frac{162\pi}{5}$ ii) $\frac{27\pi}{2}$ **11** b) $\frac{\pi}{10}$ **12** $\frac{1250\pi}{3}$ **13** a) $P(-2, 4), Q(2, 4)$ b) i) $\frac{256\pi}{3}$ ii) 16π

14 a) $\frac{32\pi}{5}$ b) $\frac{3\pi}{2}$

Exercise 5F

1 a) $(3, 3)$ b) $\frac{15}{2} - 8\ln 2$ **2** a) $(3, 9)$ b) $30 - 32\ln 2$ **3** a) $P(\pi, 0), Q(2\pi, 0)$ b) $\pi, 3\pi$ **4** $\frac{e^2+1}{4}$

5 a) $A(2, 2), B(3, 0)$ b) $3\frac{3}{5}$ c) $\frac{65\pi}{12}$ **6** a) $A\left(1, \frac{1}{e}\right)$ b) $1 - \frac{5}{2e}$ c) $\frac{(3e^2 - 19)\pi}{12e^2}$ **7** b) $\frac{11}{30}, \frac{1}{30}$

8 a) $A(1, \frac{1}{2}), B(2, \frac{2}{3})$ c) $1 - \ln(\frac{8}{3})$ **9** b) $e - 2$ c) $\frac{(e^2-5)\pi}{4}$

Check out

1 a) $\frac{1}{9}(x + 5)^9 - \frac{5}{8}(x + 5)^8 + c$ b) 0.804 (3 d.p.) **2** a) $\frac{e^{3x}}{9}(3x - 1) + c$ b) 0.734 (3 d.p.) **3** a) $\frac{1}{6}\tan^{-1}\left(\frac{x}{6}\right) + c$

3 b) 1.56 (3 s.f.) **4** a) $\frac{1366\pi}{15} = 286$ (3 s.f.) b) 16.8 (3 s.f.)

Revision exercise 5

1 a) $\frac{\pi}{2} - 1$ **2** a) $\frac{2x}{x^2+9}$ **3** a) $4x\ln x$ b) 2.94 **4** $\frac{1}{4}$ **5** $\frac{5\pi}{6}$ **6** a) $\frac{4}{4-x^2}$ c) ii) 0.524 (3 s.f.)

7 a) 3.77 (3 s.f.) b) $2y = 5x - 1$ c) i) $3x\sqrt{x^2+3}$ ii) 0.935 (3 s.f.) **8** a) $-4e^{-3}$ b) $y = -4e^{-3}x + 17e^{-3}, x = \frac{17}{4}$ c) $(-1, e)$

8 d) $3(1 - 2e^{-3})$ **9** a) $-\frac{1}{2}x\cos 2x + \frac{1}{4}\sin 2x + c$ b) i) $\frac{\pi}{4}$ **10** $a = 1, b = 2, c = -2, I = \frac{3}{2} - 2\ln 2$

11 a) $\frac{1}{4}x^4\ln x - \frac{x^4}{16} + c$ b) 46.4

Chapter 6

Check in

1 a) i) -3 ii) 3 b) i) 1.176 ii) -0.625 c) i) 2 ii) -18.1 (3 s.f.) **2** a) i) $x = \frac{1}{3-2y}$ ii) $x = \sqrt{\frac{1-4y}{3}}$

2 b) i) $4x + 5 = 0$ ii) $4x^2 + 20x - 6 = 0$ **3** 0.227 (3 s.f.)

Exercise 6A

7 d) $-5.19, 0.19$

Exercise 6B

1 a) 0.1716 b) 0.2915 c) 0.1835 d) diverges e) oscillates **2** a) $3, 2.\dot{6}, 2.875, 2.739\,130\,435$

3 a) $0.2, 0.238\,095\,238\,1, 0.235\,955\,056\,2, 0.236\,074\,270\,6$ **4** a) $0.877\,582\,561, 0.639\,012\,494, 0.802\,685\,1, 0.694\,778\,026$

5 a) 2.718 281 828, 15.154 262 24, 3,814,279.105 **6** a) 1.732 050 808, 1.706 072 567, 1.703 532 855, 1.703 284 362, 1.703 260 047

6 b) $3x^2 - x - 7 = 0$ **7** a) $x^3 - 2x^2 - 1 = 0$ b) 2.2056 **8** 1.15

9 $x_0 = 0.5$, 0.629 960 527 9, 0.615 442 002 4, 0.618 561 656 6; 0.62 **10** 2.22 **11** 0.22 **12** b) i) 2.646 ii) 3.873 iii) 6.164

Exercise 6C

1 1.14 **2** 2.13 **3** 9.58 **4** 0.730 **5** 7.71 **6** 1.33 **7** 80.7 **8** 1.41 **9** 28.0 **10** 15.1 **11** 0.656

12 7.21 **13** b) $\frac{617}{315}$ c) 0.659% **14** a) i) 2.052 ii) 2.094 b) 2 c) i) 2.6% ii) 4.7%

Check out

2 b) 1.23 **3** a) 5.64 (3 s.f.) b) 9.914 (3 s.f.)

Revision exercise 6

1 b) ii) 3, $2.\dot{5}$, 2.469 299, 2.466 216 iv) $\sqrt[3]{15}$ **2** a) 3.742, 3.968, 3.996 b) ii) 4 **3** 1.434

4 a) $A\left(1, \frac{\pi}{2}\right), B\left(-1, -\frac{\pi}{2}\right)$ b) 0.565 **5** a) i) $-x\sin x + \cos x$ iii) 0.86 b) $\frac{\pi}{2} - 1$ **6** b) 4.055

7 a) 0.381 82 b) 0.381 70 **9** a) $x = 0.737$ b) 0.660, 0.633, 0.645, 0.640 c) $2^{-x} = x$

C3 Practice Paper

1 a) i) $3\cos 3x$ ii) $e^{2x}(3\cos 3x + 2\sin 3x)$ **2** a) i)

x	1	3	5
f(x)	1.5811	1.1785	0.8575

ii) 7.23 (3 s.f.) b) 29

3 a) $\frac{1}{2}\sin 2x + c$ b) 0.102 (3 s.f.) **4** a) Reflection in y-axis; stretch parallel to x-axis, scale factor 3 b) ii) min.

5 a) i) $fg(x) = \frac{1}{2 - 5x}$ ii) domain: all real x, $x \neq \frac{2}{5}$; range: all real f(x), f(x) $\neq 0$ c) $x = -5$ or $x = -1$

6 a) $\frac{2}{3}\sqrt{3x^2 + 2} + c$ b) $\frac{1}{3}\ln(\frac{29}{5})$ **7** a) $\frac{1}{5}$ c) 0.896 (3 d.p.) d) ii) 0.896, 0.045

Chapter 7

Check in

1 b) $(x + 1)$ is a factor **2** a) −5 b) 7 **3** a) $3(2 - 3x)$ b) $5x(2 - x)$ c) $(2x - 1)(x + 3)$ d) $(x + 1)(x - 2)(x + 3)$

Exercise 7A

1 a) −1 b) −23.5 c) 6 d) $6\frac{1}{3}$ e) $-4\frac{1}{8}$ f) 14 **2** −9 **3** $\frac{10}{3}$ **4** 20 **5** 7 **6** 34 **7** $a = -12, b = 9$

8 $a = 3, b = -2$ **9** $b = 11, c = 1$.

Exercise 7B

1 a) $(x - 3)(x + 1)(x - 2)$ b) $(x + 1)(x - 2)(x - 4)$ c) $(x - 1)(x - 2)(x - 3)$ d) $(x + 1)(x - 2)(x - 5)$ e) $(x + 5)(x - 1)^2$

1 f) $(x - 2)^3$ **2** a) $(x - 1)(2x - 1)(x + 1)$ b) $(x - 2)(2x + 1)(x + 1)$ c) $(x - 1)(3x - 1)(x + 2)$ d) $(2x - 1)(x + 1)(x + 3)$

2 e) $(x^2 + 1)(3x - 1)$ f) $(5x - 1)(x + 1)(x + 2)$ **3** a) $-\frac{3}{2}, 1, 2$ b) $-\frac{1}{3}, 3, 4$ c) $-\frac{5}{2}, -1, 3$ d) $-\frac{1}{5}, 1, 6$ e) −2, −6.16, 0.162

3 f) 3, 1.70, 5.30 **4** a) $-3, \frac{1}{2}, 2$ b) $\frac{1}{3}, 2, 3$ c) $-1\frac{1}{2}, 1, 4$ d) −6 e) $-\frac{1}{2}, \frac{1}{5}, 1$ f) $-\frac{2}{3}, \frac{1}{2}, 3$ **5** 5

6 $a = 5, b = 6, c = -1$; −5, −3, 2 **7** $c = -2, d = 5, e = 10$ (or $c = 5, d = -2, e = 10$); −5, 1, 2

8 $a = -1, b = 14, c = -8$; 1, 2, 4 **9** $a = 5, b = -4, c = 5$; 4 **10** $b = 5, c = -3$; 3

11 $(2x + 1)(x - 2)(x - 5), x = -\frac{1}{2}$ or $x = 2$ or $x = 5$ **12** $(x + 2)^2(x - 3), x = -2$ or $x = 3$

13 $a = 3, b = -2, c = 60, x = -5$ or $x = 4$ or $x = 3$ **14** $c = 3, d = 1, e = -27, x = 4$ or $x = -3$ or $x = -\frac{3}{2}$

15 $a = 4, b = 2, c = -3, x = -\frac{1}{3}$ or $x = \frac{1}{4}$ or $x = \frac{1}{2}$

Exercise 7C

1 a) $\frac{5x+7}{(x+3)(x-1)}$ b) $\frac{x-10}{(x+4)(x-3)}$ c) $\frac{2(x+13)}{(x+4)(x-5)}$ d) $\frac{x^2+6}{(x+2)(x-3)}$ e) $\frac{2x^2-5x-2}{(x+2)(x-2)}$ f) $\frac{4x+34}{(2x+5)(2x-3)}$

1 g) $\frac{5x^2+3x+4}{(x+4)(x-2)}$ h) $\frac{16x^2-25x+18}{(x^2+3)(2x-5)}$ **2** a) $\frac{3x+8}{(x+4)(x-1)}$ b) $-\frac{4x+11}{(x+2)(x+4)}$ c) $\frac{2x-1}{(x-1)(x-2)}$ d) $\frac{7x-6}{(x-2)^2}$

2 e) $\frac{-3x-4}{(x+2)(x+3)}$ f) $\frac{x-3}{(2x+5)(x-1)}$ **3** $\frac{7}{5}$ **4** $\frac{5x-1}{(2x+1)(x-3)}, -\frac{9}{4}$ **5** $\frac{x^2+x+3}{(x+1)(x-2)}, -\frac{3}{2}, \frac{17}{4}$ **6** $a=2, b=3$

7 $A=1, B=1$ **8** $P=2, Q=8$ **9** $a=1, b=2$

Exercise 7D

1 a) $\frac{1}{x+2}$ b) $\frac{2}{x+5}$ c) $\frac{x}{x-5}$ d) $\frac{3x}{x-3}$ e) $\frac{x+5}{x+3}$ f) $\frac{x-7}{x-2}$ g) $\frac{x+2}{x-6}$ h) $\frac{x-5}{x+2}$ **2** a) $\frac{1}{x+2}$ b) $\frac{3}{2x+3}$ c) $\frac{x}{x+5}$

2 d) $\frac{2x}{x-1}$ e) $\frac{3x+1}{2x-1}$ f) $\frac{2x-1}{2x+1}$ g) $\frac{5x+2}{4x-3}$ h) $\frac{2x-5}{x-1}$ **3** a) $\frac{5x}{x+4}$ b) 6 **4** a) $\frac{5x-1}{3x+5}$ b) -11

5 a) $\frac{3x+2}{x-3}$ b) $5\frac{3}{4}$

Exercise 7E

1 a) $A=2, B=7$ b) $A=3, B=-7$ c) $A=5, B=-13$ d) $A=2, B=5$
2 a) $A=2, B=3, C=14$ b) $A=1, B=1, C=-11$ c) $A=4, B=-12, C=69$
3 a) $A=1, B=-2, C=8$ b) $A=2, B=-6, C=36$ c) $A=4, B=24, C=-20$
4 a) $A=2, B=3, C=3, D=7$ b) $A=3, B=-2, C=6, D=3$ c) $A=1, B=-2, C=-13, D=-35$
5 a) $A=4, B=8, C=12$ b) $A=1, B=0, C=4, D=11$ c) $A=5, B=-9, C=19$ d) $A=4, B=-17$
5 e) $A=3, B=1, C=6, D=4$ f) $A=-7, B=39$

Exercise 7F

1 a) $\frac{2}{x+3}+\frac{1}{x-2}$ b) $\frac{2}{x+4}+\frac{3}{x-3}$ c) $\frac{3}{x+2}-\frac{1}{x+1}$ d) $\frac{2}{2x-1}-\frac{3}{3-x}$ e) $\frac{2}{x+4}+\frac{5}{x-2}$ f) $\frac{3}{2x+5}+\frac{1}{x-2}$

2 a) $\frac{1}{x+3}-\frac{2}{x+2}+\frac{1}{x+1}$ b) $\frac{2}{x+1}+\frac{3}{x-1}-\frac{4}{x-2}$ c) $\frac{1}{3(x+2)}+\frac{1}{6(x-1)}-\frac{1}{2(x+3)}$ d) $\frac{1}{2x+1}+\frac{2}{3(2x-1)}-\frac{1}{3(x-2)}$

2 e) $\frac{4}{x+4}-\frac{2}{x+3}+\frac{3}{2x+5}$ f) $\frac{1}{x}-\frac{1}{2(x+3)}+\frac{1}{2(3x+1)}$ **3** a) $\frac{2}{x+2}-\frac{1}{(x+2)^2}$ b) $\frac{4}{x-3}+\frac{3}{(x-3)^2}$

3 c) $\frac{3}{x-4}-\frac{2}{(x-4)^2}$ d) $\frac{3}{x+1}+\frac{2}{(x+1)^2}-\frac{3}{x+2}$ e) $\frac{2}{x-2}+\frac{3}{(x-2)^2}+\frac{4}{x+3}$ f) $\frac{3}{2x+1}+\frac{2}{(2x+1)^2}-\frac{3}{x+2}$

4 a) $1+\frac{3}{x+5}+\frac{2}{x-3}$ b) $2-\frac{1}{x+2}-\frac{4}{x+1}$ c) $2+\frac{3}{2(2x+1)}+\frac{3}{2(2x-3)}$ d) $2-\frac{1}{3x-1}+\frac{3}{2x-3}$ e) $x+3+\frac{1}{x+3}+\frac{1}{x-2}$

4 f) $x+\frac{1}{2(3x-1)}-\frac{1}{2(x-3)}$ **5** $\frac{5}{2(x+3)}+\frac{3}{2(x+1)}$ **6** $\frac{1}{2x+1}-\frac{4}{3x+5}+\frac{1}{x-3}$ **7** $\frac{3}{(x+1)^2}-\frac{2}{x+1}$

8 $\frac{2}{(x-3)^2}+\frac{3}{x-3}+\frac{1}{x+2}$ **9** $\frac{3}{x-1}-\frac{2}{x^2+3x+3}$ **10** $1+\frac{3}{x+5}-\frac{2}{x-3}$ **11** $\frac{2}{x-4}-\frac{3}{2x-1}$

12 $\frac{3}{(x+4)^2}+\frac{2}{x+4}-\frac{1}{x+3}$ **13** $\frac{1}{x-5}-\frac{1}{x-4}$ **14** $\frac{1}{x-2}-\frac{1}{x+5}$

Check out

1 a) 0; $(x - \frac{1}{2})$ or $(2x - 1)$ is a factor of $P(x)$ c) $P(x) = (2x - 1)(4x + 3)(x + 2)$ **2** 4 **3** $\dfrac{3x}{(2x-1)(2x+1)}$

4 a) $5 - \dfrac{2}{x+1}$ b) $2 + \dfrac{7}{2x+1} - \dfrac{1}{x-3}$

Revision exercise 7

1 $\frac{1}{4}$ **2** b) $(x - 4)(2x - 1)(3x + 2)$ **3** $2 - \dfrac{3}{x+2}$ **4** b) $p(\frac{1}{2}) = 0$ c) $(x + 2)(2x - 1)(3x - 1)$ d) $\dfrac{\pi}{6}, \dfrac{5\pi}{6}$, 0.34, 2.80

5 c) $(x - 2)(2x - 1)(2x + 5)$ d) 1.107, 0.464, −1.190, −2.035, −2.678, 1.951 **6** b) 1 c) $(2x + 1)(x - 1)(4x - 1)$

6 d) 2.09, 4.19, 0, 1.32, 4.97, $2\pi = 6.28$ **7** $\dfrac{2}{x-1} - \dfrac{1}{x} - \dfrac{1}{x^2}$ **8** a) i) $\dfrac{3}{x-2} - \dfrac{2}{x-4}$ ii) $\dfrac{6}{x-4} - \dfrac{6}{x-2}$

Chapter 8

Check in

1 a) 35 b) −70 000 **2** a) $1 + 5x + 10x^2 + 10x^3 + 5x^4 + x^5$ b) $16 - 160x + 600x^2 - 1000x^3 + 625x^4$

3 a) $\dfrac{3}{x+2} - \dfrac{2}{2x-1}$ b) A = 1, B = 2, C = −1

Exercise 8A

1 a) $1 - 2x + 3x^2 - 4x^3$, $|x| < 1$ b) $1 + \frac{1}{2}x - \frac{1}{8}x^2 + \frac{1}{16}x^3$, $|x| < 1$ c) $1 - 6x + 24x^2 - 80x^3$, $|x| < \frac{1}{2}$

1 d) $1 + 6x + 27x^2 + 108x^3$, $|x| < \frac{1}{3}$ e) $1 + x + 2x^2 + \frac{14}{3}x^3$, $|x| < \frac{1}{3}$ f) $1 - 3x^2 + 6x^4 - 10x^6$, $|x| < 1$

1 g) $1 + 3x + 9x^2 + 27x^3$, $|x| < \frac{1}{3}$ h) $1 - 3x - \frac{9}{2}x^2 - \frac{27}{2}x^3$, $|x| < \frac{1}{6}$ **2** a) $\frac{1}{2} - \frac{1}{4}x + \frac{1}{8}x^2 - \frac{1}{16}x^3$, $|x| < 2$

2 b) $2 + \frac{1}{4}x - \frac{1}{64}x^2 + \frac{1}{512}x^3$, $|x| < 4$ c) $\frac{1}{3} + \frac{2}{27}x + \frac{2}{81}x^2 + \frac{20}{2187}x^3$, $|x| < \frac{9}{4}$ d) $4 + x - \frac{1}{15}x^2 + \frac{1}{96}x^3$, $|x| < \frac{8}{3}$

2 e) $\frac{1}{8} + \frac{3}{16}x + \frac{3}{16}x^2 + \frac{5}{32}x^3$, $|x| < 2$ f) $2 - \frac{1}{4}x - \frac{1}{64}x^2 - \frac{1}{512}x^3$, $|x| < 4$ g) $\frac{1}{6} - \frac{1}{36}x + \frac{1}{216}x^2 - \frac{1}{1296}x^3$, $|x| < 6$

2 h) $\frac{1}{9} + \frac{4}{27}x + \frac{4}{27}x^2 + \frac{32}{243}x^3$, $|x| < \frac{3}{2}$ **3** a) $1 + 2x + 2x^2$ b) $\frac{1}{2} + \frac{5}{4}x - \frac{5}{8}x^2$ c) $\frac{1}{4}x + \frac{1}{16}x^2$ d) $2 + 17x + 84x^2$ e) $8 - x - \frac{33}{16}x^2$

3 f) $7 + \frac{1}{2}x + \frac{9}{8}x^2$ g) $16 - 72x + 9x^2$ h) $\frac{1}{16} - \frac{3}{64}x - \frac{5}{128}x^2$ **4** a) i) $1 + 8x + 24x^2$ ii) $1 - \frac{1}{2}x - \frac{1}{8}x^2$ b) $1 + \frac{15}{2}x + \frac{159}{8}x^2$

5 a) i) $1 + x - \frac{1}{2}x^2$ ii) $1 + 4x + 10x^2$ b) $1 + 5x + \frac{27}{2}x^2$, $|x| < \frac{1}{2}$ **6** a) i) $1 + x + x^2 + x^3$ ii) $1 + 2x + 4x^2 + 8x^3$

6 b) $1 + 3x + 7x^2 + 15x^3$ **7** a) i) $1 - x - x^2 - \frac{5}{3}x^3$ ii) $1 + 4x + 16x^2 + 64x^3$ b) $1 + 3x + 11x^2 + \frac{127}{3}x^3$, $|x| < \frac{1}{4}$

8 $1 + \frac{1}{2}x - \frac{1}{8}x^2 + \frac{1}{16}x^3$, 1.004 987 56 **9** $1 - \frac{1}{2}x - \frac{3}{8}x^2$, 0.999 500 **10** $\frac{1}{2} + \frac{3}{16}x + \frac{27}{256}x^2$, 0.5019

12 a) $1 + \frac{1}{200}x - \frac{1}{80\,000}x^2 + \frac{1}{16\,000\,000}x^3$ **14** a) $1 + \frac{1}{64}x - \frac{3}{8192}x^2$

Exercise 8B

1 a) $\dfrac{1}{1+x} + \dfrac{1}{1-2x}$, $2 + x + 5x^2 + 7x^3$, $|x| < \frac{1}{2}$ b) $\dfrac{2}{1-x} - \dfrac{1}{1+3x}$, $1 + 5x - 7x^2 + 29x^3$, $|x| < \frac{1}{3}$

1 c) $\dfrac{2}{1+5x} - \dfrac{1}{1+3x}$, $1 - 7x + 41x^2 - 223x^3$, $|x| < \frac{1}{5}$ d) $\dfrac{2}{1+x} + \dfrac{3}{2-x}$, $\frac{7}{2} - \frac{5}{4}x + \frac{19}{8}x^2 - \frac{29}{16}x^3$, $|x| < 1$

1 e) $\dfrac{1}{2+x} - \dfrac{1}{3+x}$, $\frac{1}{6} - \frac{5}{36}x + \frac{19}{216}x^2 - \frac{65}{1296}x^3$, $|x| < 2$ f) $\dfrac{1}{1+x^2} + \dfrac{3}{1-x}$, $4 + 3x + 2x^2 + 3x^3$, $|x| < 1$

2 a) $\dfrac{1}{1-2x} - \dfrac{1}{1-x} + \dfrac{1}{(1-x)^2}$, $1 + 3x + 6x^2$, $|x| < \frac{1}{2}$ b) $\dfrac{4}{1-3x} + \dfrac{1}{(1+x)^2}$, $5 + 10x + 39x^2$, $|x| < \frac{1}{3}$

2 c) $\dfrac{1}{1+x} - \dfrac{1}{1+2x} + \dfrac{1}{(1+2x)^2}$, $1 - 3x + 9x^2$, $|x| < \frac{1}{2}$ d) $\dfrac{2}{2+x} + \dfrac{4}{1-2x} - \dfrac{3}{(1-2x)^2}$, $2 - \frac{9}{2}x - \frac{79}{4}x^2$, $|x| < \frac{1}{2}$

3 a) $A = 2, B = -2, C = 1, D = 3$ b) $\frac{5}{2} + \frac{31}{4}x - \frac{133}{8}x^2$, $|x| < \frac{1}{3}$ **4** a) $A = 3, B = -15, C = 0, D = 1$

4 b) $-1 - \frac{2}{3}x +$ [illegible] x^2, $|x| < \frac{1}{2}$

Check out

1 a) $1 - \frac{1}{2}x - \frac{1}{8}x^2 - \frac{1}{16}x^3 \ldots$ b) $\frac{1}{27}(1 - 2x + \frac{8}{3}x^2 - \frac{80}{27}x^3 \ldots)$; $|x| < \frac{3}{2}$ **2** a) $1 - \frac{1}{3}x - \frac{1}{9}x^2 \ldots$ b) $\frac{1458}{701}$ **3** $\frac{5}{2} - \frac{13}{4}x + \frac{23}{8}x^2$

Revision exercise 8

1 a) $1+\frac{x}{3}-\frac{x^2}{9}$ b) $2+\frac{x}{3}-\frac{x^2}{18}$ **2** a) i) $1-x+x^2-x^3$ ii) $1+2x-2x^2+4x^3$ b) $k=4$ **3** a) $\frac{3}{1+3x}+\frac{2}{2-x}$
3 b) $1-3x+9x^2$ d) $4-\frac{17}{2}x+\frac{109}{4}x^2$ e) $|x|<\frac{1}{3}$ **4** a) $1-4x+12x^2-32x^3\ldots$ b) $|x|<\frac{1}{2}$ **5** a) $1+\frac{1}{2}x-\frac{1}{8}x^2\ldots$
5 b) i) $2+\frac{1}{2}x-\frac{x^2}{16}\ldots$ ii) $|x|<2$ **6** a) $\frac{1}{1-x}+\frac{2}{2+x}$ b) ii) $1+x+x^2\ldots$ c) $2+\frac{x}{2}+\frac{5}{4}x^2\ldots$ **7** $2x-12x^2+48x^3\ldots$
8 a) $\frac{2}{1+2x}+\frac{1}{4-x}$ b) ii) $1-2x+4x^2\ldots$ iii) $\frac{9}{4}-\frac{63}{16}x+\frac{513}{64}x^2\ldots$ iv) $|x|<\frac{1}{2}$ **9** a) $1+\frac{3}{10}x+\frac{3}{50}x^2+\frac{x^3}{100}\ldots$ b) $K=\frac{1}{2}$
9 c) 1.030 610 152 128 36 (14 d.p.)

Chapter 9

Check in

1 a) 50° b) 0.262 (3 s.f.) **3** a) $-\frac{1}{2}\cos 2x+c$ b) $\frac{1}{2}$

Exercise 9A

1 a) $\frac{63}{65}$ b) $\frac{56}{65}$ c) $-\frac{33}{56}$ **2** a) $\frac{16}{65}$ b) $\frac{56}{65}$ c) $\frac{16}{63}$ **3** a) $\frac{31\sqrt{2}}{50}$ b) $\frac{17\sqrt{2}}{50}$ c) $-\frac{17}{31}$ **4** 1 **5** $\frac{3}{11}$ **6** $-\frac{5}{3}$
7 $\frac{1+2\sqrt{3}}{2-\sqrt{3}}(=8+5\sqrt{3})$ **8** $\frac{1+3\sqrt{3}}{3-\sqrt{3}}\left(=\frac{6+5\sqrt{3}}{3}\right)$ **9** a) $\sqrt{3}$ b) $\sqrt{2}-1$ c) $\frac{1}{2+\sqrt{3}}(=2-\sqrt{3})$ d) 1
9 e) $\frac{2\sqrt{3}-1}{2-\sqrt{3}}(=4+3\sqrt{3})$ f) $\frac{\sqrt{3}-\sqrt{2}}{\sqrt{2}-1}$

Exercise 9B

1 a) $\frac{24}{25}$ b) $-\frac{7}{25}$ c) $-\frac{24}{7}$ **2** a) $-\frac{119}{169}$ b) $\frac{120}{169}$ c) $\frac{169}{120}$ **3** a) $-\frac{4}{3}$ b) $\frac{4}{5}$ c) $-\frac{5}{3}$ **4** a) $\frac{\sqrt{3}}{2}$ b) $\frac{1}{2}$ c) $\frac{1}{\sqrt{3}}$
5 a) $-\frac{625}{527}$ b) $\frac{336}{625}$ c) $-\frac{527}{336}$ **6** a) $-\frac{12}{5}$ b) $\frac{13}{12}$ c) $-\frac{5}{13}$ **7** a) 0, 180°, 360°, 80.4°, 279.6° b) 90°, 270°, 41.8°, 138.2°
7 c) 90°, 270°, 210°, 330° d) 60°, 300°, 109.5°, 250.5° e) 48.6°, 131.4° f) 90°, 194.5°, 345.5° g) 104.5°, 255.5°
7 h) 0, 90°, 180°, 270°, 360°, 45°, 135°, 225°, 315° **8** (decimal answers) a) 0, 3.14, 6.28, 1.05, 5.24 b) 1.57, 4.71, 1.05, 2.09
8 c) 0, 3.14, 6.28, 2.09, 4.19 d) 1.57, 3.67, 5.76 e) 2.09, 4.19 f) 0, 6.28, 1.05, 5.24 g) 0, 6.28, 2.09, 4.19
8 h) 1.57, 4.71, 0.79, 2.36, 3.93, 5.50 **8** (as mutiples of π) a) $0, \pi, 2\pi, \frac{\pi}{3}, \frac{5\pi}{3}$ b) $\frac{\pi}{2}, \frac{3\pi}{2}, \frac{\pi}{3}, \frac{2\pi}{3}$ c) $0, \pi, 2\pi, \frac{2\pi}{3}, \frac{4\pi}{3}$
8 d) $\frac{\pi}{2}, \frac{7\pi}{6}, \frac{11\pi}{6}$ e) $\frac{2\pi}{3}, \frac{4\pi}{3}$ f) $0, 2\pi, \frac{\pi}{3}, \frac{5\pi}{3}$ g) $0, 2\pi, \frac{2\pi}{3}, \frac{4\pi}{3}$ h) $\frac{\pi}{2}, \frac{3\pi}{2}, \frac{\pi}{4}, \frac{3\pi}{4}, \frac{5\pi}{4}, \frac{7\pi}{4}$

Exercise 9C

3 a) $(2y+1)(2y-1)(y-1)$ c) 30°, 90°, 150°, 210°, 330° **4** a) $(y+1)(y-1)(4y-1)$ c) 0°, 180°, 360°, 75.5°, 284.5°

Exercise 9D

The constant of integration is omitted in the answers to these questions.

1 a) $\frac{1}{2}x-\frac{\sin 2x}{4}$ b) $4x+2\sin 2x$ c) $3x+\frac{3}{2}\sin 2x$ d) $\frac{5}{2}x-\frac{5}{4}\sin 2x$ e) $\frac{1}{2}x-\frac{\sin 6x}{12}$ f) $8x+\sin 8x$
2 a) $\sin x-\frac{1}{3}\sin^3 x$ b) $-16\cos x+\frac{16}{3}\cos^3 x$ c) $6\sin x-2\sin^3 x$ d) $5\sin x-\frac{5}{3}\sin^3 x$ e) $-\frac{1}{2}\cos 2x+\frac{1}{6}\cos^3 2x$
2 f) $12\sin 3x-4\sin^3 3x$ **3** a) $\frac{2+\pi}{2}$ b) $2\pi-4$ c) $\frac{3\pi}{2}$ d) $2\pi-4$ e) $\frac{\pi-2}{8}$ f) 3π

Exercise 9E

1 a) $5, \frac{4}{3}$ b) $13, \frac{12}{5}$ c) $\sqrt{29}, \frac{5}{2}$ d) $\sqrt{29}, \frac{2}{5}$ e) $\sqrt{2}, 1$ f) $25, \frac{3}{4}$ g) $2, \frac{1}{\sqrt{3}}$ h) $2\sqrt{5}, 2$ **2** a) $13\sin(\theta+22.6°)$ b) 7.4°, 127.4°

3 a) $5\sin(\theta - 36.9°)$ b) 60.4°, 193.3° **4** a) 1.57, 5.76 or $\frac{\pi}{2}, \frac{11\pi}{6}$ b) 1.83, 6.02 or $\frac{7\pi}{12}, \frac{23\pi}{12}$ c) 1.40, 5.68 d) 0.90, 3.28
4 e) 4.15, 5.92 f) 1.11, 5.93 **5** a) $\sqrt{97}\cos(2\theta + 24.0°)$ b) 14.2°, 141.8° **6** b) 1.51, 5.70 **7** a) 105°, 165°
7 b) 236.3°, 326.3° c) 7.3°, 140.2° **8** a) $10\sin(\theta + 36.9°)$ b) greatest: 10 at $\theta = 53.1°$; least: -10 at $\theta = 233.1°$
9 a) $3 + \sqrt{5}\cos(\theta - 0.46)$ b) greatest: $3 + \sqrt{5}$ at $\theta = 0.46$; least: $3 - \sqrt{5}$ at $\theta = 3.61$
10 a) greatest: 13 at $\theta = 67.4°$; least: -13 at $\theta = 247.4°$ b) greatest: $\sqrt{5}$ at $\theta = 26.6°$; least: $-\sqrt{5}$ at $\theta = 206.6°$
10 c) greatest: 12 at $\theta = 143.1°$; least 2 at $\theta = 323.1°$ d) greatest: $2 + \sqrt{2}$ at $\theta = 225°$; least: $2 - \sqrt{2}$ at $\theta = 45°$
10 e) greatest: $10 + \sqrt{5}$ at $\theta = 296.6°$; least: $10 - \sqrt{5}$ at $\theta = 116.6°$ f) greatest: 10 at $\theta = 228.2°$; least: 4 at $\theta = 48.2°$

Check out

1 b) $-\cos x$ **2** b) 41.8°, 138.2°, 194.5°, 345.5° (1 d.p.) **3** a) $\sqrt{34}\cos(x - 0.54)$ b) i) $-\sqrt{34} = -5.83$ ii) 3.68
3 c) 1.35, 6.01 (3 s.f.)

Revision exercise 9

1 a) $10\cos(x + 53.1°)$ b) 19.4°, 234.4° **2** a) 1.176 (3 d.p.) b) $26\sin(\theta + 1.176)$ c) i) 26 ii) 0.395 rad (3 d.p.)
3 a) i) $L = 2\sin\theta + 4\cos\theta$ ii) $4.472\sin(\theta + 1.107)$ b) i) 4.472 ii) 0.46 (2 d.p.) **4** a) $6\cos 2x - 2\sin 2x$
4 b) $y = -2x + \frac{\pi}{2} + 3$ c) i) $3\sin 4x - 4\cos 4x + 5$ ii) 17 **5** a) $5\sin(\theta - 36.9°)$ b) 60.4°, 193.3° **6** b) 14°, 194°, 90°, 270°
7 a) $3 - 3\cos 2\theta$ b) $\frac{\pi - 3}{24}$ c) 0.766, 2.375 (3 d.p.) **8** a) $13\cos(\theta - 1.176)$ b) 0.129, 2.223 (3 d.p.) **9** b) $2 - \sqrt{3}$
10 a) $\frac{5}{13}$ b) $\frac{63}{65}$ **11** $2 + \sqrt{2} + \pi = 6.5558$ **12** b) 114.3, 335.7 **13** a) $2 + \frac{\pi^2}{2}$ c) 57.1 (3 s.f.) **14** 48.2°, 120°, 311.8°, 240°

Chapter 10

Check in

1 a) i) 2.236 ii) 0.015 iii) 0.149 b) i) 0.333 ii) 1900 iii) 1.51 **2** a) i) 1.236 ii) 1.513 iii) 1.259
2 b) i) 2.431 ii) 0.307 iii) 0.653

Exercise 10A

1 a) 500 b) 1 093 500 c) 6.92 days **2** a) 80 °C b) 0.3125 °C c) 4 min **3** a) £15 000 b) 0.934 c) 3 yrs 6 months
4 a) 6 b) 0.168 c) 19 months **5** a) €618 b) €695.56 c) $600 \times (1.03)^n$ d) 18 **6** a) 2.14 m b) 3.93 m
6 c) $2 \times (1.07)^n$ d) 35 **7** a) 1.04 b) 21.7% c) $(1.04^n - 1)\%$ d) 14 **8** a) initial population b) 323
8 c) no bacteria die; all reproduce at same rate d) some bacteria die

Exercise 10B

1 a) $12 000 b) $1983.59 c) 14 months **2** a) 20 b) 66 c) 7 months **3** a) 80 °C b) 29.4 °C c) 9 min
4 a) 100 °C b) 74 °C c) 6.9 min **5** a) 0.006 93 b) 3.47 yrs

Check out

1 a) 1.076 c) £9000 d) 61.4 years **2** a) i) 5000 ii) 1433 b) 16 min 26 sec c) 0.0186 … All bacteria destroyed

Revision exercise 10

1 a) £100 b) £121.55 c) 8.31 **2** a) i) 50 ii) 100 b) 2.8 min **3** a) 0.040 546 … b) 27.1 **4** a) 1000
4 c) i) $t = \frac{\ln\left(\frac{N}{1000}\right)}{\ln 1.0423}$ ii) 167 min **5** b) 11.9 g **6** a) i) 15 000 ii) 0.155 b) 2009

Chapter 11

Check in

1 a) i) $12x^3 + \frac{12}{x^4}$ ii) $-18\sin 6x$ iii) $-8e^{-4x}$ b) i) $\frac{1}{2}\ln x - 8\sqrt{x} + c$ ii) $-8\cos\frac{1}{2}x + c$ iii) $-\frac{4}{e^{\frac{3}{2}x}} + c$

2 a) $-10x(4 - x^2)^4$ b) $-e^{-x}(2\sin 2x + \cos 2x)$ c) $\frac{1}{2}e^{-3x}(2x - 3x^2)$ **3** a) $-\sqrt{4 - x^2} + c$ b) $\frac{1}{2}x\sin 2x + \frac{1}{4}\cos 2x + c$

3 c) $e^{3x}\left(\frac{x^2}{3} - \frac{2x}{9} + \frac{2}{27}\right) + c$ **4** $\frac{2}{x - 3} - \frac{3}{2x + 1}$

Exercise 11A

1 $\frac{2x}{3y^2}$ b) $\frac{3y}{2y - 3x}$ c) $-\frac{2xy + y^2}{2xy + x^2}$ d) $\frac{2 - 3y}{3(x + y^2)}$ e) $\frac{4x^3 - y^2 - 6}{2xy}$ f) $\frac{6x^5 - 5y^3 - 9y}{3x(5y^2 + 3)}$ g) $\frac{x - 2y}{x}$ h) $x - 1$ i) $\frac{7x^4(x^2 - 5y^3)}{1 + 21x^5y^2}$

2 a) $\frac{1}{10}$ b) $\frac{4}{3}$ c) $\frac{1}{4}$ d) $\frac{4}{27}$ e) -6 f) 0 g) -15 h) $-\frac{1}{2}$ **3** $5y - 3x + 18 = 0$ **4** $y + 2x = 5, 2y = x$

5 $18y + x = 12, 3y - 54x + 323 = 0$ **6** $15x - 8y = 36, 7y - 15x = 36$ **7** (2, 1), (2, 5) **8** $x = -4, x = 1$

Exercise 11B

1 a) $y = (x - 3)^2$ b) $y = 2\sqrt{x}$ c) $y = 3x^2 + 6x - 2$ d) $y = \frac{4}{x}$ e) $y = \frac{3}{x}$ f) $y = \frac{12}{x} - 1$ **2** a) $9x^2 + 4y^2 = 36$

2 b) $4x^2 + y^2 = 4$ c) $16x^2 + 25y^2 = 400$ d) $x^2 - y^2 = 12$ e) $y^2 - x^2 = 8$ f) $y^2 - x^2 = 24$ **3** a) $\frac{2}{t}$ b) $\frac{1}{3t^2}$ c) $5\sqrt{t}$

3 d) $(2t - 1)^2$ e) $\frac{3(t - 1)}{8}$ f) $\frac{2\sqrt{t^3} - 1}{2 - \sqrt{t}}$ g) $-2t^2(t + 2)$ h) $\frac{-(3 + \sqrt{t})^2}{2}$ i) $-\tan t$ j) $2\cot t$ k) $\frac{3\cos 6t}{\cos 2t}$ l) $\frac{2(t + \sin 2t)}{1 - \sin t}$

4 a) 2 b) $-\frac{1}{9}$ c) -36 d) -6 e) $-\frac{3}{50}$ f) 1 g) $-\frac{1}{3}$ h) $\frac{8}{\pi - 3}$ **5** $y = x + 19$ **6** $y = 3x - 4, y + 3x + 16 = 0$

7 $y = x - \frac{1}{2}$ **8** $4y + 2x + 1 = 0$ **9** $12y - x = 30, y + 12x = 75$ **10** $y + x = 2, 4y - x + 4 = 0$ **11** (−1, 0), (15, −4)
12 $4y - x = 19, 4y + x + 43 = 0$ **13** (−2, 2) **14** a) $2y - x + 5 = 0, y + 2x + 30 = 0$ b) $\frac{1}{2} \times -2 = -1$, (−11, −8)

15 b) $(\frac{3}{4}, \frac{9}{4}), (\frac{5}{4}, -\frac{25}{4})$ **16** b) $\frac{4}{19}, -1$ **17** b) $\left(\frac{\pi + 1}{2}, \frac{\pi + 6\sqrt{3}}{6}\right), \left(\frac{5\pi + 1}{2}, \frac{5\pi - 6\sqrt{3}}{6}\right)$ **18** b) $(1 + \sqrt{2}, \sqrt{2}), (1 - \sqrt{2}, -\sqrt{2})$

Exercise 11C

The constant of integration is omitted from these answers.

1 a) $\frac{1}{2}x^2 + 2x - 3\ln(x - 2)$ b) $\frac{1}{2}x^2 - 3x + \ln(x + 4)$ c) $x^2 + 5x + 6\ln(x - 3)$ d) $x^2 + x + 2\ln(2x - 1)$ e) $\frac{1}{3}x^3 - x - 2\ln(x - 2)$
1 f) $x^3 - x^2 + \ln(x + 1)$ g) $x^3 + 2x^2 - 6x + \ln(x - 3)$ h) $\frac{1}{4}x^4 + \frac{1}{3}x^3 + \frac{1}{2}x^2 + x + \ln(x - 1)$ **2** a) $2\ln(x - 1) - \ln(x - 2)$

2 b) $\ln(x + 3) + \frac{1}{2}\ln(2x - 1)$ c) $\frac{1}{4}\ln\left(\frac{x - 2}{x + 2}\right)$ d) $2x + \ln(x - 1) + 3\ln(x - 4)$ e) $8x - 25\ln(x + 5) + 9\ln(x - 3)$

2 f) $x + \ln\left(\frac{2x - 3}{x - 1}\right)$ g) $x - \frac{2}{x + 1} + \ln(x + 1)$ h) $\ln\left(\frac{x - 2}{x + 1}\right) - \frac{2}{x - 2}$ i) $\ln\left(\frac{2 - x}{5 - x}\right) + \frac{1}{5 - x}$ j) $\ln\left(\frac{x - 1}{x + 1}\right) - \frac{2}{x - 1}$

2 k) $3\ln\left(\frac{2x + 1}{x + 1}\right) + \frac{1}{x + 1}$ l) $\ln\left(\frac{3x - 1}{x + 2}\right) - \frac{2}{x + 2}$ **3** a) $\ln\left(\frac{3}{2}\right)$ b) $\frac{1}{2}\ln 7 - \frac{6}{7}$ c) $14 + \ln\left(\frac{4}{3}\right)$ d) $\ln\left(\frac{3}{2}\right) - \frac{1}{6}$ e) $\frac{1}{2}\ln\left(\frac{4}{3}\right)$

3 f) $6 - \ln\left(\frac{4}{3}\right)$ g) $\frac{1}{12} + \frac{1}{4}\ln 2$ h) $28 + \ln 10$ **4** a) $\ln(1 + x)$ b) $-x - \ln(1 - x)$ c) $\frac{1}{2}\ln(1 + x^2)$ d) $\sin^{-1}x$

4 e) $\ln(1 - x) + \frac{1}{1 - x}$ f) $2\sqrt{1 + x}$ g) $\tan^{-1}x$ h) $\frac{1}{1 - x}$ i) $x - \ln(1 + x)$ j) $\frac{1}{2}\ln\left(\frac{1 + x}{1 - x}\right)$ k) $-\sqrt{1 - x^2}$ l) $\frac{2}{3}(x - 2)\sqrt{(1 + x)}$

5 a) $A(-2, \frac{1}{5}), B(2, \frac{1}{5})$ b) $\frac{4}{5} - \frac{1}{3}\ln 5$ **6** a) $P(3, 3)$ b) $\frac{3}{2} + \ln 4$ **7** b) $2\ln 3$ **8** a) $(-\frac{25}{4}, -25), Q(4, 16)$

Exercise 11D

The constant of integration is omitted in these answers.

1 a) $\frac{1}{4}x^2(x^2+8x+18)$ b) $2\ln x - \ln(x+1)$ c) $-x\cos x + \sin x$ d) $\dfrac{4}{2-x}$ e) $\sin^{-1}x - \sqrt{1-x^2}$ f) $\frac{1}{2}\ln(x^2-4)$ g) $-e^{\cos x}$

1 h) $\ln x - \dfrac{1}{x^2}$ i) $3\ln(x+1) - 2\ln(x-1)$ j) $x\ln x - x$ k) $3 - x - 3\ln(3-x)$ l) $\frac{1}{3}\sqrt{(x^2+1)^3}$ m) $\dfrac{e^{2x}}{4}(2x-1)$

1 n) $\dfrac{x}{4}(x^3-20x^2+150x-500)$ o) $-\frac{1}{4}\cos^4 x$ p) $x + \ln(x-2) - 2\ln(x+3)$ q) $\ln(e^x+1)$ r) $-\frac{2}{15}(4+3x)\sqrt{(2-x)^3}$

2 a) $x - 3\ln(x+2)$ b) $\ln\left(\dfrac{x+3}{x+4}\right)$ c) $\frac{1}{20}(x^4-5)^5$ d) $-2\ln(5-x)$ e) $-\dfrac{e^{-2x}}{4}(2x+1)$ f) $\frac{1}{3}x^3 - \frac{1}{2}x^2 - 6x$

2 g) $3\ln(x+2) - 2\ln(x-1) + \dfrac{1}{(x-1)}$ h) $x - 4\ln(x+4)$ i) $\frac{1}{9}\tan^3 3x$ j) $3\sin^{-1}x - \sqrt{1-x^2}$ k) $\frac{1}{2}e^{2x}$ l) $\dfrac{x}{2}\sin 2x + \dfrac{1}{4}\cos 2x$

2 m) $\frac{2}{15}\sqrt{(5x-1)^3}$ n) $\ln x - \tan^{-1}x$ o) $\ln(x^3+5x)$ p) $\ln(1-\cos x)$ q) $\frac{2}{15}(3x-4)\sqrt{(x+2)^3}$ r) $\dfrac{x^6}{36}(6\ln x - 1)$

3 a) $x + 5\ln(x-5)$ b) $\ln(1+\tan x)$ c) $\frac{1}{9}(x^3+3x-2)^3$ d) $\dfrac{1}{6}\ln\left(\dfrac{x-3}{x+3}\right)$ e) $\dfrac{e^{5x}}{25}(5x-1)$ f) $-\dfrac{1}{2(x^2+1)}$

3 g) $3\ln(x+1) + \ln x + \dfrac{2}{x}$ h) $\frac{1}{2}e^{x^2}$ i) $\frac{1}{80}(8x-3)(2x+3)^4$ j) $\frac{2}{3}\sqrt{x}(x-3)$ k) $-\dfrac{x}{3}\cos 3x + \dfrac{1}{9}\sin 3x$ l) $\frac{1}{2}\ln(x^2+1) + \tan^{-1}x$

3 m) $-\frac{1}{2}\cos(x^2)$ n) $\frac{1}{2}\ln(2x+1)$ o) $\ln(x+1) + 3\ln(x-3)$ p) $3\sqrt{2x+1}$ q) $\ln(2-x) + \dfrac{2}{2-x}$ r) $\dfrac{x}{2}(\ln x - 1)$

Exercise 11E

1 a) $y = \pm\sqrt{(x+1)^2+c}$ b) $y = \pm\sqrt{x^3-2x+c}$ c) $y = \sqrt[3]{3x^2+9x+c}$ d) $y = (\frac{1}{2}x^3+c)^2$ e) $y = \left(\dfrac{1}{x-1}+c\right)^2$

1 f) $y = \frac{1}{2}[1+(x^2+c)^2]$ g) $y = Ae^{x^3}$ h) $y = \ln(\frac{1}{2}x^2+c)$ i) $y = -\ln(\cos x + c)$ j) $\sin y = x + \cos x + c$

2 a) $y = 2 + 4x - x^3$ b) $y = \sqrt{x^2+2x+9}$ c) $y^2 - 6y = x^2 + 6x - 8$ d) $y = \sqrt[3]{20+x-x^2}$ e) $y = \dfrac{x^2}{1-x^2}$ f) $y = (1+\sqrt{x+1})^2$

2 g) $y = 7e^{x^2}$ h) $\cos y = \dfrac{1}{4} + \dfrac{1}{x}$ i) $y = 2 - \dfrac{2}{x}$ j) $y = \ln\left(\dfrac{1+2\cos x}{2}\right)$ **3** $y = e^{x^2-9} - 1$ **4** $y = \sin^2 x + 2\sin x - 2$

Exercise 11F

1 b) 12 500 c) population cannot grow indefinitely – restrictions on space, food, … and rats will die.

2 b) 6 hours 40 minutes **3** b) 4 days 14 hours **4** b) 46 hours 40 minutes **5** b) 0.277 s

Exercise 11G

1 c) 250 **2** c) 9 **3** c) 4 years 11 months **6** c) 921 000

Check out

1 a) $-\dfrac{y(2y+6x)}{x(4y+3x)}$ b) at $(1, \frac{1}{2})$ gradient $= -\frac{7}{10}$, at $(1, -2)$ gradient $= -\frac{4}{5}$ **2** a) $xy^2 = y^2 - 16$ b) $y = -4x - 10$

3 $\frac{2}{5}\ln(x+3) - \frac{1}{15}\ln(3x-1) + c$ **4** $y = 1 + e^{-\frac{1}{x}-1}$ **5** a) $x = \dfrac{25t}{5t+4}$ b) $\frac{8}{15}$ days c) at $x = 5$, $\dfrac{dx}{dt} = 0$

Revision exercise 11

1 a) $-\dfrac{1}{4t^2}$ b) $y = 4x - \frac{7}{2}$ **2** a) $\dfrac{2}{x+4} + \dfrac{4}{7-2x}$ b) $2\ln\frac{49}{4}$ **3** a) $(2, -1)$ $(2, -3)$ b) $\frac{4}{9}, -\frac{4}{9}$ **4** a) $-\dfrac{3}{x+4} + \dfrac{4}{2x+1}$

4 b) $4\ln 3 - 3\ln 2$ **5** a) i) $\frac{1}{6}$ b) $y^2 = \ln(x+2) + 1 - \ln 3$ **6** b) $1.8\ \text{m s}^{-1}$ **8** a) $-\dfrac{1}{3t^2}$ b) $y = 3x - 5$

9 $\frac{1}{5}t = \ln 9 - \ln(10-x)$; $t = 0.589$ (3 d.p.) **10** b) $y = -\frac{1}{3}x + \sqrt{2}$ **11** a) ii) $\dfrac{2}{x-4} - \dfrac{2}{x+4}$ b) $3 + 2\ln 3$

12 b) 1:23.3 pm (83.3 minutes) **13** a) ii) $A = 6, B = -3, C = 8$ b) $3\ln 2 + \frac{4}{3}$ **14** b) $3y = -4x + 7$

15 a) $xe^x - e^x$ b) $\ln y = e^x(x-1)+1$ **16** a) $\pm\frac{5}{3}\sqrt{5}$ b) $-1.5, 1.5$ (2 s.f.) **17** $-\dfrac{1}{2(1+2x)} + \dfrac{1}{4(1+2x)^2} + c$

18 a) $(3, \frac{7}{3})$ $(3, -\frac{1}{3})$ b) $1, -1$ **19** a) i) $\dfrac{dh}{dt} = -k\sqrt{h}$ iii) 0.293 (3 d.p.) b) 6 hours 50 minutes

20 a) $1 + \dfrac{4}{2x-3} - \dfrac{3}{x+2}$ b) $4 + 4\ln 3 - 3\ln 2$ **21** b) $2\sqrt{A} = 2t + 1$ or $t = \sqrt{A} - \frac{1}{2}$ c) ~17 days

22 b) i) $\ln P = kt + c, k = \frac{1}{30}\ln 2$ ii) 22 **23** a) $\dfrac{1}{3+\cos x} + c$ b) $y = \dfrac{1}{3+\cos x} + \frac{3}{4}$

Chapter 12

Check in

1 a) 3 b) $-\frac{1}{2}$ **2** a) 10 b) $7\sqrt{2}$ **3** $AB = 3\sqrt{2}, BC = 10, AC = \sqrt{106}, \hat{A} = 74.05°, \hat{B} = 81.87°, \hat{C} = 24.08°$

Exercise 12A

1 a) 5 b) $\sqrt{74}$ c) 13 d) $2\sqrt{5}$ e) 3 f) $\sqrt{61}$ g) $\sqrt{155}$ h) $\sqrt{83}$ **2** ± 7 **3** ± 4 **4** ± 3 **5** $\begin{pmatrix}5\\2\\3\end{pmatrix}$ **6** $\begin{pmatrix}3\\3\\-13\end{pmatrix}$

7 $\begin{pmatrix}8\\1\\-8\end{pmatrix}$ **8** $\begin{pmatrix}-3\\10\\0\end{pmatrix}$ **9** $\begin{pmatrix}10\\-11\\-1\end{pmatrix}$ **10** $\begin{pmatrix}-7\\3\\2\end{pmatrix}$ **11** a) $\begin{pmatrix}2\\-11\\18\end{pmatrix}$ b) $\sqrt{449}$ **12** a) $\begin{pmatrix}5\\17\\-8\end{pmatrix}$ b) $\sqrt{378}$

13 a) $\begin{pmatrix}20\\16\\-8\end{pmatrix}$ b) $\begin{pmatrix}8\\16\\20\end{pmatrix}$

Exercise 12B

1 a) $\overrightarrow{PQ} = \begin{pmatrix}3\\-2\\2\end{pmatrix}$ b) $\overrightarrow{RS} = \begin{pmatrix}6\\-4\\4\end{pmatrix}$ **2** a) $\overrightarrow{AB} = \begin{pmatrix}-1\\2\\-3\end{pmatrix}$ b) $\overrightarrow{CD} = \begin{pmatrix}-3\\6\\-9\end{pmatrix}$ **3** a) $\overrightarrow{AB} = \begin{pmatrix}2\\-3\\1\end{pmatrix}$ b) $\overrightarrow{CD} = \begin{pmatrix}-8\\12\\-4\end{pmatrix}$

4 a) $2\sqrt{29}$ b) $(-2, 1, 2)$ **5** a) $5\sqrt{2}$ b) $(2.5, 3, 0.5)$ **6** a) $(4, -1, -2)$ b) $(-1, -3, 6)$ c) $\sqrt{93}$

7 a) $(2, 2, -1)$ b) $(-5, -4, 4)$ c) $\sqrt{110}$ **8** a) $\overrightarrow{OP} = \begin{pmatrix}5\\-1\\3\end{pmatrix}, \overrightarrow{OQ} = \begin{pmatrix}5.5\\-2\\3.5\end{pmatrix}, \overrightarrow{OR} = \begin{pmatrix}1\\1.5\\1\end{pmatrix}, \overrightarrow{OS} = \begin{pmatrix}0.5\\2.5\\0.5\end{pmatrix}$

Exercise 12C

1 a) $\begin{pmatrix}2\\1\\3\end{pmatrix}, \begin{pmatrix}4\\2\\6\end{pmatrix}$ b) 1 : 2 **4** 8, 3

Exercise 12D

1 a) −9 b) −9 c) 26 d) −13 e) 25 f) 13 **2** a) −3 b) −3 c) 12 d) 21 e) 9 f) 14 **3** a) 1 b) 3 c) 3
3 d) 2 e) −13 f) 17 **4** a) −45 b) −43 c) 24 d) 2 e) −19 f) 91 **5** a) Perpendicular b) Parallel
5 c) Parallel d) Neither e) Parallel f) Perpendicular g) Parallel h) Neither **6** a) 75.7° b) 45° c) 53.1°
6 d) 115.3° e) 119.2° f) 66.4° g) 95.3° h) 69.4° **7** a) 10 b) $1\frac{1}{4}$ **8** a) −12 b) 12 **9** $2\frac{4}{7}$ **10** 18
11 2 or −5 **12** 1 or 3

Exercise 12E

1 $\begin{pmatrix}x\\y\end{pmatrix} = \begin{pmatrix}4\\3\end{pmatrix} + t\begin{pmatrix}1\\-2\end{pmatrix}$ **2** $\begin{pmatrix}x\\y\\z\end{pmatrix} = \begin{pmatrix}5\\-2\\3\end{pmatrix} + t\begin{pmatrix}4\\-3\\1\end{pmatrix}$ **3** $\begin{pmatrix}x\\y\end{pmatrix} = \begin{pmatrix}5\\-1\end{pmatrix} + t\begin{pmatrix}1\\-1\end{pmatrix}$ **4** $\begin{pmatrix}x\\y\end{pmatrix} = \begin{pmatrix}2\\6\end{pmatrix} + t\begin{pmatrix}3\\-8\end{pmatrix}$

5 $\begin{pmatrix} x \\ y \\ z \end{pmatrix} = \begin{pmatrix} -1 \\ 2 \\ -3 \end{pmatrix} + t\begin{pmatrix} 7 \\ 1 \\ 3 \end{pmatrix}$ **6** a) $\begin{pmatrix} x \\ y \\ z \end{pmatrix} = \begin{pmatrix} -4 \\ 1 \\ -3 \end{pmatrix} + t\begin{pmatrix} 2 \\ -1 \\ 3 \end{pmatrix}$ **7** a) $\begin{pmatrix} x \\ y \\ z \end{pmatrix} = \begin{pmatrix} 3 \\ 6 \\ -1 \end{pmatrix} + t\begin{pmatrix} -1 \\ 2 \\ 2 \end{pmatrix}$ **8** a) $\begin{pmatrix} x \\ y \\ z \end{pmatrix} = \begin{pmatrix} 12 \\ 19 \\ 15 \end{pmatrix} + t\begin{pmatrix} 2 \\ 3 \\ 5 \end{pmatrix}$

9 a) $\begin{pmatrix} x \\ y \\ z \end{pmatrix} = \begin{pmatrix} -3 \\ 2 \\ 6 \end{pmatrix} + t\begin{pmatrix} 5 \\ 3 \\ -6 \end{pmatrix}$

Exercise 12F

1 $\begin{pmatrix} 10 \\ 4 \end{pmatrix}$ **2** $\begin{pmatrix} 2 \\ -3 \end{pmatrix}$ **3** a) $\begin{pmatrix} 2 \\ 5 \\ 9 \end{pmatrix}$ b) 15.6° **4** a) $\begin{pmatrix} 1 \\ 3 \\ 5 \end{pmatrix}$ b) 58.5° **5** 39.1° **6** 64.3° **7** 151.8°

8 a) $\mathbf{r} = \begin{pmatrix} -2 \\ 0 \\ 4 \end{pmatrix} + \mu\begin{pmatrix} 3 \\ 1 \\ -1 \end{pmatrix}$ **9** b) $\mathbf{r} = \begin{pmatrix} 3 \\ 5 \\ -2 \end{pmatrix} + \mu\begin{pmatrix} -6 \\ 15 \\ -3 \end{pmatrix}$

Exercise 12G

1 a) (8, 5, 1) b) $3\sqrt{2}$ **2** a) (11, −1, 16) b) $3\sqrt{3}$ **3** a) (−10, −1, 0) b) $5\sqrt{14}$ **4** a) $\mathbf{r} = \begin{pmatrix} 0 \\ 5 \end{pmatrix} + \lambda\begin{pmatrix} 3 \\ 1 \end{pmatrix}$

5 a) $\mathbf{r} = \begin{pmatrix} -1 \\ -2 \end{pmatrix} + \lambda\begin{pmatrix} 1 \\ 2 \end{pmatrix}$ **6** a) $\mathbf{r} = \begin{pmatrix} -1 \\ 1 \end{pmatrix} + \lambda\begin{pmatrix} 1 \\ 1 \end{pmatrix}$ **7** a) i) $\frac{\sqrt{3}}{3}$ **8** a) i) $\frac{\sqrt{2}}{3}$ **9** a) $\begin{pmatrix} 6 \\ 3 \\ -9 \end{pmatrix}$ b) $\begin{pmatrix} 6\lambda - 5 \\ 3\lambda - 4 \\ -9\lambda \end{pmatrix}$

9 c) (1, 4, 2) **10** a) $\begin{pmatrix} 4 \\ 8 \\ -4 \end{pmatrix}$ b) $\begin{pmatrix} -5 \\ -8 \\ -3 \end{pmatrix} + \lambda\begin{pmatrix} 4 \\ 8 \\ -4 \end{pmatrix}$ c) (6, 4, 2)

Check out

1 a) $\begin{pmatrix} 7 \\ -1 \\ -2 \end{pmatrix}$ b) i) $\begin{pmatrix} 3 \\ 5 \\ -4 \end{pmatrix}$ ii) **p**, **r** **2** a) $\mathbf{r} = \begin{pmatrix} 2 \\ -3 \\ 1 \end{pmatrix} + \lambda\begin{pmatrix} 3 \\ 5 \\ -4 \end{pmatrix}$ b) ii) (8, 7, −7) **3** 129.2° **4** $\frac{\sqrt{2530}}{11}$

Revision exercise 12

1 a) $\mathbf{r} = \begin{pmatrix} 3 \\ -1 \\ 2 \end{pmatrix} + \lambda\begin{pmatrix} -1 \\ 1 \\ 0 \end{pmatrix}$ b) $\begin{pmatrix} 1 \\ 1 \\ 2 \end{pmatrix}$ d) (5, −3, 2) **2** a) (3, 8, 4) b) 77°(nearest degree) **3** a) (7, 2, 4) **4** b) i) −1

5 a) 130° **6** a) 79.1° **7** b) $\mathbf{r} = \mathbf{a} + \lambda\mathbf{b}$ **8** a) $p = -9$, $\begin{pmatrix} 15 \\ 2 \\ -3 \end{pmatrix}$ b) 42°

C4 Practice Paper

1 a) $2\sin x\cos x$ b) $\frac{3}{4}$ **2** a) iii) $(x + 2)(2x - 3)(2x - 1)$ b) $A = 1, B = -2, C = \frac{3}{4}, D = 0$ **3** a) i) $1 + \frac{1}{3}x - \frac{1}{9}x^2$

3 b) 1.995 825 (6 d.p.) **4** a) i) $\sqrt{29}\cos(\theta - 68.2°)$ ii) 147.5°, 348.9° **6** b) i) $2x^2\frac{dy}{dx} + 4xy + 2(y - 1)\frac{dy}{dx} = 0$

7 a) i) $\begin{pmatrix} 2 \\ -1 \\ 2 \end{pmatrix}$ ii) 3 iii) $\mathbf{r} = \begin{pmatrix} 1 \\ 4 \\ 3 \end{pmatrix} + \lambda\begin{pmatrix} 2 \\ -1 \\ 2 \end{pmatrix}$ **8** a) i) $\frac{1}{x} + \frac{1}{(1 - x)}$ b) i) $\frac{1}{9}$ iii) $t = 48$ (24 hours later)

Formulae

This section lists formulae which relate to the Core modules C3 and C4, and which candidates are expected to remember. These formulae will **not** be included in the AQA formulae book.

Candidates may use relevant formulae included in the formulae booklet without proof.

Trigonometry

$\sec^2 A = 1 + \tan^2 A$

$\operatorname{cosec}^2 A = 1 + \cot^2 A$

$\sin 2A = 2 \sin A \cos A$

$$\cos 2A = \begin{cases} \cos^2 A - \sin^2 A \\ 2\cos^2 A - 1 \\ 1 - 2\sin^2 A \end{cases}$$

$$\tan 2A = \frac{2 \tan A}{1 - \tan^2 A}$$

$a \cos\theta + b \sin\theta = R \sin(\theta + \alpha)$, where $R = \sqrt{a^2 + b^2}$ and $\tan\alpha = \dfrac{a}{b}$

$a \cos\theta - b \sin\theta = R \cos(\theta + \alpha)$, where $R = \sqrt{a^2 + b^2}$ and $\tan\alpha = \dfrac{b}{a}$

Differentiation

function	derivative
e^{kx}	ke^{kx}
$\ln x$	$\dfrac{1}{x}$
$\sin kx$	$k \cos kx$
$\cos kx$	$-k \sin kx$
$f(x)g(x)$	$f'(x)g(x) + f(x)g'(x)$
$f(g(x))$	$f'(g(x))g'(x)$

Integration

function	integral
$\cos kx$	$\dfrac{1}{k}\sin kx + c$
$\sin kx$	$-\dfrac{1}{k}\cos kx + c$
e^{kx}	$\dfrac{1}{k}e^{kx} + c$
$\dfrac{1}{x}$	$\ln\lvert x\rvert + c \qquad (x \neq 0)$
$f'(g(x))g'(x)$	$f(g(x)) + c$

Volumes

Volume of solid of revolution:

About the x-axis: $$V = \int_a^b \pi y^2 \, dx$$

About the y-axis: $$V = \int_c^d \pi x^2 \, dy$$

Vectors

$$\begin{bmatrix} x \\ y \\ z \end{bmatrix} . \begin{bmatrix} a \\ b \\ c \end{bmatrix} = xa + yb + zc = (\sqrt{x^2 + y^2 + z^2})(\sqrt{a^2 + b^2 + c^2}) \cos \theta$$

Mathematical Notation

Set notation

$\in$	is an element of
$\notin$	is not an element of
$\{x_1, x_2, \ldots\}$	the set with elements $x_1, x_2, \ldots$
$\{x: \ldots\}$	the set of all x such that …
$\mathrm{n}(A)$	the number of elements in set A
$\mathbb{N}$	the set of natural numbers, $\{1, 2, 3, \ldots\}$
$\mathbb{Z}$	the set of integers, $\{0, \pm1, \pm2, \pm3, \ldots\}$
$\mathbb{Z}^+$	the set of positive integers, $\{1, 2, 3, \ldots\}$
$\mathbb{Q}$	the set of rational numbers, $\left\{\frac{p}{q}: p \in \mathbb{Z}, q \in \mathbb{Z}^+\right\}$
$\mathbb{Q}^+$	the set of positive rational numbers, $\{x \in \mathbb{Q}, x > 0\}$
$\mathbb{R}$	the set of real numbers
$\mathbb{R}^+$	the set of positive real numbers, $\{x \in \mathbb{R}, x > 0\}$
(x, y)	the ordered pair, x, y

Miscellaneous symbols

$=$	is equal to
$\neq$	is not equal to
$\equiv$	is identical to or is congruent to
$\approx$	is approximately equal to
$\propto$	is proportional to
$<$	is less than
$\leqslant, \ngtr$	is less than or equal to, is not greater than
$>$	is greater than
$\geqslant, \nless$	is greater than or equal to, is not less than
∞	infinity
$p \Rightarrow q$	p implies q (if p then q)
$p \Leftarrow q$	p is implied by q (if q then p)
$p \Leftrightarrow q$	p implies and is implied by q (p is equivalent to q)
$\exists$	there exists
$\forall$	for all

Operations

$a + b$	a plus b
$a - b$	a minus b
$a \times b, ab, a.b$	a multiplied by b
$a \div b, \frac{a}{b}, a/b$	a divided by b
$\sum_{i=1}^{n} a_i$	$a_1 + a_2 + \ldots + a_n$
$\sqrt{a}$	the positive square root of a

$n!$	n factorial
$\binom{n}{r}$	the binomial coefficient $\frac{n!}{r!(n-r)!}$ for $n \in \mathbb{Z}^+$

Functions

$\mathrm{f}(x)$	the value of the function f at x
$\mathrm{f}: A \to B$	f is a function under which each element of set A has an image in set B
$\mathrm{f}: x \to y$	the function f maps the element x to the element y
f^{-1}	the inverse function of the function f
$\lim_{x \to a} \mathrm{f}(x)$	the limit of $\mathrm{f}(x)$ as x tends to a
$\frac{\mathrm{d}y}{\mathrm{d}x}$	the derivative of y with respect to x
$\frac{\mathrm{d}^n y}{\mathrm{d}x^n}$	the nth derivative of y with respect to x
$\mathrm{f}'(x), \mathrm{f}''(x), \ldots, \mathrm{f}^{(n)}(x)$	first, second, …, nth derivatives of $\mathrm{f}(x)$ with respect to x
$\int y \, \mathrm{d}x$	the indefinite integral of y with respect to x
$\int_a^b y \, \mathrm{d}x$	the definite integral of y with respect to x between the limits $x = a$ and $x = b$

Exponential and logarithmic functions

e	base of natural logarithms
e^x, $\exp x$	exponential function of x
$\log_a x$	logarithm to the base a of x
$\ln x$, $\log_e x$	natural logarithm of x
$\log_{10} x$	logarithm of x to base 10

Circular functions

sin, cos, tan, cosec, sec, cot	the circular functions
$\sin^{-1}$, $\cos^{-1}$, $\tan^{-1}$, cosec^{-1}, $\sec^{-1}$, $\cot^{-1}$	the inverse circular functions

Vectors

$\mathbf{a}$	the vector $\mathbf{a}$
$\overrightarrow{AB}$	the vector represented in magnitude and direction by the directed line segment AB
$\mathbf{a}$	a unit vector in the direction of $\mathbf{a}$
$\mathbf{i}, \mathbf{j}, \mathbf{k}$	unit vectors in the directions of the Cartesian coordinate axes
$\lvert \mathbf{a} \rvert$, a	the magnitude of $\mathbf{a}$
$\lvert \overrightarrow{AB} \rvert$, AB	the magnitude of AB
$\mathbf{a}.\mathbf{b}$	the scalar product of $\mathbf{a}$ and $\mathbf{b}$

Index